Communications
in Computer and Information Science   **1039**

Commenced Publication in 2007
Founding and Former Series Editors:
Phoebe Chen, Alfredo Cuzzocrea, Xiaoyong Du, Orhun Kara, Ting Liu,
Krishna M. Sivalingam, Dominik Ślęzak, Takashi Washio, and Xiaokang Yang

Editorial Board Members

More information about this series at http://www.springer.com/series/7899

Piotr Gaj · Michał Sawicki ·
Andrzej Kwiecień (Eds.)

Computer Networks

26th International Conference, CN 2019
Kamień Śląski, Poland, June 25–27, 2019
Proceedings

 Springer

Editors
Piotr Gaj 🆔
Silesian University of Technology
Gliwice, Poland

Michał Sawicki 🆔
Silesian University of Technology
Gliwice, Poland

Andrzej Kwiecień
Silesian University of Technology
Gliwice, Poland

ISSN 1865-0929   ISSN 1865-0937   (electronic)
Communications in Computer and Information Science
ISBN 978-3-030-21951-2   ISBN 978-3-030-21952-9   (eBook)
https://doi.org/10.1007/978-3-030-21952-9

This Springer imprint is published by the registered company Springer Nature Switzerland AG
The registered company address is: Gewerbestrasse 11, 6330 Cham, Switzerland

Preface

Computer networks are one of the most important part of the contemporary digital world. They are like veins, very important but not visible unless they are damaged. What would happen with all of modern devices around us if the computer networks stopped working? The result in most cases would be the same. They would become useless or their functionalities would be significantly reduced. Computer networks and communication technologies are the key part of current digital world. Networking of almost everything is constantly growing and everything indicates that it will still grow.

Currently, almost every area of human activity is based on fast and effective communication and reliable access to information. As a result, it becomes indispensable to have an in-depth knowledge of how to manage, model, and design networked systems. Owing to the high dynamics and the multiplicity of emerging technologies, it is necessary to constantly expand and exchange knowledge and gain experiences in this field. Conferences are the events where independent points of view are presented and where experts and users can spontaneously hold discussions. This book contains the proceedings of one such event.

The conference was established at the Faculty of Automatic Control, Electronics and Computer Science at the Silesian University of Technology in Gliwice. The cycle of these initial editions was approximately annual but not strictly constant. It depended on various conditions, e.g., the fact that the event was new and not widely known. From year to year, the conference grew in popularity and recognition and its range gradually evolved from local to nationwide. In 1999, the conference was organized outside the university for the first time. This year, the venue was the historical Palace in Kamień Śląski, where currently a sanctuary of St. Jack Odrowąż is also located.

In 2009, the conference proceedings were published in Springer's CCIS series for the first time, as volume 39. Since then, the conference proceedings have been published in this series each year. All these books were well accepted by readers and highly rated by external entities.

For many years, among co-organizers, technical co-sponsors, and partners was the Section of Computer Networks and Distributed Systems belonging to the Committee on Informatics of the Polish Academy of Sciences (PAN), IEEE Poland Section, and the International Network for Engineering Education and Research (iNEER).

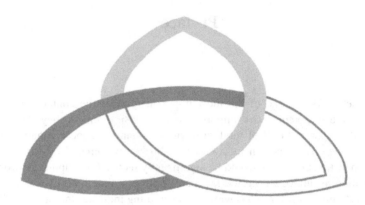

26th International Science Conference *Computer Networks*

This year, the 26th edition of the conference took place. Consequently, the Computer Networks conference has been present on the conference market for over a quarter of a century, giving participants a chance to meet people, discuss valid and important issues, as well as disseminate research and application results in the proceedings published in a significant and recognized book series. Over the past 26 years of the conference's history, all important topics related to the development of computer networks have been discussed at the conference and major breakthroughs in this area have been deliberated. Thus, we believe that the event has had a significant contribution to the global pool of achievements in this domain.

Computer networks are still the main means for data transfer among distributed nodes of all computer systems currently available. Thus, this area of research and application is up-to-date and very important for current industrial and social activities, as well as for the needs of the future. Computer network-related techniques and technologies are not spectacular at their base, but without them many spectacular services would be unavailable.

For this edition of the CN conference, approximately 70 papers were submitted. To maintain the high quality of the conference proceedings, only 29 carefully selected papers are published in this book. Each paper was reviewed by three independent reviewers in a double-blind process. This book collates the research work of scientists from numerous notable research centers. Each of three chapters includes highly stimulating studies of the wide spectrum of both science and practical-oriented issues regarding the computer networks and communication domain, which may interest a wide readership.

– Computer Networks
  This chapter comprises 14 papers. They refer to general computer networks problems, such as architecture design, analyzing, modeling and programming, as well as issues related to networked systems, their operations and management, Internet networks, wireless solutions, cybersecurity, real-time approach, security, quality, and reliability.

- Communications

  This chapter contains eight papers related to the general communication architectures, mobile and wireless communication, sensor network usage in vehicles, spectrum sharing in cognitive femtocells, indoor location based on ultra-wideband technology, problems of communication quality in sparse channels, problems with resilience of virtualization containers in cloud services, analysis of communication models and protocols in smart grid environment, among others.
- Queueing Theory and Queuing Networks

  This chapter contains seven papers. The papers refer to the analysis of energy consumption within cloud architecture, analysis of HTTP performance in industrial control systems, performance models of cluster-based Web systems, and some important issues of networks quality and reliability, all in the context of queueing theory.

On behalf of the Program and Organizing Committee of the CN Conference, we would like to express our gratitude to all authors for sharing their research results and for their assistance in producing this volume, which we believe is a reliable reference in the computer networks domain.

We also want to thank the members of the Technical Program Committee and all reviewers for their involvement and participation in the reviewing process.

If you would like to help us make the CN conference more attractive and interesting, please send us your opinions and proposals at cn@polsl.pl.

April 2019

Piotr Gaj
Michał Sawicki
Andrzej Kwiecień

Organization

CN 2019 was organized by the Institute of Informatics of the Faculty of Automatic Control, Electronics and Computer Science, Silesian University of Technology (SUT) and supported by the Committee on Informatics of the Polish Academy of Sciences (PAN), the Section of Computer Networks and Distributed Systems in technical co-operation with the IEEE and consulting support of the iNEER organization.

The company Bombardier Inc. was the conference industrial partner.

Executive Committee

All members of the Executive Committee are from the Silesian University of Technology, Poland.

Honorary Member	Halina Węgrzyn
Organizing Chair	Piotr Gaj
Organizing Support	Jacek Stój
Technical Volume Editor	Michał Sawicki
Technical Support	Aleksander Cisek
Technical Support	Ireneusz Smołka
Office	Małgorzata Gładysz
Web Support	Piotr Kuźniacki

Co-ordinators

PAN Co-ordinator	Tadeusz Czachórski
IEEE PS Co-ordinator	Jacek Izydorczyk
iNEER Co-ordinator	Win Aung

Program Committee

Program Chair

Andrzej Kwiecień	Silesian University of Technology, Poland

Honorary Members

Win Aung	iNEER, USA
Adam Czornik	Silesian University of Technology, Poland
Bogdan M. Wilamowski	Auburn University, USA

Technical Program Committee

Davide Adami	University of Pisa, Italy
Wessam Ajib	University of Quebec at Montreal, Canada

Olumide Akinwande	Imperial College London, UK
Tülin Atmaca	Institut National de Télécommunication, France
Rajiv Bagai	Wichita State University, USA
Sebastian Bala	University of Opole, Poland
Jiří Balej	Mendel University in Brno, Czech Republic
Alexander Balinsky	Cardiff University, UK
Zbigniew Banaszak	Koszalin University of Technology, Poland
Thomas Bauschert	Chemnitz University of Technology, Germany
Robert Bestak	Czech Technical University in Prague, Czech Republic
Tomasz Bilski	Poznań University of Technology, Poland
Grzegorz Bocewicz	Koszalin University of Technology, Poland
Leoš Bohac	Czech Technical University in Prague, Czech Republic
Leszek Borzemski	Wrocław University of Technology, Poland
Juan Felipe Botero	Universidad de Antioquia, Colombia
Markus Bregulla	University of Applied Sciences Ingolstadt, Germany
Maria Carla Calzarossa	University of Pavia, Italy
Berk Canberk	Istanbul Technical University, Turkey
Valeria Cardellini	University of Rome Tor Vergata, Italy
Emiliano Casalicchio	Blekinge Institute of Technology, Sweden
Amlan Chatterjee	California State University, USA
Ray-Guang Cheng	National University of Science and Technology, Taiwan
Andrzej Chydziński	Silesian University of Technology, Poland
Tadeusz Czachórski	Silesian University of Technology, Poland
Dariusz Czerwiński	Lublin University of Technology, Poland
Adam Czubak	University of Opole, Poland
Andrzej Duda	INP Grenoble, France
Alexander N. Dudin	Belarusian State University, Belarus
Peppino Fazio	University of Calabria, Italy
Max Felser	Bern University of Applied Sciences, Switzerland
Holger Flatt	Fraunhofer IOSB-INA, Germany
Jean-Michel Fourneau	Versailles University, France
Janusz Furtak	Military University of Technology, Poland
Rosario G. Garroppo	University of Pisa, Italy
Natalia Gaviria	Universidad de Antioquia, Colombia
Roman Gielerak	University of Zielona Góra, Poland
Mariusz Głabowski	Poznan University of Technology, Poland
Agustín J. González	Federico Santa María Technical University, Chile
Anna Grocholewska-Czuryło	Poznań University of Technology, Poland
Daniel Grzonka	Cracow University of Technology, Poland
Sebastien Harispe	IMT Mines Alés, France
Artur Hłobaż	University of Lodz, Poland
Edward Hrynkiewicz	Silesian University of Technology, Poland
Zbigniew Huzar	Wrocław University of Technology, Poland
Mauro Iacono	University of Campania Luigi Vanvitelli, Italy

Jacek Izydorczyk	Silesian University of Technology, Poland
Sergej Jakovlev	University of Klaipeda, Lithuania
Agnieszka Jakóbik	Cracow University of Technology, Poland
Jürgen Jasperneite	Ostwestfalen-Lippe University of Applied Sciences, Germany
Krzysztof Juszczyszyn	Wrocław University of Science and Technology, Poland
Anna Kamińska-Chuchmała	Wrocław University of Science and Technology, Poland
Jerzy Klamka	IITiS Polish Academy of Sciences, Gliwice, Poland
Wojciech Kmiecik	Wrocław University of Science and Technology, Poland
Grzegorz Kołaczek	Wrocław University of Science and Technology, Poland
Ivan Kotuliak	Slovak University of Technology in Bratislava, Slovakia
Zbigniew Kotulski	Warsaw University of Technology, Poland
Demetres D. Kouvatsos	University of Bradford, UK
Stanisław Kozielski	Silesian University of Technology, Poland
Henryk Krawczyk	Gdańsk University of Technology, Poland
Udo R. Krieger	Otto-Friedrich-University Bamberg, Germany
Michał, Kucharzak	Wrocław University of Technology, Poland
Mirosław Kurkowski	Police Academy in Szczytno, Poland
Piotr Lech	West-Pomeranian University of Technology, Poland
Piotr Lechowicz	Wrocław University of Technology, Poland
Ricardo Lent	University of Houston, USA
Jerry Chun-Wei Lin	Western Norway University of Applied Sciences, Norway
Zbigniew Lipiński	University of Opole, Poland
Wolfgang Mahnke	ascolab GmbH, Germany
Francesco Malandrino	Polytechnic University of Turin, Italy
Aleksander Malinowski	Bradley University, USA
Marcin Markowski	Wrocław University of Science and Technology, Poland
Przemysław Mazurek	West-Pomeranian University of Technology, Poland
Agathe Merceron	Beuth University of Applied Sciences, Germany
Jarosław Miszczak	IITiS Polish Academy of Sciences, Poland
Vladimir Mityushev	Pedagogical University of Cracow, Poland
Jolanta Mizera-Pietraszko	Opole University, Poland
Evsey Morozov	Petrozavodsk State University, Russia
Włodzimierz Mosorow	Lodz University of Technology, Poland
Sasa Mrdovic	University of Sarajevo, Bosnia and Herzegovina
Mateusz Muchacki	Pedagogical University of Cracow, Poland
Gianfranco Nencioni	University of Stavanger, Norway
Sema F. Oktug	Istanbul Technical University, Turkey
Remigiusz Olejnik	West Pomeranian University of Technology, Poland

Michele Pagano	University of Pisa, Italy
Nihal Pekergin	University Paris-Est Créteil, France
Maciej Piechowiak	University of Kazimierz Wielki in Bydgoszcz, Poland
Piotr Pikiewicz	College of Business in Dąbrowa Górnicza, Poland
Jacek Piskorowski	West Pomeranian University of Technology, Poland
Bolesław Pochopień	Silesian University of Technology, Poland
Oksana Pomorova	Khmelnitsky National University, Ukraine
Sławomir Przyłucki	Lublin University of Technology, Poland
Tomasz Rak	Rzeszow University of Technology, Poland
Stefan Rass	University of Klagenfurt, Austria
Stefano Rovetta	University of Genoa, Italy
Przemysław Ryba	Wrocław University of Science and Technology, Poland
Vladimir Rykov	Russian State Oil and Gas University, Russia
Wojciech Rząsa	Rzeszow University of Technology, Poland
Dariusz Rzońca	Rzeszow University of Technology, Poland
Alexander Schill	TU Dresden, Germany
Olga Siedlecka-Lamch	Czestochowa University of Technology, Poland
Artur Sierszeń	Lodz University of Technology, Poland
Mirosław Skrzewski	Silesian University of Technology, Poland
Adam Słowik	Koszalin University of Technology, Poland
Pavel Smolka	University of Ostrava, Poland
Tomas Sochor	University of Ostrava, Czech Republic
Maciej Stasiak	Poznań University of Technology, Poland
Ioannis Stylios	University of the Aegean, Greece
Grażyna Suchacka	Opole University, Poland
Wojciech Sułek	Silesian University of Technology, Poland
Zbigniew Suski	Military University of Technology, Poland
Bin Tang	California State University, USA
Kerry-Lynn Thomson	Nelson Mandela Metropolitan University, South Africa
Oleg Tikhonenko	Cardinal Stefan Wyszynski University, Poland
Ewaryst Tkacz	Silesian University of Technology, Poland
Homero Toral Cruz	University of Quintana Roo, Mexico
Mauro Tropea	University of Calabria, Italy
Leszek Trybus	Rzeszów University of Technology, Poland
Kurt Tutschku	Blekinge Institute of Technology, Sweden
Selda Uyanik	Istanbul Technical University, Turkey
Adriano Valenzano	National Research Council of Italy, Italy
Bane Vasic	University of Arizona, USA
Peter van de Ven	CWI – Centrum Wiskunde & Informatica, The Netherlands
Miroslaw Voznak	VSB-Technical University of Ostrava, Czech Republic
Sylwester Warecki	Broadcom Ltd., USA
Jan Werewka	College of Economics and Computer Science, Poland

Tadeusz Wieczorek	Silesian University of Technology, Poland
Lukasz Wisniewski	OWL University of Applied Sciences, Germany
Przemysław Włodarski	West Pomeranian University of Technology, Poland
Józef Woźniak	Gdańsk University of Technology, Poland
Hao Yu	Xilinx, USA
Grzegorz Zareba	University of Arizona, USA
Krzysztof Zatwarnicki	Opole University of Technology, Poland
Zbigniew Zieliński	Military University of Technology, Poland
Liudong Zuo	California State University, USA
Piotr Zwierzykowski	Poznań University of Technology, Poland

Referees

Davide Adami	Jacek Izydorczyk	Sławomir Przyłucki
Tülin Atmaca	Sergej Jakovlev	Tomasz Rak
Rajiv Bagai	Agnieszka Jakóbik	Stefano Rovetta
Jiří Balej	Jürgen Jasperneite	Przemysław Ryba
Zbigniew Banaszak	Anna	Dariusz Rzońca
Robert Bestak	Kamińska-Chuchmała	Alexander Schill
Tomasz Bilski	Ivan Kotuliak	Olga Siedlecka-Lamch
Leoš Bohac	Zbigniew Kotulski	Mirosław Skrzewski
Maria Carla Calzarossa	Stanisław Kozielski	Pavel Smolka
Valeria Cardellini	Henryk Krawczyk	Tomas Sochor
Emiliano Casalicchio	Udo R. Krieger	Maciej Stasiak
Amlan Chatterjee	Mirosław Kurkowski	Ioannis Stylios
Andrzej Chydziński	Andrzej Kwiecień	Grażyna Suchacka
Tadeusz Czachórski	Piotr Lech	Wojciech Sułek
Dariusz Czerwiński	Piotr Lechowicz	Oleg Tikhonenko
Adam Czubak	Ricardo Lent	Ewaryst Tkacz
Alexander N. Dudin	Marcin Markowski	Homero Toral Cruz
Peppino Fazio	Przemysław Mazurek	Leszek Trybus
Max Felser	Agathe Merceron	Kurt Tutschku
Holger Flatt	Jarosław Miszczak	Adriano Valenzano
Jean-Michel Fourneau	Vladimir Mityushev	Peter van de Ven
Rosario G. Garroppo	Jolanta Mizera-Pietraszko	Sylwester Warecki
Natalia Gaviria	Evsey Morozov	Jan Werewka
Roman Gielerak	Włodzimierz Mosorow	Tadeusz Wieczorek
Mariusz Głąbowski	Mateusz Muchacki	Lukasz Wisniewski
Anna	Gianfranco Nencioni	Przemysław Włodarski
Grocholewska-Czuryło	Remigiusz Olejnik	Józef Woźniak
Daniel Grzonka	Michele Pagano	Krzysztof Zatwarnicki
Sebastien Harispe	Maciej Piechowiak	Zbigniew Zieliński
Artur Hłobaż	Piotr Pikiewicz	
Mauro Iacono	Jacek Piskorowski	

Sponsoring Institutions

Organizer

Institute of Informatics, Faculty of Automatic Control, Electronics and Computer Science, Silesian University of Technology

Co-organizer

Committee on Informatics of the Polish Academy of Sciences, the Section of Computer Networks and Distributed Systems

Official, Industrial Partner

Bombardier Transportation

Technical Partner

Technical Co-sponsor

IEEE Poland Section

Conference Partner

iNEER

Contents

Communications

Computer Networks

Performance Modeling of the Consensus Mechanism in a Permissioned Blockchain

Udo R. Krieger[1(✉)], Michael H. Ziegler[1], and Hendrik L. Cech[2]

[1] Fakultät WIAI, Otto-Friedrich-Universität,
An der Weberei 5, 96047 Bamberg, Germany
udo.krieger@ieee.org
[2] Fakultät Informatik, Technische Universität München,
85748 Garching, Germany

Abstract. We consider a permissioned blockchain and analyze the dissemination and commitment processes of blocks among its corresponding miner nodes in the underlying peer-to-peer network. We propose a Markovian non-purging (n, k) fork-join queueing model to analyze the delay performance of the synchronization process among these miner nodes that apply a vote-based consensus procedure. We determine the impact of the most influential design and load parameters on the resulting commitment delay of new blocks that are appended to the blockchain after successful commitment decisions and the approval by the fully distributed consensus procedure. The proposed analysis of a permissioned blockchain is illustrated by means of a simple example of a fully interconnected P2P graph applying mean-value analysis techniques.

Keywords: Blockchain · Consensus mechanism ·
Performance modeling · Fork-join queueing network ·
Mean-value analysis

1 Introduction

The *Internet-of-Things* (IoT) describes the fundamental paradigm shift of enhancing previously analog devices and their associated gateways to the Internet with effective computing, storage and networking capabilities (cf. [1,2,15]). Combined with the paradigm of fog and edge computing, the Internet-of-Things will provide the fast growing basis for new, rapidly evolving application scenarios of the blockchain technology (cf. [5,11,12,15,22]). A fog computing architecture integrating the basic functionality of a scalable blockchain framework like Plasma or Multichain [27] constitutes the major motivation of our research regarding the performance analysis of the blockchain technology (cf. [9,25]).

In recent years, major research efforts have been devoted to public-permissionless blockchains such as Bitcoin [17] or Ethereum [6] and the basic functionality of the involved consensus protocols (cf. [20]). The latter process is

P. Gaj et al. (Eds.): CN 2019, CCIS 1039, pp. 3–17, 2019.
https://doi.org/10.1007/978-3-030-21952-9_1

governed by a block validation. It is influenced by the properties of the underlying public-key cryptographic system that is controlling the pseudonymous interactions among the clients based on their emitted transactions. It depends also on the executed functionality of the communicating partners that are associated with these cryptographically signed transactions and smart contracts in the related peer-to-peer (P2P) overlay network of a blockchain. This distributed dissemination and synchronization processes regarding the aggregation of transactions into so-called blocks and the associated validation of transactions and their blocks by the subset of all mining nodes, called the miners or validators, constitute core elements of the distributed database functionality. It is applied to store the non-immutable transaction history in the blockchain.

Considering the application of blockchain technology in advanced Internet-of-Things scenarios, a permissioned blockchain such as Hyperledger Fabric [3] or Multichain [27] that use different variants of a lightweight vote-based consensus scheme constitutes a more adequate alternative to a permissionless blockchain with its heavyweight protocol (cf. [12,23]). Regarding the performance of the blockchain dynamics and thereby induced vulnerabilities due to long-tailed communication delays, only a few studies have applied a queueing-theoretical framework to analyze the underlying consensus mechanism and its implications on the blockchain (see [13,18,21]). The study of Göbel et al. [14] represents the most prominent realization of such a theoretical approach. We intend to investigate this response time issue studied in [14] in a more detailed manner. For this purpose we analyze the dissemination and commitment processes of blocks among the authenticated miner nodes in the peer-to-peer network associated with the fundamental transaction layer of a permissioned blockchain. It is assumed that the latter applies a vote-based consensus procedure which is inspired by the Practical Byzantine Fault Tolerance (PBFT) scheme [8].

We propose a Markovian non-purging (n, k) fork-join queueing model to analyze the response time of the block validation and synchronization processes among all mining nodes that apply such a vote-based procedure. We determine some relevant design and load parameters influencing the resulting commitment delay of new blocks that are appended to the blockchain view of the transaction history after a successful approval by the distributed consensus procedure. The proposed analysis is illustrated by means of a simple example of a fully interconnected P2P graph in a permissioned blockchain applying mean-value analysis techniques.

The paper is organized as follows. In Sect. 2 the fundamental properties of public-permissionless and permissioned blockchains applying a Proof-of-Work or vote-based procedure for issued transactions are briefly discussed. Section 3 presents the performance modeling and analysis of the consensus mechanism in a vote-based blockchain. In Subsects. 3.1 and 3.2 we derive a fork-join queueing network to describe the distributed decision processes associated with blocks and analyze its performance. In Subsect. 3.3 a simple example of a P2P network of miners is used to illustrate our analysis approach. In Sect. 4 the discussion is finalized by some conclusions and a brief outlook on future work.

2   Fundamental Properties of a Blockchain

A blockchain is a decentralized database which is distributed via replication among its contributing peers. Data are not directly updated, but a collection of change records represented by transactions are appended as aggregated entities called blocks (see Fig. 1). All participating peers can validate the state changes and must agree on the state of those data stored in the associated data plane. Thus, a blockchain represents the immutable history of state changes triggered by the transactions of the peers up to the most recent block. The blockchain technology has been introduced a decade ago to realize a truly decentralized digital currency called *Bitcoin* in a peer-to-peer interaction mode among its users without the need of a centralized authority (cf. [17]). In recent years the blockchain concept has been extended enormously and applied to diverse areas such as logistics, manufactoring, or the aforementioned, rapidly evolving Internet-of-Things (IoT) applications (cf. [1,2,6,7,11]).

From a technical point of view, the distributed control model of a blockchain like Bitcoin [17] or Ethereum [6] organizes the interworking of a set of identifiable, interconnected peers that belong to the underlying peer-to-peer (P2P) network of this transaction-oriented system architecture according to three basic features:

- *Decentralized architecture:* The blockchain functionality is executed by a set of independent peers that can dynamically join and leave the underlying P2P network. These peers can be maintained by different entities that do not even need to know their identities or intentions.
- *Fundamentally anonymous interaction:* The functionality of a permissionless blockchain allows peers to participate in the P2P network without identification, while a permissioned blockchain requires an authentication. Read access is not recorded and writing data to the storage plane of a blockchain is supported by a pseudonymity concept to protect the privacy of the interacting peers.
- *Stability guarantees:* Peers participating in the interactions can store data in the storage plane of this P2P network and can be assured that these data will not be manipulated, even if they do not trust other peers.

Peers of a blockchain share new cryptographically signed transactions with the whole P2P overlay. For this purpose an aggregation of a finite set of transactions into a block is applied using hashed links and a Merkle root tree technique, see Fig. 1 (cf. [6,17]). After a verification step to validate the information of a generated block each peer stores all transactions recently created by that block and then restarts to gather transactions for the next block, see Fig. 2. The goal of negotiating a common history of blocks among the peers realizes a complex *consensus* issue in the associated distributed agreement model.

The choice of an efficient consensus mechanism coincides with the access model of a peer in a blockchain. Bitcoin [17] is the prime example of an open, *unpermissioned*, i.e. *permissionless*, P2P network. Access to the data in the network or participation in the consensus process are not restricted. In the alternative *permissioned* blockchain, e.g. Hyperledger Fabric [3], the access to the

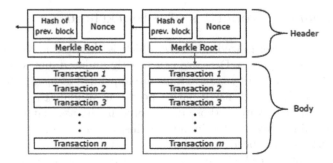

Fig. 1. A block header is indirectly dependent on the transactions by a Merkle root, and including the hash of each previous block in the current block header, a linking among the blocks is achieved.

P2P network requires authentication and authorization. The latter can be realized by well-known techniques, e.g., public-key cryptography, shared secrets, or white listed IP addresses. Regarding many IoT application scenarios, e.g. certain health-care applications, this approach is more appropriate to reduce the complexity of the transaction management in the blockchain.

The consensus mechanisms of a blockchain can be divided into *proof-based* and *vote-based* schemes (cf. [7,20]). Bitcoin [17], for instance, applies an expensive proof-based consensus mechanism called Proof-of-Work (PoW), whereas the Byzantine Fault Tolerant (BFT) replication, i.e., a solution to the *Byzantine*

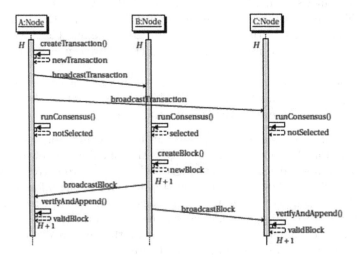

Fig. 2. Sequence diagram of an interaction between peers A, B, C where new transactions are created. A new block is created by a selected mining peer $B \equiv j \in V$ that sends it to all connected peers, which in turn verify the block and append it to their blockchain replica such that the block height H is increased, respectively.

generals problem subject to malicious nodes, constitutes a vote-based scheme (cf. [16,23]). The algorithms Practical Byzantine Fault Tolerance (PBFT) [8], BFT-SMaRt [4], and Delegated Byzantine Fault Tolerance (dBFT) are widely applied representatives of this latter class. Crash-tolerant protocols that protect the distributed system of a blockchain only against crashed but honest nodes constitute a related weaker class of consensus protocols. An extensive overview of these different types of the consensus mechanisms can be found in [20, Table 3].

Here we restrict our attention to the consensus scheme of a permissioned, vote-based blockchain in an IoT scenario. We focus on the transaction processing by those authenticated voting peers of the overlay, called mining peers, miner nodes or validators, that are selected and authorized to propose the adherence of a new block to the data plane of the blockchain, see Fig. 3.

3 Performance Analysis of the Consensus Mechanism

3.1 A Fork-Join Queueing Model of the Consensus Protocol

We consider a permissioned blockchain that applies a vote-based consensus process for the transactions emitted by the peers of the blockchain. The associated P2P network of miners comprises authenticated nodes of the blockchain that process the incoming transactions and execute the validation function to decide on a block index for the next block based on cryptographic functions. The latter include a set of validated transactions and their hashes aggregated into a Merkle tree. Finally, a validated new block is generated and appended to the blockchain, see Figs. 1 and 2. In the following we assume that the associated P2P network of authenticated miners comprises n nodes governing fog cells, see Fig. 3.

We model the associated P2P network of these n miners by an undirected, weighted graph $G = (V, E, c)$ with nodes $V = \{0, 1, \ldots, n - 1\}$ and describe

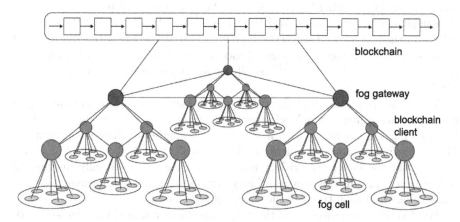

Fig. 3. P2P structure of a permissioned blockchain in a fog computing architecture with miner nodes (blue) running at fog gateways and blockchain clients (red) managing IoT devices (green) in fog cells. (Color figure online)

the edges $e = (i, j) \in E$ in terms of the corresponding symmetric adjacency matrix $A = A(G) \in \mathbb{R}^{n \times n}$ of the miner graph G. We assume that the graph is strongly connected and A is an irreducible matrix. The entry $A_{(i,j)} = c(i, j) \geq 0$, $i, j \in V$, indicates the length of the underlying direct path between two miners $i \in V, j \in V$ according to the number of the logical links at the IP-network layer (or physical links) in the P2P network.

Considering a starting node $i \in V$, e.g. $i = 0$, we construct the corresponding shortest paths $p(i, j)$ of lengths

$$d(i, j) = (A^{k(i,j)})_{(i,j)}, \quad k(i, j) = \mathrm{argmin}_{k \geq 1}\{k \in \mathbb{N} : (A^k)_{(i,j)} > 0\}$$

to all its reachable neighbors $j \neq i, j \in V$, in the transitive closure of the reachability relation E among these nodes generated by A. We assume that $0 < \varepsilon(i) \leq d$ is the eccentricity of node $i \in V$, i.e. the greatest distance between i and any other miner $j \neq i$, and $0 < d = \max_{i \in V} \varepsilon(i)$ is the diameter of the P2P network.

Without loss of generality, we arrange the resulting associated reachability tree $\tau(i)$ of a considered miner $i = 0$ in such a way that we order all paths $p(i, j)$ from i to all nodes $j \neq i$ in ascending order of the length $d(i, j)$ starting with an initial path (i, i) of length zero as leftmost entry. It means that we construct the partition of V according to the level structure of its mining graph G subject to the selected initial miner node $i = 0$ which can be determined by an enhanced breadth-first search procedure in G. We assume that equal length paths to a certain peer j are randomly resolved and the resulting tree $\tau(i)$ has $n - 1$ paths of monotone increasing orders $0, 1, \ldots, \varepsilon(i) \leq d$ to all $j \in V \setminus \{i\}$. It means we assume a geodetic graph with a selection of a unique shortest path from an initial miner $i = 0$ to all other mining peers $j \neq i$ in the blockchain.

We describe the transactions generated by node $i = 0$ that are sent to a considered miner $j \neq i$ and aggregated by it into a cluster structure of blocks in terms of a point process $\{B_i^{(j)} : j \geq 0\}$ of an associated class $C_i = \{i\} \subset V$. We assume that the resulting block traffic is a Poisson process with rate $\lambda_i = \lambda$ with class-dependent mean interarrival time $1/\lambda_i$. The latter traffic of these blocks $B_i^{(j)}$ and their inherent transactions $\{T_i^{(k(j),l)} : 1 \leq l \leq k(j)\}$ is validated and approved by a vote-based consensus process in each miner after a broadcast of the related transactions $T_i^{(k(j),l)}$ to all corresponding miner nodes $j \in V$, see Fig. 2.

These disseminations and the associated validation processes from an initial node $i = 0$ according to the level structured mining tree $\tau(i)$ are described by a fork-join queueing network with $n = \varepsilon(i) \leq d$ branches of single-server stations $P_1, \ldots P_n$ with station-dependent service rates $\mu_k, 1 \leq k \leq n$, see Fig. 4. We assume that the latter are arising from independent, exponentially distributed service times. This model captures the aggregated transmission, propagation and transaction validation times of blocks along the tree $\tau(i)$.

We describe the processes of the cryptographically signed transfer and verification from the initial miner i to a node $j \neq i$ on the forward path and on the backward path by an infinite server queue with exponentially distributed service times $S_{(i,j)}, S_{(j,i)}$ with common mean $\mathbb{S}_{(i,j)} = d(i, j)/\nu = \mathbb{S}_{(j,i)} = d(j, i)/\nu$, respectively. The latter are accumulated to the exponentially distributed service time for blocks $B_i^{(j)}$ of the transaction traffic from i with mean $\mathbb{S}_{(j,j)} = \chi_{ji}/\eta_j$ and the related waiting time for processing, i.e. the resulting sojourn time $R_{(j,j)}$ in the queue P_j of miner j. This mean $\mathbb{S}_{(j,j)}$ of the block traffic of class C_i generated by a mining peer i is arising from the local validation and block creation procedures applied by the addressed miner node j. The sharing coefficient of block traffic class C_i at node j is given by $\chi_{ji} = \chi_j = 1/n$ and reflects the uniform proportion $\chi_j = 1/n$ of the service rate η_j of node j that is assigned to class C_i among all n traffic classes. We can model the time difference between competing consensus strategies by a real scaling term $s \in (0,1)$ and then replace η_j in terms of a scaling dependent service rate $\eta_j(s) = \widehat{\eta}_j/s$.

Using Norton's theorem (cf. [10]) or related mean-value approximations, we can approximate the resulting Erlang distribution $\widehat{R}_{(i,j)} = S_{(i,j)} + R_{(j,j)} + S_{(j,i)}$ including the transfer-verification delays $S_{(i,j)}, S_{(j,i)}$ of the forward and backward paths from i to j and the response time $R_{(j,j)}$ of the miner j by an exponential distribution with the same mean $1/\mu_j = 1/\mu$. We approximate this three station network of two infinite server queues of the combined transfer-verification delays and the single-server queue of the block stream processing at miner j by a (λ, μ)-equivalent single server with identical mean response time $\widehat{\mathbb{R}}_{(i,j)} = \mathbb{E}(\widehat{R}_{(i,j)})$, i.e.

$$\widehat{\mathbb{R}}_{(i,j)} = 2\frac{d(i,j)}{\nu} + \frac{\chi_j \cdot s/\widehat{\eta}_j}{1 - n \cdot \lambda \cdot s/\widehat{\eta}_j} = \frac{1}{\mu - \lambda}, \quad \mu = 1/\widehat{\mathbb{R}}_{(i,j)} + \lambda \qquad (1)$$

subject to a scaling of the block service rate $\eta_j = \widehat{\eta}_j/s, s \in (0,1)$.

In the following we consider a vote-based consensus process of the blocks by the involved P2P network G with $n = |V|$ miner nodes. We assume that $k = n - \lfloor (n-1)/3 \rfloor \approx 2/3 \cdot n < n$ nodes must agree before a block is approved by the miners. This approval is modelled by a non-purging fork-join process, see Fig. 5 regarding $n = 3, k = 2$ (cf. [24]).

3.2 Analysis of the Fork-Join Model of a Vote-Based Consensus Protocol

We intend to analyze the impact of broadcasting transactions within the P2P network of authenticated miners and the effect of a vote-based consensus protocol on the approval of blocks in the blockchain. To study the effect of the design and load parameters on the performance metrics of the consensus protocol, such as the scalability in terms of the number of mining nodes n and the delay spreading during the dissemination of transactions and block approval messages within the P2P network due to the distance structure, we assume here that all miners $j \in V$ have to handle the same total load of block arrivals as result of the aggregated shared information on the transactions, see Fig. 2. The latter is given in terms of a uniform arrival rate $\widehat{\lambda}_j = \lambda_S(n) = \sum_{i=1}^{n} \lambda_i = |V| \cdot \lambda$ due to broadcasting. We conclude that the block arrival rate on each node j is given by $\lambda_S(n) = n \cdot \lambda$. Moreover, we assume a common service rate $\eta_j = \widehat{\eta}$ for the block generation and validation processes of each miner node j. Then the proportion dedicated at miner j to each service class C_i determined by the n block streams of the different miners is given by $\chi_{ji} = \chi_j = 1/n$ and yields an individual service rate $\eta_{ji} = 1/\chi_j \cdot \eta_j = \hat{\eta}/n > 0$ per block stream $\{B_i^{(j)} : j \geq 1\}$ of a selected miner node i.

We further assume that the P2P mining network is designed in such a way that a homogeneous connectivity structure of the underlying graph G is established which yields a uniform level structure of all associated trees $\tau(i)$ with $n-1$ branches of equal length $\varepsilon(i) = d(i,j) = d \geq 1$ to all nodes $j \neq i$.

We follow Wang et al. [24] to analyze the delay performance of the derived non-purging (n,k) fork-join model. Therein pending jobs are not flushed when the consensus is reached by a completion of k out of n jobs which are traveling along the n different queues P_i of the model, see Figs. 4 and 5. As discussed in the previous Subsect. 3.1, this model is derived from the vote-based consensus mechanism which is applied to the block traffic of a considered miner $i = 0$ along its mining tree $\tau(i)$.

The special case of a classical (n,n) fork-join queue is symmetrical in the sense that all n queues P_i in its branches are interchangeable, see Fig. 4 (cf. [24]). Therefore, any $k < n$ arbitrarily chosen queues $\{P_{i_1}, \ldots, P_{i_k}\}$ generate the same joint probability distribution and their resulting stable sojourn times are jointly-identical random variables (cf. [24]). The basic fork-join queue and the non-purging (n,k) fork-join queue under the same job arrival process of rate λ and with the same P_i-queue's service time distribution with uniform rate μ are called (λ, μ)-equivalent queues, see Fig. 5 (cf. [24]). We use these analytic means to compute the delay metrics of the (n,k) fork-join queue of the consensus mechanism.

We compute the first passage time Σ_i until $k = \lfloor 2/3 \cdot n \rfloor \in \mathbb{N}$ miners $j \in V \setminus \{i\}$ have returned their affirmative approval result to the initiating node $i = 0$ in the fork-join queueing network with its n single-server queues P_i of common service rate μ. Due to Wang et al. [24], this time Σ_i coincides with the sojourn time $t_{(n,k)}$ of a general non-purging (n,k) fork-join queue. Applying results on a

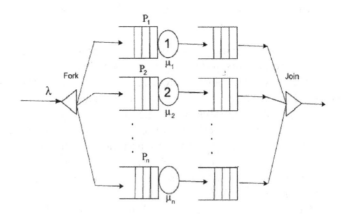

Fig. 4. Fork-join queueing model.

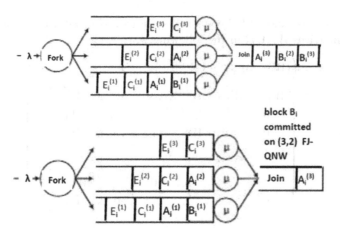

Fig. 5. Non-purging fork-join queueing model with a uniform service rate μ and an approval by $k = 2$ out of $n = 3$ nodes for block sequences $(A_i), (B_i), (C_i), (E_i)$.

linear transformation of the kth order statistics $X_{(n,k)}$ of independent, identically distributed random variables $\mathcal{X} = \{X_1, \ldots, X_n\}$ and the representation of the distribution function of the maximum $M_k = \max\{X_1, \ldots, X_k\}$ by means of coefficients $\{A_i^{(n,k)} : i = 1, \ldots, n\}$, it can be represented by a linear combination of the sojourn times $t_{(j,j)}$ of the underlying (λ, μ)-equivalent basic fork-join queues in the following way (cf. [24]):

$$t_{(n,k)} = \sum_{j=k}^{n} W_j^{(n,k)} \cdot t_{(j,j)} \quad 1 \le k \le n \tag{2}$$

$$W_i^{(n,k)} = \sum_{j=k}^{i} \binom{n}{j} \cdot A_i^{(n,j)} \quad 1 \le k \le n, \quad 1 \le i \le n \tag{3}$$

$$A_i^{(n,k)} = \begin{cases} 1, & i = k \\ -\sum_{j=1}^{i-k} \binom{n-i+j}{j} \cdot A_{i-j}^{(n,k)} & k+1 \le i \le n \end{cases} \tag{4}$$

The recursion (4) of the coefficients $A_i^{(n,k)}, 1 \le k < n$, can be simplified in the following manner:

$$A_{k+l}^{(n,k)} = \begin{cases} 1, & l = 0 \\ -\sum_{j=0}^{l-1} \binom{n-k-j}{n-k-l} \cdot A_{k+j}^{(n,k)} & 1 \le l \le n - k \end{cases} \tag{5}$$

Using the $W^{(n,\cdot)}$- and $A^{(n,\cdot)}$-coefficients in (3), (5), we can define upper-triangular matrices W, A, B in the following way:

$$W = \left((W^{(n,\cdot)})_{k,i} \right) = \left(W_i^{(n,k)} \right)$$

$$= \begin{pmatrix}
W_1^{(n,1)} & W_2^{(n,1)} & W_3^{(n,1)} & \cdots & \cdots & W_n^{(n,1)} \\
0 & W_2^{(n,2)} & W_3^{(n,2)} & \cdots & \cdots & W_n^{(n,2)} \\
0 & 0 & W_3^{(n,3)} & \cdots & \cdots & W_n^{(n,3)} \\
\vdots & \cdots & \ddots & \ddots & \cdots & \vdots \\
\vdots & \cdots & \cdots & 0 & W_{n-1}^{(n,n-1)} & W_n^{(n,n-1)} \\
0 & \cdots & \cdots & \cdots & 0 & W_n^{(n,n)}
\end{pmatrix} \tag{6}$$

$$A = \left((A^{(n,\cdot)})_{k,i} \right) = \left(A_i^{(n,k)} \right)$$

$$= \begin{pmatrix}
A_1^{(n,1)} & A_2^{(n,1)} & A_3^{(n,1)} & \cdots & \cdots & A_n^{(n,1)} \\
0 & A_2^{(n,2)} & A_3^{(n,2)} & \cdots & \cdots & A_n^{(n,2)} \\
0 & 0 & A_3^{(n,3)} & \cdots & \cdots & A_n^{(n,3)} \\
\vdots & \cdots & \ddots & \ddots & \cdots & \vdots \\
\vdots & \cdots & \cdots & 0 & A_{n-1}^{(n,n-1)} & A_n^{(n,n-1)} \\
0 & \cdots & \cdots & \cdots & 0 & A_n^{(n,n)}
\end{pmatrix} \tag{7}$$

$$B = \left((B^{(n,\cdot)})_{k,i} \right) = ((B_i)_{i \geq k}) = \begin{pmatrix}
B_1 & B_2 & B_3 & \cdots & \cdots & B_n \\
0 & B_2 & B_3 & \cdots & \cdots & B_n \\
0 & 0 & B_3 & \cdots & \cdots & B_n \\
\vdots & \cdots & \ddots & \ddots & \cdots & \vdots \\
\vdots & \cdots & \cdots & 0 & B_{n-1} & B_n \\
0 & \cdots & \cdots & \cdots & 0 & B_n
\end{pmatrix} \tag{8}$$

$$B_i = \binom{n}{i} \tag{9}$$

Then the $W^{(n,\cdot)}$-coefficients can be computed in terms of the B_i- and $A^{(n,\cdot)}$-coefficients by a simple matrix multiplication of two upper-triangular matrices:

$$W = B \cdot A \tag{10}$$

The expected sojourn time $\mathbb{T}_{(n,k)} = \mathbb{E}(t_{(n,k)})$ of a general non-purging (n,k) fork-join queue can be represented by a linear combination of the expected sojourn times of the (λ, μ)-equivalent basic fork-join queues $P_i, 1 \leq i \leq n$, in the following way (cf. [24, Theorem 4]):

$$\mathbb{T}_{(n,k)} = \sum_{i=k}^{n} W_i^{(n,k)} \mathbb{T}_i \tag{11}$$

Here $\mathbb{T}_i = \mathbb{E}(t_{(i,i)})$ is the expected sojourn time of the (λ,μ)-equivalent basic (i,i) fork-join queue and $W_i^{(n,k)}$ are the corresponding W-coefficients given by (3).

In the special case of an exponentially distributed service time distribution with a uniform service rate μ in each branch of the fork-join network, we can apply Nelson's approximation $\mathbb{T}^*_{(n,k)}$ of the expected mean sojourn time $\mathbb{E}(t_{(n,k)})$ (cf. [24, Theorem 5], [19]).

$$\mathbb{T}^*_{(n,k)} = \max\{\widehat{\mathbb{T}}_{(n,k)}, 0\} \tag{12}$$

$$W_S^{(n,k)} = \sum_{i=\max\{2,k\}}^{n} W_i^{(n,k)} \left[\frac{11H_i + 4\rho(H_2 - H_i)}{H_2}\right] k \in \{1,\ldots,n\} \tag{13}$$

$$\widehat{\mathbb{T}}_{(n,k)} = \begin{cases} n \cdot \frac{1/\mu}{1-\rho} + \frac{12-\rho}{88} \cdot \frac{1/\mu}{1-\rho} \cdot W_S^{(n,1)} & k = 1 \\ \frac{12-\rho}{88} \cdot \frac{1/\mu}{1-\rho} \cdot W_S^{(n,k)} & k \geq 2 \end{cases} \tag{14}$$

to the (n,k) fork-join queue. Here $H_i = \sum_{k=1}^{i} \frac{1}{k}$ are the harmonic numbers, λ is the single arrival rate of all queues P_i and $\rho = \lambda/\mu$ is the corresponding load. This approximation is based on the mean response time $\mathbb{E}(R_{M/M/1}) = \frac{1/\mu}{1-\rho} = (\mu - \lambda)^{-1}$ of an M/M/1 queue.

3.3 Performance Evaluation of an Illustrative P2P Network

To investigate the scalability and commitment delay of the vote-based consensus protocol applied in a permissioned blockchain, we consider a fully interconnected P2P network of n miner nodes. They are integrated into a fog computing architecture as fog gateways controlling corresponding fog cells and their blockchain clients (cf. Fig. 3). To illustrate the mean-value analysis approach, we consider here only a small, simple configuration scenario of two interconnected routers R_1, R_2 that couple several fog cells arranged in two clusters with a population of n miners where $\lfloor n/2 \rfloor$ peers in a fog cell cluster are attached to each router R_i. It is assumed that there is a negligible link delay between the latter routers in the P2P network compared to the access delay by the peers.

Given a simple example with $n = 6$ peers, three attached nodes Φ_1, Φ_2, Φ_3 and Φ_4, Φ_5, Φ_6, respectively, interact as mining peers of this blockchain in each cell cluster. The related miner graph $G = (V, E)$, $V = \{0, \ldots, 5\}$, has an associated irreducible, block-structured adjacency matrix:

$$A(G) = \begin{pmatrix} 0 & 2 & 2 & 2 & 2 & 2 \\ 2 & 0 & 2 & 2 & 2 & 2 \\ 2 & 2 & 0 & 2 & 2 & 2 \\ 2 & 2 & 2 & 0 & 2 & 2 \\ 2 & 2 & 2 & 2 & 0 & 2 \\ 2 & 2 & 2 & 2 & 2 & 0 \end{pmatrix} = \begin{pmatrix} A_1 & 2E_3 \\ 2E_3 & A_1 \end{pmatrix}, \quad A_1 = \begin{pmatrix} 0 & 2 & 2 \\ 2 & 0 & 2 \\ 2 & 2 & 0 \end{pmatrix}, \quad E_3 = \begin{pmatrix} 1 & 1 & 1 \\ 1 & 1 & 1 \\ 1 & 1 & 1 \end{pmatrix}$$

We consider the representative traffic of blocks generated by the peer $\Phi_0 \equiv 0$ to all other peers $\{\Phi_1, \ldots, \Phi_5\} \equiv \{1, \ldots, 5\} = V \setminus \{0\}$. We assume that the rate of this generated block traffic is given uniformly by $\lambda_i = \lambda > 0$ for each peer $i \in V$. Due to the broadcasting of the blocks of class i from the considered peer $i = 0$ to all others $j \neq i$, a representative miner, e.g. $j = 2$, will get additionally block traffic with a rate $\lambda_S(n-1) = (n-1)\lambda$, i.e. in total it has to process $\lambda_S(n) = n\lambda$ blocks per time unit t_u.

We assume a block service rate $\mu_j = \mu$ at each node $j \in V$ related to $\widehat{\mathbb{R}}_{(i,j)} = 2d(i,j)/\nu + \frac{\chi_j/\eta_j}{1-\lambda_S(n)/\eta_j}, \chi_j = 1/n$, which captures the mean link delay $\mathbb{S}_{(i,j)} = \mathbb{E}(S_{(i,j)}) = d(i,j)/\nu$ and link distance $d(i,j) = d(0,2) = 2$ between the sender $i = 0$ of the transaction set and the receiving miner $j = 2$ processing these items in terms of related blocks (see Fig. 2).

Inspired by previous measurements (cf. [13]), we assume a block traffic rate $\lambda < 1/500$ per time unit $t_u = 1\,\mathrm{s}$ and a mean block service time of $\mathbb{E}(S_{(j,j)}) = \chi_j/\eta_j(s) = \chi_j \cdot s/\widehat{\eta}_j = s/n \cdot 20\ t_u$'s. It includes a scaling term $s \in (0,1) \subset \mathbb{R}$ that can model, for instance, the time difference between competing consensus strategies. The mean delay on a link including transmission, propagation, and transaction validation times is supposed to be $\mathbb{S}_{(i,j)} = d(i,j)/\nu = 500\,\mathrm{s}$.

We consider an example with $n \in \{6, 12, 18, 24, 27\}$ peers in the miner network and the initial peer $i = 0$ where $n/2$ miners are attached to each router R_1, R_2 with a negligible transfer delay between them. Then a uniform arrival rate $\lambda_0 = 1/634$ blocks/sec, the mean block service time $1/\eta_j(s) = 20\,\mathrm{s}$ for $s = 1$, as well as the link delay value $1/\nu = 250\,\mathrm{s}$ between the miners and the unidirectional distance $d(0,j) = 2$ are applied as uniform load model to all peers $j \neq i$ in (1).

We can calculate the corresponding mean block synchronization delay $\mathbb{T}_{(n,k)}$ of the consensus mechanism by Taylor's approximation $\mathbb{T}^*_{(n,k)}$ in (12) to (14) for a required feedback by $k = 2/3 \cdot n \in \{4, 8, 12, 16, 18\}$ peers. The outcome is depicted in Fig. 6 where a parametrization in terms of the scaling factor $s \in (0,1)$ of the block processing time and a variable load $\rho = \lambda/\mu$ in the case $n = 12, k = 8, s = 1$ is used. It reveals the robustness w.r.t. scaling and a linear behavior of the mean consensus delay, as expected, due to the low utilization levels of the (λ, μ)-equivalent queues in the fork-join network of the considered examples.

An indispensable validation of the sketched performance model in a fully virtualized IoT scenario requires an implementation of a vote-based consensus protocol in a permissioned blockchain such as the Multichain [27] framework. Such a blockchain system that is integrated into a virtualized fog computing environment is currently under development in a LINUX test bed. It is based on a cluster of Raspberry Pi's with their 64-bit ARM processor architecture and Hypriot Cluster Lab (HCL) realizing a master-slave clustering by Docker Swarm as software environment (cf. [9,15,26]). However, only a few preliminary performance results regarding the basic Docker Swarm operation mode of the blockchain clients that are calling a developed Multichain Docker image running on a master node are available so far (cf. [9]). An extended measurement study regarding the realized commitment delay of the consensus protocol and its comparison with the developed performance model is a subject of our future research.

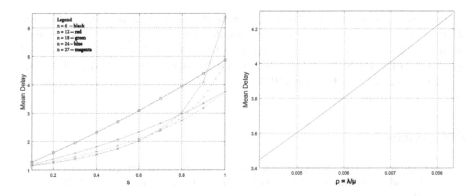

Fig. 6. Approximation $\mathbb{T}^*_{(n,k)}$ of the mean consensus delay by a (n, k)-fork-join model for $n \in \{6, 12, 18, 24, 27\}$ parametrized by the scaling factor $s \in (0, 1)$ (left-hand side) and for $n = 12, k = 8, s = 1$, parametrized by the load $\rho = \lambda/\mu$ (right-hand side).

4   Conclusions

A blockchain is a decentralized database which is distributed via replication among its contributing clients. It is governed by a dissemination of signed transaction sets and a consensus-based validation of the derived aggregated objects called blocks by the mining peers. In recent years public-permissionless blockchains such as Bitcoin [17] or Ethereum [6] and the basic functionality of the involved Proof-of-Work based consensus protocol have been intensively studied (cf. [20]). Regarding the application of blockchain technology in advanced Internet-of-Things scenarios, permissioned blockchains such as Hyperledger Fabric [3] or Multichain [27] that use different variants of a lightweight vote-based consensus protocol constitute a more adequate alternative to public-permissionless blockchains.

Considering such a permissioned blockchain with a vote-based consensus algorithm embedded in a fog computing environment, we have analyzed the dissemination and commitment processes of blocks among the corresponding mining nodes in the underlying peer-to-peer network. Inspired by Wang et al. [24], we have proposed a Markovian, non-purging (n, k) fork-join queueing model to analyze the response time performance of the block synchronization process among these mining nodes that apply a vote-oriented consensus procedure. We have determined the influencing parameters of the commitment delay of new blocks that are appended to the blockchain after a successful approval by the distributed consensus procedure.

It has been a major goal of our performance study to investigate the scalability issues of the consensus strategy and its inherent limitations by means of the derived analytic fork-join queueing model and to gain insights on the impact of all basic load parameters of the model. The proposed mean-value analysis of the fundamental delay performance metric has been illustrated by means of a simple example of a fully interconnected P2P graph arising from the interconnected clients in several fog cells.

Regarding the accuracy of the proposed approach, an indispensable validation of our performance model by extended simulations and adequate measurements in an IoT setting that is fully virtualized in terms of Docker [26] containers and operating in Docker Swarm mode such as our Raspberry Pi test bed HCL-BaFog based on Hypriot Cluster Lab (HCL) (cf. [9,15,25]) will constitute important items of our future research.

References

1. Al-Fuqaha, A., et al.: Internet of Things: a survey on enabling technologies, protocols, and applications. IEEE Commun. Surv. Tutorials **17**(4), 2347–2376 (2015). Fourth Quarter
2. Atzori, L., et al.: Internet of Things: a survey. Comput. Netw. **54**, 2787–2805 (2010)
3. Androulaki, E., et al.: Hyperledger fabric: a distributed operating system for permissioned blockchains. In: Oliveira, R., Felber, P., Hu, Y.C. (eds.) Proceedings of the Thirteenth EuroSys Conference, EuroSys 2018, Porto, Portugal, 23–26 April 2018, pp. 30:1–30:15. ACM (2018)
4. Bessani, A., Sousa, J., Alchieri, E.: State machine replication for the masses with BFT-SMART. In: 44th Annual IEEE/IFIP International Conference on Dependable Systems and Networks (DSN), pp. 355–362. IEEE (2014)
5. Brody, P., Pureswaran, V.: Device democracy: saving the future of the Internet of Things. IBM, September 2014
6. Buterin, V.: Ethereum White Paper: A Next-Generation Smart Contract and Decentralized Application Platform (2013). https://www.blockchainresearchnetwork.org/research/whitepapers/. Accessed 22 Nov 2017
7. Cachin, C., Vukolić, M.: Blockchains consensus protocols in the wild. arXiv preprint arXiv:1707.01873 (2017)
8. Castro, M., Liskov, B.: Practical Byzantine fault tolerance. In: Seltzer, M.I., Leach, P.J. (eds.) Proceedings of the Third USENIX Symposium on Operating Systems Design and Implementation (OSDI), New Orleans, Louisiana, USA, 22–25 February 1999, pp. 173–186. USENIX Association (1999)
9. Cech, H.L., Großmann, M., Krieger, U. R.: A fog computing architecture to share sensor data by means of blockchain functionality. In: 2019 IEEE International Conference on Fog Computing (ICFC 2019) (2019, accepted paper)
10. Chandy, K.M., Herzog, U., Woo, L.: Parametric analysis of queuing networks. IBM J. Res. Dev. **19**(1), 36–42 (1975)
11. Christidis, K., Devetsikiotis, M.: Blockchains and smart contracts for the Internet of Things. IEEE Access **4**, 2292–2303 (2016)
12. Conoscenti, M., Vetrò, A., De Martin, J.C.: Blockchain for the Internet of Things: a systematic literature review. In: IEEE/ACS 13th International Conference of Computer Systems and Applications (AICCSA) (2016)
13. Decker, C., Wattenhofer, R.: Information propagation in the Bitcoin network. In: 13th IEEE Conference on Peer-to-Peer Computing, pp. 1–10 (2013)
14. Göbel, J., Keeler, H.P., Krzesinski, A.E., Taylor, P.G.: Bitcoin blockchain dynamics: the selfish-mine strategy in the presence of propagation delay. Perform. Eval. **104**, 23–41 (2016)
15. Großmann, M., Eiermann, A., Renner, M.: Hypriot Cluster Lab: An ARM-powered cloud solution utilizing Docker. In: 23rd International Conference on Telecommunications (ICT 2016), Thessaloniki, Greece, 16–18 May 2016 (2016)

16. Lamport, L., Shostak, R., Pease, M.: The Byzantine generals problem. ACM Trans. Program. Lang. Syst. **4**(3), 382–401 (1982)
17. Nakamoto, S.: Bitcoin: A Peer-to-Peer Electronic Cash System (2008). https://bitcoin.org/bitcoin.pdf. Accessed 30 Nov 2018
18. Natoli, C., Gramoli, V.: The Balance Attack Against Proof-Of-Work Blockchains: The R3 Testbed as an Example, 30 December 2016. arXiv:1612.09426v1
19. Nelson, R.D., Tantawi, A.N.: Approximate analysis of fork/join synchronization in parallel queues. IEEE Trans. Comput. **37**(6), 739–743 (1988)
20. Nguyen, G.-T., Kim, K.: A survey about consensus algorithms used in blockchain. J. Inf. Process. Syst. **14**(1), 101–128 (2018)
21. Ron, D., Shamir, A.: Quantitative analysis of the full bitcoin transaction graph. In: Sadeghi, A.-R. (ed.) FC 2013. LNCS, vol. 7859, pp. 6–24. Springer, Heidelberg (2013). https://doi.org/10.1007/978-3-642-39884-1_2
22. Shafagh, H., Hithnawi, A., Burkhalter, L., Duquennoy, S.: Towards blockchain-based auditable storage and sharing of IoT Data. arXiv Preprint arXiv:1705.08230 (2017)
23. Vukolić, M.: The quest for scalable blockchain fabric: proof-of-work vs. BFT replication. In: Camenisch, J., Kesdoğan, D. (eds.) iNetSec 2015. LNCS, vol. 9591, pp. 112–125. Springer, Cham (2016). https://doi.org/10.1007/978-3-319-39028-4_9
24. Wang, H., et al.: Approximations and bounds for (n, k) fork-join queues: a linear transformation approach. In: 18th IEEE/ACM International Symposium on Cluster, Cloud and Grid Computing (CCGRID) (2018)
25. Ziegler, M.H., Großmann, M., Krieger, U.R.: Integration of Fog Computing and Blockchain Technology Using the Plasma Framework. Technical report, University of Bamberg (2019, submitted)
26. Docker Inc.: Docker Overview (2018). https://docs.docker.com/engine/docker-overview/. Accessed 28 Aug 2018
27. MultiChain: Multichain 1.0 Beta 2 and 2.0 Roadmap (2017). https://www.multichain.com/blog/2017/06/multichain-1-beta-2-roadmap/. Accessed 2 Sept 2018

An Architectural Framework Proposal for IoT Driven Agriculture

Godlove Suila Kuaban[1]([✉]), Piotr Czekalski[2], Ernest L. Molua[3], and Krzysztof Grochla[1]

[1] Institute of Theoretical and Applied Informatics, Polish Academy of Sciences, Baltycka 5, 44–100 Gliwice, Poland
{gskuaban,kgrochla}@iitis.pl
[2] Institute of Informatics, Silesian Technical University, Akademicka 16, 44–100 Gliwice, Poland
piotr.czekalski@polsl.pl
[3] Faculty of Argriculture and Veterinary Medicine, University of Buea, P.O. Box 63, Buea, SW Region, Cameroon
molua.ernest@ubuea.cm

Abstract. The Internet of Things is paving the way for the transition into the fourth industrial revolution with the mad rush of connecting physical devices and systems to the internet. IoT is a promising technology to drive the agricultural industry, which is the backbone for sustainable development especially in developing countries like those in Africa that are experiencing rapid population growth, stressed natural resources, reduced agricultural productivity due to climate change, and massive food wastage. In this paper, we assessed challenges in the adoption of IoT in developing countries in agriculture. We propose a cost effective, energy efficient, secure, reliable and heterogeneous (independent of the IoT protocol) three layer architecture for IoT driven agriculture. The first layer consists of IoT devices and it is made up of IoT driven agriculture systems such as smart poultry, smart irrigation, theft detection, pest detection, crop monitoring, food preservation, and food supply chain systems. The IoT devices are connected to the gateways by low power LoRaWAN network. The gateways and local processing servers co-located with the gateways create the second layer. The cloud layer is the third layer, which exploits the open source FIWARE platform to provide a set of public and free-to-use API specifications that come along with open source reference implementations.

Keywords: IoT · FIWARE · LoRaWAN

1 Introduction

The Internet of Things (IoT), also known as the Internet of Everything, has moved from hype to reality with the continuous rush of connecting physical devices or things to the Internet. The first industrial revolution was marked

© Springer Nature Switzerland AG 2019
P. Gaj et al. (Eds.): CN 2019, CCIS 1039, pp. 18–33, 2019.
https://doi.org/10.1007/978-3-030-21952-9_2

by mechanization with the invention of the steam engine, the second industrial revolution by the discovery of electrical and other forms of energy to power factories for mass production, the third industrial revolution by the combination of advances in electronics and the digital revolution that introduced computers to automate production. Now the Internet of Things is paving the way for the fourth industrial revolution where every component of the factory is being connected to each other and to the Internet to permit real time management and control of factories which has led to the so-called Industry 4.0.

A similar evolution is taking place in the agricultural sector, from the first agricultural revolution that was labour intensive, to the second agricultural revolution marked by mechanized farming which resulted from the invention of the steam engine in the first industrial revolution and the discovery of electrical and other forms of energy in the second industrial revolution, to the third agricultural revolution motivated by advances in biotechnology, genetic engineering and the need to feed the rapidly increasing population. The recent use of IoT and other information and communication technologies to design smart farms and to automate a lot of farming processes in order to boost food production can be regarded as setting the pace for another agricultural revolution. The Industry 4.0 trend, which is powered by IoT is transforming the production capabilities of all industries, including the agricultural industry resulting in agriculture 4.0 [38].

The major challenges to the deployment of IoT solutions in farms located in rural areas, especially in developing countries, have limited Internet connectivity, insufficient (or no) energy supply, and the high cost of the infrastructure relative to the income of an average farmer. In this paper we survey agricultural challenges that could be addressed with the use of IoT and the challenges to the adoption of IoT in developing countries and then propose an architectural framework for IoT driven agriculture. This study was carried out based on the African context but can be adapted to address agricultural challenges in other developing countries.

The rest of the paper is organised into six sections. Section 2 contains a short review of related literature. Section 3 contains the assessment of agricultural challenges while in Sect. 4, key challenges to the adoption of IoT in developing countries are discuses with focus on Africa. Section 5 contains the proposed reference architecture for IoT driven agriculture and conclusions and future research work is in Sect. 6.

2 Review of Related Literature

Like other technological innovations, there are relatively negligible efforts toward IoT research and innovation in Africa as the continent has always waited for other continents such as Europe, North America, and Asia to develop technologies based on their needs and the continent end up adopting technologies that do not fit its realities and challenges [42]. Recent findings [47] revealed that the majority of the information about IoT in Africa is found in the categories of news, magazines, and other trade publications with relatively very little scientific publications on IoT in Africa. It is very important that more scientific research should be conducted to develop IoT solutions that will address some of the important challenges of the continent, the state of deployment, innovation progress, and other social and economic aspects of IoT in Africa.

The key IoT enablers, the state of IoT deployment and a proposal of possible fields including agriculture where IoT can be used to address some of the challenges in Africa were discussed in [43]. A preliminary research studies in [13] assessed the level of preparedness of the economies in Sub Saharan Africa to adopt IoT and their findings revealed that the Sub Saharan African (SSA) region is lagging behind other regions in four out of the five indices that were considered which include Network Readiness Index, ICT Development Index, Global Innovation Index, Global Competitiveness Index, and Knowledge Economic Index (KEI). Areas, where IoT can play a significant role in increasing prosperity and reducing poverty in Africa, through enhancing basic services and sectors such as agriculture and healthcare, were discussed in [47]. The potential of IoT to reduce poverty in rural communities in South Africa and Zambia were investigated in [12] with emphasis on agriculture while some few areas where smart farming can be applied to create a direct impact on farmers in SSA region were evaluated in [11].

According to the 2016 report on the demographic profile of African countries by the United Nations Economic Commission for Africa, the population of African countries has nearly trebled from 478 million in 1980 to 1.2 billion in 2015 and it is expected to reach 2.4 billion by 2050 [40]. There is a need to adopt innovative and efficient methods to meet up with the challenges introduced by the rapid population growth. Due to the slow adoption of mechanized agriculture similar to what is happening in other industries in Sub-Saharan Africa, land productivity is among the lowest in the world, with over 60% of farm power provided mostly by women, the elderly and children [41].

Fog Computing is based on the Cloud Computing paradigm and extends it to the edge of the network, adding a middle layer of communication and data processing between the end devices and the servers located in the cloud. It has been proved as an architecture well suited for IoT applications [50] and it has been shown to have good performance in terms of latency and reliability [48]. It has been proposed to describe this network of interactions [49], and is well suited for lightweight edge processing in IoT applications such as smart cities, health, agriculture etc. The fog layer is, therefore, used to shift some of the regular lightweight processing from the cloud closer to the IoT devices (IoT layer). A FIWARE based fog computing architecture for smart cities that moves stream processing tasks to the edge of the network to reduce latency was proposed in [5].

FIWARE (where FI stands for Future Internet) is an open source platform that makes it easy to build and deploy sophisticated and innovative internet applications with relatively reduced costs and complexity of serving large numbers of users (especially IoT devices) globally and handling data at a large scale. It provides a set of public and free-to-use API specifications that come along with open source reference implementations. It is made up of seven main parts called "generic enablers" and domain-specific enablers for certain domains like smart cities, energy [6], agriculture etc. The basic functioning of the FIWARE is based on the Next Generation Service Interface (NGSI) open API which defines the data model (context entities, context attributes, and context metadata), context data interface and context availability [3]. The NGSI enables openness and interoperability due to its powerful and simple RESTful API which enables access to the IoT context information [5] regardless of the IoT communication protocol used.

3   Assessment of the Agricultural Challenges in Africa

Despite its potential, the agricultural sector in Africa is facing some challenges that include but are not limited to climate fluctuation, pests control, storage problems, theft, and farm monitoring. Majority of the agricultural challenges in the continent can be addressed using cost effective, energy efficient and reliable IoT solutions in order to increase agricultural productivity.

Climate fluctuation is a global challenge, which has led to the occurrence of extreme temperatures; increasing aridity; and unpredictable rainfall. These fluctuations in environmental variables have led to an increase in malnutrition and infectious diseases in SSA as it is gradually disrupting the rainfed agricultural systems on which the majority of the population rely on [22]. Rapid and extreme fluctuations in temperature and humidity is also a threat to livestock farming such as poultry farming as poultry birds are very sensitive to temperature, humidity, and air quality. The combination of climatic and non-climatic drivers and stressors will likely exacerbate the vulnerability of Africa's agricultural systems to climate change, coupled with the fact that some African communities do not have adequate capacity to cope with or adapt to, the negative impacts of climate change [21]. The UN Secretary General, António Guterres in his speech at the climate summit 2018 in Poland said that climate change is the most important issue that we are facing, that farmers in the Sahel are losing their livelihood due to climate change. The authors in [20] revealed the negative impact of climatic fluctuations on agriculture in Nigeria and recommended that a sustained increase in agricultural productivity can be achieved by using innovative technologies to control the climate fluctuations. The IoT technology could be leveraged to automate the control of environmental variables such as temperature, humidity, lighting, soil moisture etc in order to optimize the resources used and also to improve agricultural productivity.

According to findings from the Climate Change, Agriculture and Food Security (CCAFS) research program, one-sixth of global food production is lost to pest and that climate change may bring a greater risk of pests and diseases to African agricultural systems, affecting crop, livestock, and fisheries productivity [19]. The tropical climate in Africa creates a good breeding ground for pests such as insects, birds, giant rats, and other animals that damage crops and prey on organisms needed in the farming ecosystem. Pests, such as e.g. wheat weevil, can cause great damage to the harvested grains. The challenges in pest management are detecting the presence of pests, predicting a possible outbreak of pests and diseases, and to have good response strategies to keep pests damage under control. IoT could be used to detect the presence of pests, predict the possible outbreak of pests and advice farmers on possible techniques to keep its effects under control.

Theft of farm products, tools, and other assets is a serious threat to the agricultural systems in developing countries. This is as a result of the rising level of unemployment and poverty, especially in poorly developed rural communities. Livestock farming is seriously affected by theft wherein one successful theft operation, the time, energy and investments of the farmer can be rendered useless. It is even more complicated in areas where the farms are located far away from the human settlement as intruders can easily get away unnoticed. Therefore, it is important to develop anti-theft systems that will be able to detect intruders, scare them away and inform the farm owner in real time and possibly ensure timely intervention of local security agents or neighbors. With the use of IoT, it is possible to develop reliable and affordable anti-theft systems based on the context of the given environment in order to reduce the chances of farmers losing their investments on a single theft operation.

The inability to monitor the crops or livestock in real time in order to schedule farming activities and reduce the number of visits to the farms makes farming relatively expensive and time demanding to small scale farmers, who may have other jobs to sustain their families. Most of the crops cultivated in the tropics require careful monitoring to determine when to remove weed that may overshadow crops, when to apply fertilizer and when to harvest. Weeds are uncultivated plants that are competing with cultivated plants [18], they can have negative effects on the crops if not removed on time, such as reduced crop yield; compete with cultivated crops for soil water thus increasing irrigation demands during the dry season, reduce chances of cross-pollination, and damage plant health. It is also challenging to know exactly when the crops are ready for harvesting as some crops may get ready unexpectedly and may be damaged before the estimated time. Livestock farmers face similar problems of time-consuming and labor intensive activities such as giving food and water to the animals, monitoring the well being of the animals and cleaning the environment where the animals are kept. IoT can be used to automate some farming activities which will reduce the number of visits to the farm and therefore, save the time and cost of running the farms.

Farmer and traders dealing with agricultural products in Africa face a serious challenge of storing products like fruits, vegetables, tubers, and cereals. As the agricultural systems in Africa are rainfed, the majority of the crops such as cereals and tubers cultivated during the rainy season must be preserved and marketed throughout the year even during the non-harvesting season. Monitoring the state and quality of farm products along the food supply chain challenge that needs to be tackled by all the stakeholders in the agricultural supply chain to ensure that appropriate actions can be taken to avoid food wastage. IoT sensors can be used to monitor environmental conditions such as temperature and humidity under which the farm products are stored, to detect any abnormalities and alert the owner to take appropriate actions.

4 Key Challenges to the Adoption of IoT Driven Agriculture

There are a lot of research and innovation projects going on in Europe, America and Asia to develop, standardize and deploy IoT technologies but very little is being done in the African continent just as in the past technological revolutions [42,44]. Due to very little research and innovation in Africa, most of the technologies and use cases developed outside of the African continent, some times do not fully address the challenges of the continent and their capabilities not fully exploited. The adoption of IoT can trigger innovation in other sectors such as agriculture, health, transportation and heavy industries, which gives the hope that the continent can leapfrog technologically to catchup with the rest of the world.

Insufficient power supply is one of the challenges to the adoption of IoT in Africa and SSA in particular that constitute 46 of Africa's 54 countries. According to a series of reports from Oxfam and the Renewable and Appropriate Energy Laboratory at the University of California, Berkeley, the region has the lowest energy supply capacity with the most acute forms of energy poverty in the world. Poor management of energy utility services and the high cost of energy [17], makes it difficult to automate some farming processes which require energy to power machines, sensors, communications devices, and local data processing. However, African nations are progressively upgrading their energy capacity and also embarrassing renewable energy initiatives which indicate a promising future for IoT in Africa. Due to limited supply of energy and high cost of energy we advocate the development of secure, reliable and energy efficient IoT systems in order to reduce the cost of managing IoT system and increase the rate of adoption.

Limited internet connectivity and the high cost of internet connection in some African countries is a hindrance to the adoption of IoT in the African continent. Despite the steady growth in mobile (2G, 3G and 4G) and internet connectivity, Africa is still far behind the rest of the world, with a penetration rate of 35.2% representing 10.9% of the total world internet users [16]. Due to insufficient experts, little competition among operators, poor regulation of ICT and telecommunication sectors in most African countries, mobile coverage is low, the quality of internet connectivity is poor and cost of internet connectivity is relatively high compared with other continents [15]. Poor and high cost of internet connectivity will make IoT solutions relatively difficult to manage and unaffordable for low income earners especially farmers in rural communities. Despite the significant effort by must nations to develop modern telecommunication infrastructure, there is still need to develop IoT solutions to addresses the needs of the continent while optimising the cost of energy, connectivity and provisioning but ensuring an acceptable level of security and quality of service.

Insufficient skilled labour in the area of IoT, data science, and agriculture which could be as a result of mass movements of graduates from STEM (Science, Technology, Engineering and Mathematics) fields for greener pastures. Majority of farmers in African rural communities are small scale farmers and a lot of them little or no formal or informal training on agricultural practices that can improve yields. The problem of insufficient IoT skills is because many universities have not yet adopted IoT as a formal course or program. In order to tackle the problem of insufficient IoT skills, we are currently developing free IoT courses with remote laboratory infrastructure under the IOT-OPEN.EU (Innovative Open Education on IoT: improving higher education for European digital global competitiveness), ERASMUS+ Key Action 2 (Strategic Partnership) project. This will provide students and IoT enthusiast including those from developing countries free access to our resources for learning and research. We are building other cooperation with institutions in Africa as part of IoT Technology for Sub Saharan Africa (ITSA) project.

The rate of technology adoption in Africa is very low especially in the central African Sub-region which have the lowest technological adoption, with mechanized agriculture and IoT still being a hype. Technological adoption requires an updated educational curriculum but most developing nations do not have dynamic educational curriculum to keep them abreast with the new developments in hardware, software, and communications technologies [14]. It is also difficult to convince small and medium-size enterprises to buy technological solutions as they are sometimes reluctant to pay for technological solutions partly, due to inadequate financial capital. Adequate attention is not given to the agriculture and IT sectors by some developing countries as their main focus is on natural resources like oil, minerals, natural gas etc which are already being stressed. IoT (and IoT-driven agriculture) research and innovation initiatives based on local problems especially the energy, connectivity, security and cost constraints will enable an increase in the rate of technology adoption.

Another challenge to the adoption of IoT generally is the fact that IoT has a lot of cyber security challenges, as it has been predicted that by 2020, 25% of cyber attacks will be targeted towards IoT devices [9]. The 2016 high-profile Mirai attack conducted with the use of an army of IoT botnets was catastrophic as it disrupted internet services for more than 900,000 Deutsche Telekom customers in Germany, infected almost 2,400 TalkTalk routers in the UK [8] and disrupted the DNS services of Dyn, which further led to the disruption of services of website such as Twitter, Amazon, Reddit, and Spotify [9]. According to a research survey in [10] it was found that they are very few cybersecurity initiatives in the African continent as the few countries that had cybersecurity initiatives ended only at cybercrime legislation with little or no actions. The fact that IoT systems are highly vulnerable and can be easily exploited might scare a lot of African enterprises from adopting IoT as it may be considered as a potential threat to their systems.

5   The Proposed Architectural Framework

In most IoT solutions, the sensing devices are usually powered by batteries and some low power communications protocols have been developed to minimize energy consumption. Currently, security in IoT networks is the major preoccupation as IoT devices can easily be compromised due to their low processing power capabilities resulting from its power constraints. However, power and security are not the only constraints when deploying IoT solutions especially in developing countries but also internet connectivity and the cost of setting up and managing the IoT infrastructure. Our proposed architectural framework for IoT driven agriculture takes into consideration energy, internet connectivity, security and cost constraints. Our framework is made up of three layers which are: the sensor layer, the fog layer and the cloud layer.

5.1 The IoT Layer

The IoT layer is made up of IoT driven agriculture systems which include sensors that measure environmental variables such as temperature, humidity, soil moisture, air quality, gas concentration, vibrations and sound, cameras for crop/livestock and security monitoring, and agricultural systems to be controlled such as irrigation machines, heating systems, fans, lighting systems, and alarm systems. Some of the IoT driven agriculture use case applications that were considered include but are not limited to those shown in Fig. 1 which include: smart poultry systems, smart irrigation systems, theft detection systems in farms, pest detection systems, crop monitoring, food storage monitoring, and food supply chain. Details of cost-effective use case applications for IoT-driven agriculture based on the context of developing countries have not been presented due to limited space but will be provided in subsequent works.

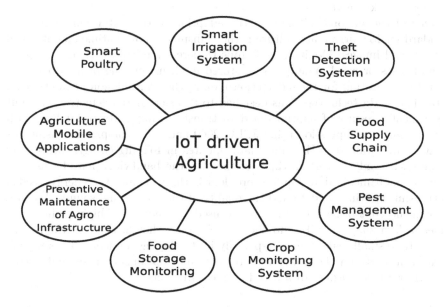

Fig. 1. Proposed IoT driven agriculture use cases for sustainable developement

Due to the limited storage and processing power of the devices in the IoT layer, the data measured by the IoT sensors is usually sent to a cloud platform for advanced processing and the feedback from the cloud platform is used to control some systems, which are systems in the farming process in this case. This has resulted in the need of low power transmission technologies and protocols for IoT such as ZigBee, 6LowPAN, BLE, NB-IoT LoRa, LoRaWAN, Sigfox, Weightless P. IoT driven agriculture use case applications require mostly LPWAN (Low Power Wide Area Network) communication technologies where sensors from many farms can be connected to one gateway. Some of the most popular LPWAN technologies include Sigfox, Weightless P, NB-IoT LoRa, and LoRaWAN.

In order to ensure low power communication between the IoT sensor devices deployed in open fields or in poultries and the gateway, the LoRa communication standard is proposed. The LoRa communication uses a proprietary spread spectrum modulation based on the Chirp Spread Spectrum modulation (CSS) and which allows achieving very good sensitivity and uses fixed channel bandwidth. The LoRa devices are very energy efficient and may operate up to a few years on a single battery. The LoRa Alliance (https://lora-alliance.org) has developed an LPWAN protocol for deployments where end-devices have limited energy and need to transmit a small amount of data (e.g. tens of bytes) at a time. The LoRa WAN provides two-way communication, with three different classes of devices (A, B and C) with different schemes of traffic transmission initiation. In addition to its energy benefits, LoRa communication standard is well suited for the application of IoT in agriculture because it provides a large radio network courage up to several kilometers.

One of the major challenges in the adoption of the LoRa communication standard for IoT driven agriculture is the transmission of multimedia data such as voice and images. Monitoring of crops to determine harvest times, the presence of pests or weed, monitoring of stored food, monitoring of livestock, and security monitoring for theft detection require the transmission of multimedia data. The LoRa technology was designed to transfer short telemetric data but experimental results of multimedia data transfer using LoRa technology with a better image compression using JPEG 2000 and voice compression using the A-law were obtained in [1]. Since data updates in IoT driven agriculture may be necessary only at certain times of the day, the bandwidth may be used to transmit multimedia data but the problem is the time and the energy needed to transmit multimedia data and the quality of the images and sounds. We also recommend that other methods may be used for delivery of high-quality multimedia data such as WiFi and the use of drones though these methods may be costly to small scale farmers especially in developing countries. Thus we will perform a study on the power requirements and optimization for multimedia data transmission using LoRa based on measurements.

5.2   The Fog Layer

In must IoT deployments, cloud platforms are used for processing of IoT data but it requires good internet connectivity. The cloud approach also have some drawbacks like high latency in IoT applications where the systems at the IoT layer requires real time control as the case of some IoT driven agriculture systems and the issue of security as the data may be compromised in the course of transmission within the internet backbone to a distant cloud server for example, attacks such as traffic reordering in the case of temperature, humidity, soil moisture etc. measurements and also traffic dropping attacks may disrupt the control process. The fog computing paradigm is a better approach to mitigate some of the drawbacks of using only the cloud for data processing by shifting some of the lightweight processing to the edge of the network which is closer to the IoT layer. To merge LoRa with Fog Computing paradigm we propose an architecture for IoT driven agriculture shown in Fig. 2. The LoRa WAN access point can be installed to provide wireless communication between the IoT layer and the fog layer alongside a local computer (e.g. a Raspberry Pi or a cheap PC). The fog/edge computing architecture is suited in this context because even if the internet is temporally unavailable for any reason the local network and the fog nodes will still keep the services working and local processing also offers energy benefits in terms of the number of communications.

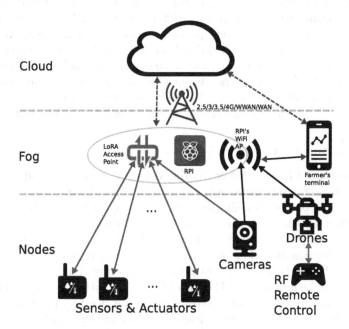

Fig. 2. Proposed architecture for IoT driven Agriculture

The fog layer in the proposed reference architecture for IoT driven agriculture shown in Fig. 3, is made up of the IoT agent, a lightweight context broker, a complex event processor (CEP), a LoRaWAN gateway and a WiFi access point. The basic functioning of FIWARE is based on the NGSI OPen API which defines the context data model, the context data interfaces and context availability but the NGSI specifications are not supported by the devices at the IoT layer.

The IoT agent between the FIWARE fog computing system and the sensors and agricultural systems at the IoT layer translates the different communication protocols such as CoAP transport, HTTP ultralight, MQTT transport, and OMA Lightweight M2M into NGSI specifications and hence the FIWARE provide open specifications for IoT deployments as it is independent of the protocols running on the different IoT devices from different vendors. IoT agents have been implemented to enable the exchange of commands between the devices at the IoT layer and the FIWARE context broker, including one that provides an interface between the LoRaWAN servers and the FIWARE context broker.

The lightweight context broker keeps track of the context entities such as: soil moisture, temperature, humidity, gas concentration, water level, light intensity etc registered by the different context data producers (sensors, actuators, drones etc.) and also keeps track of the different publish/subscribe request for the different context entities. When a new context update arrives the current value of the context attribute (moisture value, temperature value, humidity value etc.) is updated and therefore, the current context data must be forwarded to the database, the CEP and the cloud context broker for advanced processing. The CEP should be able to filter, aggregate and merge data and process the context data so that when any of the event processing rules or conditions configured is meet, the CEP generates an alert and publish to the lightweight context broker. For example, if the temperature or the soil moisture level is below or above the required level, the CEP can generate an alert and publish on the lightweight context broker which can then send notifications to the consumers who have subscribed for this context event (farmers, actuators, cloud context broker etc).

5.3   The Cloud Layer

The introduction of fog or edge computing does not eliminate cloud computing in IoT systems as there is always a need for advanced and computationally intensive processing. Computationally intensive processing such as big data analytics, advanced data visualisation, GIS (Geographic Information System) processing to generate thematic maps and statistics, AI-based knowledge discovery, data driven agriculture applications etc should be shifted to the FIWARE cloud. Details of the FIWARE cloud are not provided by our reference architecture but more details can be found in [6]. The FIWARE cloud platform contains a context broker whose data producers and consumers are the cloud CEP, the lightweight context brokers at the fog layer, advanced data analytics applications, third-party IoT driven agriculture applications and users. In order to ensure that the current context data is not overwritten by new ones, the context data is written to a database. Since the cloud and the fog layers both support the NGSI specifications, there is no need for an agent for protocol translation.

In real and large scale deployment of this kind of architecture, the cloud servers may be located far away from the fog nodes and the traffic needs to be transported over traditional internet infrastructure. As mentioned above, the IoT traffic may be attacked during the course of transportation or may experience longer delays that may affect real-time control applications in IoT driven

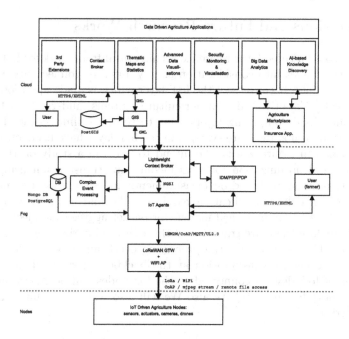

Fig. 3. Proposed reference architecture for IoT driven agriculuture based on fog computing/edge computing and fiware

agriculture. Networks operators could adopt an architecture proposed in the SerIoT (Safe and Secure Internet of Thing) project [2] for secure transportation of IoT traffic from the fog nodes to the cloud servers. The SerIoT network is a self-aware Software Defined Network (SDN) that consist of forwarding elements and a smart SDN controller that is based on Cognitive Packet Network (CPN) and its Random Neural Network (RNN) with reinforcement learning. The smart controller computes and updates the flow rules based on a goal function whose metrics are Security and Safety, Quality of Service and Energy usage. The fog nodes could be connected to a single edge SerIoT node which can then source-route the traffic through intermediate nodes towards the cloud servers. The SerIoT holistic solution for secure Internet of Things also contains security monitoring, intrusion detection, honeypots etc.

FIWARE provides a security generic enabler (GE) to ensure the security of the FIWARE platform as security is one of the must compelling requirement of IoT deployments. The FIWARE security GE components were discussed in [6]. The Identity Manager (IDM) can create users (devices in the IoT layer, other applications, and operators), roles and permission. The Policy Enforcement Point (PEP) which is a proxy server performs authentication and optional authorization checks. The Policy Decision Point (PDP) provides authorization services by deciding the actions the users are allowed to perform. It is possible to implement advanced security analysis, visualization and reporting in the cloud platform.

6 Conclusions and Future Research Works

We have proposed a three-layer reference architectural framework for IoT driven agriculture that takes into consideration energy, internet connectivity, security and cost constraints. It consists of the IoT layer which is made up of low power LoRaWAN networks of IoT driven agriculture systems, and then a FIWARE based fog and cloud computing layers. Due to the page limit, a lot of details about the IoT driven agriculture systems in the IoT layer have been left out but will be provided in future works. We intend to perform a study on the power requirements and optimization for multimedia data transmission using LoRa communication standard based on measurements and simulation because a lot of IoT driven agriculture application requires audio and image transmissions. We will equally develop analytical models for packet aggregation mechanism at the fog node before there are transported through the internet backbone to the cloud servers. This is to reduce the processing overheads of short IoT packets and to optimize energy consumption at the backbone because it is expected that billions of IoT devices connected to the fog nodes will be generating huge amounts of short IoT packets. We also We will also test this architecture with some selected use cases.

Acknowledgements. The work presented in this paper was partially supported by the ERASMUS+ Key Action 2 (Strategic Partnership) project IOT-OPEN.EU (Innovative Open Education on IoT: improving higher education for European digital global competitiveness), reference no. 2016-1-PL01-KA203-026471 and the SerIoT Research and Innovation Action, funded by the European Commission under the H2020-IOT-2016-2017 (H2020-IOT-2017) Program through Grant Agreement 780139. The European Commission support for the production of this publication does not constitute the endorsement of the contents which reflects the views only of the authors, and the Commission cannot be held responsible for any use which may be made of the information contained therein.

References

1. Kirichek, R., Pham, V.-D., Kolechkin, A., Al-Bahri, M., Paramonov, A.: Transfer of multimedia data via LoRa. In: Galinina, O., Andreev, S., Balandin, S., Koucheryavy, Y. (eds.) NEW2AN/ruSMART/NsCC -2017. LNCS, vol. 10531, pp. 708–720. Springer, Cham (2017). https://doi.org/10.1007/978-3-319-67380-6_67
2. Domanska, J., Gelenbe, E., Czachorski, T., Drosou, A., Tzovaras, D.: Research and innovation action for the security of the Internet of Things: the SerIoT project. In: Gelenbe, E., et al. (eds.) Euro-CYBERSEC 2018. CCIS, vol. 821, pp. 101–118. Springer, Cham (2018). https://doi.org/10.1007/978-3-319-95189-8_10
3. Ivan, T., Trikos, M., Navarro-Hellín, H., Lalović, K.: FIWARE: a web of things development platform. Mil. Tech. Courier **66**(4) (2018)
4. López-Riquelme, J., Pavø'n-Pulido, N., Navarro-Hellín, H., Soto-Valles, F.: A software architecture based on FIWARE cloud for Precision Agriculture. Agric. Water Manag. **183**, 123–135 (2016)

5. Rampérez, V., Soriano, J., Lizcano, D.: A multidomain standards-based fog computing architecture. Hindawi Wireless Communications and Mobile Computing Volume. Wiley (2018)
6. Salhofer, P.: Evaluating the FIWARE platform: a case-study on implementing smart application with FIWARE. In: Proceedings of the 51st Hawaii International Conference on System Sciences (2018)
7. Soto, V.E.A.: Performance evaluation of scalable and distributed IoT platforms for smart regions, Master's degree thesis (2017)
8. Mohammed, A.F.: Security issues in IoT. IJSRSET **3** (2017)
9. Ismail, N.: The security challenges with the Internet of Things, the information age. https://www.information-age.com/internet-things-security-crisis-123470475/. Accessed 29 Aug 2018
10. Chetty, M., Goodman, S., Cole, K., LaRosa, C., Rietta, F., Schmitt, D.: Cybersecurity in Africa: An Assessment, Sam Nunn School of International Affairs Georgia Institute of Technology Atlanta, GA US (2008)
11. Ishengoma, F., Athuman, M.: Internet of Things to improve agriculture in Sub Sahara Africa - a case study. Int. J. Adv. Sci. Res. Eng. **4**(6), 8–11 (2018)
12. Dlodlo, N., Kalezhi, J.: The Internet of Things in Agriculture for Sustainable Rural Development. IEEE (2015). https://doi.org/10.1109/ETNCC.2015.7184801. https://www.researchgate.net/publication/277713549. Accessed 31 Dec 2018
13. Atayero, A., Oluwatobi, S., Alege, P.O.: An assessment of the Internet of Things (IoT) adoption readiness of Sub-Saharan Africa. J. South Afr. Bus. Res. Article ID 321563 (2016). https://doi.org/10.5171/2016.321563
14. Ejiaku, S.A.: Technology adoption: issues and challenges in information technology adoption in emerging economies. J. Int. Technol. Inf. Manag. **23**(2), Article 5 (2014)
15. Alliance for Affordable Internet (A4AI): New data: What's the price of 1GB of mobile broadband across LMICs? (2018). https://a4ai.org/new-mobile-broadband-pricing-data-2018. Accessed 21 Dec 2018
16. Internet World Stats (2017). https://www.internetworldstats.com/stats1.htm. Accessed 21 Dec 2018
17. Morrissey, J.: The energy challenge in sub-Saharan Africa, OXFAM'S Research Backgrounder, Oxfam and the Renewable and Appropriate Energy Laboratory at the University of California, Berkeley (2017)
18. Stephens, R.J.: Theory and Practice of Weed Control. Springer, New York (1982)
19. Dinesh, D., et al.: Impact of climate change on African agriculture: focus on pests and diseases. Findings from CCAFS submissions to the UNFCCC SBSTA (2015). https://cgspace.cgiar.org. Accessed 5 Dec 2018
20. Ayinde, O.E., Muchie, M., Olatunji, G.B.: Effect of climate change on agricultural productivity in Nigeria: a co-integration model approach. J. Hum. Ecol. **35**(3), 189–194 (2011)
21. Pereira, L.: Climate change impacts on agriculture across Africa. Oxford Research Encyclopedia of Environmental Science (2017)
22. Serdeczny, O., et al.: Climate change impact in the Sub-Saharan Africa: from physical challenges to their social repercussions. Regional Environmental change, special issue on models for adaptive forest management-the motive project. Springer (2015)
23. Food and Agricultural Organization of the United Nations [FAO]: ICT in agriculture: connecting smallholders to Knowledge, Networks and Institutions. The State of Food and Agriculture 2010–2011: Women in Agriculture, Closing the Gender Gap for Development. FAO, Rome (2011)

24. World Bank: ICT in Agriculture: Connecting Smallholders to Knowledge, Networks and Institutions, Updated Edition. World Bank, Washington, DC (2017). https://doi.org/10.1596/978-1-4648-1002-2
25. GSMA: Understanding the Internet of Things (IoT), Connected Living (2014)
26. European Commission: Industry 4.0 in agriculture: Focus on IoT aspects, Digital Transformation Monitor (2017)
27. Courade, G., Devèze, J.C.: Des agricultures Africaines face à de difficiles transitions, Afriquecontemporaine 217 (2006)
28. Delpeuch, F.: Le systéme alimentaire mondial à un carrefour. Cahiers de l'Agriculture **16**, 161–62 (2017)
29. AfDB: Organisation for Economic Co-operation and Development [OECD], & United Nations Development Programme [UNDP] (2017)
30. Woldemichael, A., Salami, A., Mukasa, A., Simpasa, A., Shimeles, A.: Transforming Africa's agriculture through agro-industrialization. Afr. Econ. Brief **8**(7) (2017). African Development Bank, Abidjan
31. Kanu, S.B., Salami, A.O., Numasawa, K.: Inclusive Growth: An Imperative for African Agriculture. African Development Bank, Tunis (2014)
32. Verdier-Chouchane, A., Karagueuzian, C.: Moving towards a green productive agriculture in Africa: the role of ICTs. Afr. Econ. Brief **7**, 1–12 (2016)
33. Stoŏces, M., Vaněk, J., Masner, J., Pavlik, J.: Internet of Things (IoT) in agriculture - selected aspects. AGRIS On-line Papers. Econ. Inform. **8**(1), 83–88 (2016). https://doi.org/10.7160/aol.2016.080108. ISSN 1804–1930
34. Mohammed, Z.K.A., Ahmed, E.S.A.: Internet of Things applications, challenges and related future technologies. World Sci. News **67**(2), 126–148 (2017)
35. Savale, O., Managave, A., Ambekar, D., Sathe, S.: Internet of Things in precision agriculture using wireless sensor networks. Int. J. Adv. Eng. Innov. Technol. **2**, 14–17 (2015)
36. Diaz-Bonilla, E.: Macroeconomics, Agriculture and Food Security. A guide to Policy Analysis in Developing Countries, International Food Policy Research Institute, Washington, D.C. (2015)
37. Writer, G.: IoT Applications in Agriculture (2018). https://www.iotforall.com/iot-applications-in-agriculture/. Accessed 24 Sept 2018
38. Bonneau, V., Copigneaux, B.: Industry 4.0 in Agriculture: Focus on IoT Aspects. European Commission (2017). https://ec.europa.eu/growth/tools. Accessed 24 Sept 2018
39. United Nations Economic Commision for Africa: The Demographic Profile of African Countries, ISO 14001:2004 certified (2016)
40. United Nations Economic Commision for Africa: The Demographic Profile of African Countries, ISO 14001:2004 certified. https://www.uneca.org. Accessed 17 Sept 2016
41. European Agricultural Machinary: Advancing Agricultural Mechanization (AM) to promote farming & rural development in Africa. http://cema-agri.org. Accessed 17 Sept 2014
42. Masinde, M.: IoT Applications that work for the African Continent: Innovation or Adoption? IEEE (2014). https://www.researchgate.net/publication/277713549. Accessed 31 Dec 2018
43. Ndubuaku, M., Okereafor, D.: Internet of Things for Africa: challenges and opportunities. In: Proceedings of International Conference on Cyberspace Governance - CYBERABUJA 2015 (2015). https://doi.org/10.13140/RG2.1.2532.6162
44. Onyalo, N., Kandie, H., Njuki, J.: The Internet of Things, progress report for Africa: a survey. Int. J. Comput. Sci. Softw. Eng. (IJCSSE) **4**(9) (2015)

45. Dlodlo, N., Kalezhi, J.: The Internet of Things in Agriculture for Sustainable Rural Development (2015)
46. Tzounis, A., Katsoulas, N., Bartzanas, T., Kittas, C.: Internet of Things in agriculture, recent advances and future challenges. Biosyst. Eng. **164**, 31–48 (2017)
47. Isma'ili, S., Li, M., Shen, J., He, Q., Alghazi, A.: African societal challenges transformation through IoT. In: 21st Pacific Asia Conference on Information System (PACIS), pp. 1–9 (2017)
48. Slabicki, M., Grochla, K., Performance evaluation of CoAP, SNMP and NETCONF protocols in fog computing architecture. In: Network Operations and Management Symposium (NOMS), 2016 IEEE/IFIP, pp. 1315–1319. IEEE
49. Bonomi, F., Milito, R., Zhu, J., Addepalli, S.: Fog computing and its role in the internet of things. In: Proceedings of the First Edition of the MCC Workshop on Mobile Cloud Computing. ACM, pp. 13–16 (2012)
50. Hong, K., Lillethun, D., Ramachandran, U., Ottenwälder, B., Koldehofe, B.: Mobile fog: a programming model for large-scale applications on the Internet of Things. In: Proceedings of the Second ACM SIGCOMM Workshop on Mobile Cloud Computing, pp. 15–20. ACM (2013)

The Procedure of Key Distribution in Military IoT Networks

Jan Chudzikiewicz$^{(\boxtimes)}$, Tomasz Malinowski$^{(\boxtimes)}$, Janusz Furtak$^{(\boxtimes)}$, and Zbigniew Zieliński$^{(\boxtimes)}$

Faculty of Cybernetics, Military University of Technology,
ul. gen. Sylwestra Kaliski 2, 00-908 Warsaw 46, Poland
{jan.chudzikiewicz,tomasz.malinowski,janusz.furtak,
zbigniew.zielinski}@wat.edu.pl
http://www.wat.edu.pl/

Abstract. Cryptographic key distribution is a critical stage in the implementation of network security in the Military IoT (MIoT). In MIoT, due to the resource-constrained nature of the nodes, it is important to design a key management protocol with minimum resource overhead. The proposed procedure uses asymmetric and symmetric cryptography and reduces the memory requirement. The efficient Quantis Random Number Generator is used to provide symmetric keys with full entropy. The method of delivering symmetric keys (so-called session keys) to nodes wishing to implement a safety data exchange was presented. During simulation studies of the ZigBee network, the average energy consumption for the coordinator, single router and end node participating in the encryption key distribution process was estimated. Simulation tests for the ZigBee network were prepared and implemented in the Riverbed Modeler environment.

Keywords: Asymmetric and symmetric cryptography · Military IoT · Quantis Random Number Generator

1 Introduction

The Internet of Things (IoT) is rapidly emerging worldwide interdisciplinary technology which converges such areas as networking, embedded hardware, software architectures, sensing technologies, information management, and data analytics. Till now IoT includes many civilian applications as smart city initiatives, wearable devices for near real-time health monitoring, smart homes and buildings, smarter vehicles, logistics support, etc.

An immediate consequence of the success of the IoT in the civilian domain was an attempt to apply IoT also to the military domain [1,2]. Deployment of IoT related technologies by the military has primarily focused on combat applications as Command, Control, Communications, Computers, Intelligence, Surveillance and Reconnaissance (C4ISR) to provide situational awareness to

© Springer Nature Switzerland AG 2019
P. Gaj et al. (Eds.): CN 2019, CCIS 1039, pp. 34–47, 2019.
https://doi.org/10.1007/978-3-030-21952-9_3

commanders and warfighters. The adaptation of the concept of IoT for the military domain Military IoT (MIoT) emphasized the connectivity of military objects/things that can talk one to another without human interactions. Military Internet of Things is the information network of the "things" integrated according to the IoT paradigm for military purposes. By "things" we will understand here any device where remote communication, data collection, or control might be applicable. Under this broad definition, "things" could be vehicles, appliances, materials, weapons systems or its parts, medical/health devices, electric grids, transportation infrastructure, building systems, or even living roses co-opted with sensing and transmission capabilities [2]. It is expected that MIoT environments can realize benefits similar to those in industry. However, military environments, especially battlefield environments depend on tactical communications, are much more challenging than commercial environments. Some obstacles stand in the way of successful development and deployment of IoT technologies across the military where security is the most significant challenge to broader MIoT development across the military.

The network of MIoT for battlefield environment can be seen as a collection of collaborating objects each of them equipped with some sensor nodes (SN) and transmitter-receiver devices, referred to as collecting sensor nodes (CSN) [3]. Following the work [3] we assume the MIoT network is partitioned among disjoint clusters with a limited number of objects by the organization of subordination, military units and communication capabilities. Figure 1 shows the general structure of the network.

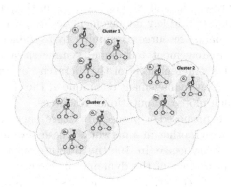

Fig. 1. The overall structure of the MIoT network

We also assume that the interested area of the MIoT is composed of a large number of self-organized sensor devices. Sensor devices usually consist of some physical sensors gathering environmental data, a microcontroller processing the data, and a radio interface to communicate with other nodes.

The mobile sensor nodes usually formed the MIoT network. In this network, wireless media with low bandwidth and small range are usually used for data exchange between CS and CSN nodes as ZigBee technology. The nodes of this network have limited power capabilities, relatively small memory resources and low computing capabilities. Communication between clusters requires considerably the use of protocols with a much greater range. Some civilian IoT communication protocols are considered to be used within MIoT [4]. The LoRaWAN protocol has gained significant adoption through support for low power, long range transmission. Narrowband IoT (NB-IoT) and Sigfox represent interesting alternatives to LoRaWAN, as they also provide long-range wireless communication with low throughput and low energy consumption. Due to the main emphasis on the demonstration of the key management procedure in the simulation tool, we assumed that communication between CSN nodes (clusters) is also carried out with the use of ZigBee technology.

In the case of such a network, it is difficult to effectively use the Certification Authority (CA) to build a trust structure. In our previous works [3,5] we proposed the use of a Trusted Platform Module (TPM) to organize security in the domain of sensor nodes. In particular, the TPM was used to build a local trust structure that was used during authentication procedures of sensor nodes in the domain. The TPM also supported all procedures for securing data stored in the nodes' resources (e.g., cryptographic keys), for all data exchange procedures between sensor nodes and procedures for detecting unauthorized interference in hardware and software resources of the nodes. In these works, the TPM in version 1.2 was used, which offers support for RSA 2048 cryptographic algorithms and the SHA-1 hash function.

The problem of key distribution in the IoT networks was recently broadly investigated by many researchers [6–8]. For instance, in the paper [6] a key management protocol with a hybrid key management technique is proposed to meet increasing security demands resource consuming by asymmetric key primitives. In the proposed key management technique, symmetric and asymmetric key primitives are used at a different level of the hierarchy of sensor nodes. When applying identity based key management technique at the cluster head level, this solution consumes much more computational resources.

In [9] authors have proposed a procedure where each ZigBee node has a shared secret key with each other in a network. These keys are stored in RAM. This causes a lot of memory usage. In [10], the authors investigate the feasibility of implementing the procedure of the dynamic distribution of encryption keys with one key distribution center. In [11] authors have proposed architecture with several Trust Centers on a large scale network. Each center manages a few hundreds of nodes.

Below we propose a procedure with one management center for several clusters and sub-centers for each cluster. This approach in our opinion will result in faster key distribution and lower energy consumption.

Resource requirement of pairing algorithm when implemented on the ARM processor using pairing functions was studied in work [7]. In work [8] the TinyPBC algorithm which provides identity-based key management without interaction between participating nodes. As was shown in the paper MICA2 sensor nodes with ATmega128L microcontroller (8 - bit/7.3828 MHz) computes pairings in 5.5 s time. In [12] a solution for trust transfer between two security domains was investigated. The main concerns of the solutions mentioned above in the context of its applications in the MIoT are scalability and time consuming by communicating nodes which belongs to different security domains.

In this paper, we propose the key management scheme which is based on the development into the network Key Distribution Nodes (KDN) equipped with a TPM. We also assume that all CSN nodes of MIoT network are equipped with the TPM. The TPM can be used in the authentication procedures of these devices, to protect the cryptographic material that has to be stored in the device's resources and to secure transmission as well. We propose to integrate the TPM and the IDQ quantum random number generator (QRNG)[1] in each of the KDN in the context of securing IoT enclaves and achieve high availability and scalability.

We see our contribution as follows. Firstly, we propose a key distribution procedure for encrypting communication between network nodes, with a particular focus on applications in the field of the Military IoT (MIoT). Secondly, we developed the KDN node by integrating the TPM and the QRNG generator to serve as trust and high-performance key distribution nodes. Next, we conduct some investigation of the proposed solution by the simulation experiments in the Riverbed Modeler environment. To achieve secure routing within the network, we adopted the solution which enforces authenticating sensor nodes to the Master node before they could start to build a routing table. It allows us to avoid several types of routing attacks, as routing table overflow attack [4].

The rest of the paper is organized as follows. In Sect. 2 overall structure of the Network for MIoT was described, and the basic terms and certain properties of the logical network structure were explained. The procedure of key distribution was presented in Sect. 3. In Sect. 4 selected components of the network implementation in the Riverbed Modeler environment are presented. In Sect. 5 some concluding remarks are given.

2   The Basic Definitions

Let us assume that our network will be described by a graph.

Definition 1. *A coherent ordinary graph describes the structure of a network* $G = \langle E, U \rangle$ *(E – set of sensor nodes, U – set of bidirectional communication link).*

Let $\hat{E} \subset E(G)$ be the set of cluster head, and $\left\{ E(G) \setminus \hat{E} \right\}$ be the set of other sensor nodes (route, and device).

---

[1] IDQ QRNG provides [15] full entropy (randomness) instantaneously and reaches 16 Mbits/s efficiency. It has successfully passed the following certifications: NIST SP800-22 Test Suite Compliance, METAS Certification, CTL Certification, Several iTech Labs individual Certificates, Compliance with the BSI's AIS31 standard and others.

Denoted by $d(e, e'|G)$ the distance between nodes e and e' in a coherent graph G, that is the length of the shortest chain (in the graph G) connecting node e with the node e'.

Definition 2. *The graph* $\langle G; \hat{E} \rangle \left(|\hat{E}| \geq 1 \right)$ *is* $(m, d|G)-$ *cluster* $\left(m \in \{1, \ldots, \right.$ $\mu(G) \}$, $d \in \{1, \ldots, D(G)\}$ $D(G)-$ *diameter of the graph* G) *if there exists the set* $\hat{E}$ *of minimum cardinality such that* [13, 14]

$$\left[\forall_{e \in \{E(G) \setminus \hat{E}\}} : \left| \left\{ e' \in \hat{E} : d(e, e'|G) \leq d \right\} \right| \geq m \right] \wedge$$
$$\wedge \left[\forall_{\{e^*, e^{**}\} \subset \hat{E}} : d(e^*, e^{**}|G) > d \right] \wedge \tag{1}$$
$$\wedge \left[(\mu(e''|G) = 1) \Rightarrow \left(e'' \notin \hat{E} \right) \right].$$

Example 1. Consider the division network G, shown in Fig. 2, into clusters satisfying (1).

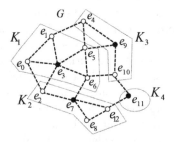

Fig. 2. An exemplary network with division into clusters

Figure 2 shown the division of the exemplary network into four clusters according to the specification $(2, 1|G)$ and satisfying (1). The set $\hat{E} = \{e_3, e_7, e_9, e_{11}\}$ of head nodes of clusters have been marked with filled circles, and the clusters have marked dotted line and were signed $\{K_1, K_2, K_3, K_4\}$. Based on the division of the structure, the communication structure of the network is created Fig. 3 shows an example of a communication structure for the network from Fig. 2.

We assume that communication between sensor nodes is carried out through the head nodes, and the head nodes can communicate with each other. The communication lines between the head nodes are not shown in Fig. 3, so as not to obscure the drawing. Communication between nodes is carried out using symmetric cryptography. The key distribution mechanism plays an important role in the process of securing communication. The procedure of key distribution is presented in Sect. 3.

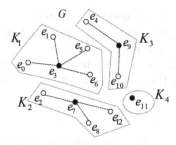

Fig. 3. Communication structure for the network from Fig. 2

3   The Procedure of Key Distribution

Figure 4 shows considered network structure. The model serves to illustrate the procedure for distributing a symmetric session key (*ses_Key*).

Assumed those sensor nodes e_{14} (node e_{ij} in procedure) and e_{24} (node e_{kl} in procedure) are nodes that want to set up secure mutual communication.

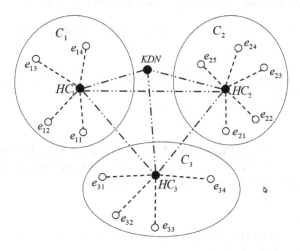

Fig. 4. The comprehensive model of communication structures

Symbols used:

C_i - symbol of a cluster (each cluster creates a security domain);

HC_i - head cluster in i-cluster (this node acts as the Gateway in the security domain);

KDN - key generator/distributing node. This node can be the head cluster node.

e_{ij} - j-sensor node in i-cluster;

pub_KDN - public key of the node KDN;

pub_e_{ij} - public key of node j-node in i-cluster;

Key_id - identifier of symmetric key (unique random number, e.g., 2 bytes long, generated by requesting sensor node);

$dest\_id$ - identifier of the destination node (it can be a mac address of e_{ij});

$source\_id$ - identifier of the source node (it can be a mac address of e_{ij}).

Key Distribution Procedure

The key distribution procedure considered in this section may be used in the sensor network only after the division of the network for the clusters has been established and after the formulation of the secure domain of the sensors in each cluster has been completed.

Assumptions:

– The sensor node e_{ij} is a member of the cluster C_i, and the sensor node e_{kl} is a member of the cluster C_k.

– The sensor node e_{ij}, whose identifier is $source\_id$, initiates the session key (ses_Key) distribution process for subsequent data exchange between the sensor node e_{ij} and sensor node e_{kl}, whose identifier is $dest\_id$.

The key distribution procedure consists of the following steps:

Step 1.

The node e_{ij} sends a request to HC_i for the pub_KDN (if it does not have this key so far).

Step 2.

If HC_i has the pu_KDN key, sends it to the node e_{ij}. Otherwise, HC_i sends the request to KDN, receives the pub_KDN, writes pub_KDN to its resources, and sends the key to the node e_{ij}.

Step 3.

The node e_{ij} saves the pub_KDN in its resources and prepares a packet containing the following data:

– pub_e_{14};
– Key_id;
– $source\_id$;
– $dest\_id$.

The packet is encrypted with the pub_KDN and sends to KDN via HC_i.

Step 4.

a. KDN after receiving the packet from node e_{ij} generates key and saves the key temporarily in its resources together with the Key_id, $dest\_id$, and $source\_id$;

b. KDN prepares packet and sends it to the node e_{ij} containing the following fields:
- Key_id;
- ses_Key;
- $dest\_id$.

The packet is encrypted using pub_e_{ij}.

Step 5.

The node e_{ij} after receiving the packet from KDN decrypts the packet and save the ses_Key and $dest\_id$ in its resources.

Step 6.

The node e_{ij} sends the packet containing Key_id to node $dest\_id$ via HC_i.

Step 7.

HC_i after receiving the packet from node e_{ij} determines to which cluster the node de_id belongs and then resends a received packet from to node e_{ij} to node $dest\_id$ via the head of its cluster.

Step 8.

a. The node e_{kl} after receiving the packet from node e_{ij} sends a request to head of its cluster C_k for the pub_KDN (if it does not have this key so far).

b. The node e_{kl} prepares a packet containing the following data:
- pub_e_{kl};
- Key_id – from received packet;
- $dest\_id$ – identifier of e_{kl}.

The packet is encrypted with the pub_KDN key and sent to KDN via HC_k.

Step 9.

a. After receiving the package from e_{kl} finds the Key_id in its resources, checks the recipient's address and prepares packet and sends it to the node e_{kl} via HC_k. The packet contains the following data:
- Key_id;
- ses_Key;
- $source\_id$;
- $dest\_id$.

The packet is encrypted using pub_e_{kl} from the received packet.

b. KDN removes data corresponding to the Key_id from its resources.

Step 10.

> The node e_{kl} after receiving the package from KDN, decrypts the packet and save the session key Key_id in its resources together with the $source\_id$. The node e_{kl} sends to the node e_{ij}, whose identifier is $source\_id$ the confirmation of receipt the session key Key_id.

It is worth noting that the described key distribution procedure concerns the secure transfer of one symmetric key between the parties. If there is a need to renew the session key, e.g., due to exceeding the key validity time, transferring the assumed portion of data volume using one key or due to loss of trust in the key in use, each party using the key can initiate the described procedure for the distribution of the new session key.

4 The Simulation Setup and Results of Energy Consumption Analysis

The purpose of our simulation tests was to estimate the energy consumption of ZigBee nodes involved in the process of determining and distributing the encryption key for secure transmission between two end devices. The simple simulation model described in [16] was used. In particular, it assumes a simplified method of calculating energy consumption, depending only on the amount of data transmitted and received by the node. The research was carried out in the Riverbed Modeler environment [16].

The simulation model implements physical and medium access control layers of the IEEE 802.15.4-2003 standard. The model supports the cluster-tree topology with PAN Coordinator, router, end devices and hierarchical routing between clusters. This implementation enables the cluster-tree network topology with proposed by authors [16,17] specific time division beacon scheduling mechanism.

The physical layer consists of a wireless radio transmitter (Tx) and receiver (Rx) compliant to the IEEE 802.15.4 specification, operating at the 2.4 GHz frequency band. A data rate is equal to 250 kbps. The transmission power is set to 1 mW, and the modulation technique is Quadrature Phase Shift Keying. The MAC layer implements the slotted CSMA/CA and GTS mechanisms [18]. MAC layer is also responsible for generating beacon frames and synchronizing the network when a given node acts as a PAN Coordinator. The application layer has a traffic generator, which generates unacknowledged and acknowledged data frames transmitted using slotted CSMA/CA.

The battery module is responsible for computing the consumed and the remaining energy during the active and inactive periods. The default values of the current draws are specified in Table 1.

The initial amount of battery energy before any activity is 34 560 joules (two AA 1,5 V 1600 mAh) and a power supply is 3 V. Configuration window with batteries parameters is shown in Fig. 5a.

Table 1. Node's energy leakage in different states

Energy consumption	Value
Receive mode	27.7 mA
Transmission mode	17.4 mA
Idle mode	35 µA
Sleep mode	16 µA

The traffic model is very simple and does not give the possibility of linking many profiled applications to the node. Configuration window of the traffic generator is shown in Fig. 5b.

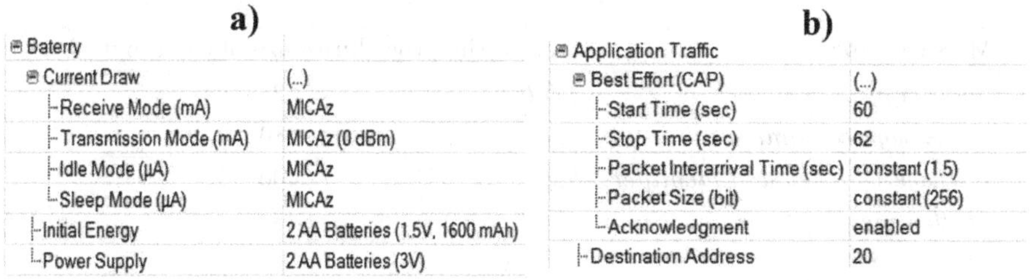

a)

⊟ Baterry	
⊟ Current Draw	(...)
⊢Receive Mode (mA)	MICAz
⊢Transmission Mode (mA)	MICAz (0 dBm)
⊢Idle Mode (µA)	MICAz
└Sleep Mode (µA)	MICAz
⊢Initial Energy	2 AA Batteries (1.5V, 1600 mAh)
└Power Supply	2 AA Batteries (3V)

b)

⊟ Application Traffic	
⊟ Best Effort (CAP)	(...)
⊢Start Time (sec)	60
⊢Stop Time (sec)	62
⊢Packet Interarrival Time (sec)	constant (1.5)
⊢Packet Size (bit)	constant (256)
└Acknowledgment	enabled
⊢Destination Address	20

Fig. 5. Node's batteries configuration window (a) and traffic generator of ZigBee node (b)

The study of energy consumption by individual nodes of the ZigBee network was carried out using a simple model, shown in Fig. 6. We intended to estimate the maximum possible energy consumption when two nodes establish a shared encryption key.

It contains two selected nodes for which the correct encryption key is determined before the basic transmission. These nodes are e_{14} and e_{24}. HC_1 router KDN and the HC_2 router are mediating in the process of generating keys for traffic encryption. Router HC_1 and HC_2 play the role of head clusters, and KDN (PAN Coordinator) functions as a key generator, as described in Sect. 3.

The simulation model includes data volumes exchanged between nodes in subsequent steps of the cryptographic key distribution procedure. Before the simulation test, it was estimated that the nodes would transmit short messages and longer ones, as in Table 2.

Taking into account the data volumes and the transmission rate of 250 kbps, it was estimated that the procedure for determining encryption key takes about 5.2 simulation seconds (sims) (the end-to-end delay in application layer was crucial here). The simulation showed an average end-to-end latency between two nodes at 0.65 sims. We do not take into account the time of information processing by the node and the network load caused by other transmissions.

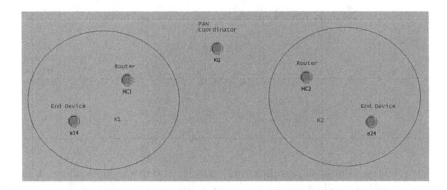

Fig. 6. Simulation model

Table 2. The size of the data generated in the application layer

Message type	Step of the procedure	Size of data [bytes]
Short request	*1,5,6*	32
A message containing a key	*2,7*	380
Package with key and identifiers	*3,4,8*	400
"hello" message	*9*	48

The next step was to determine energy consumption. For a clear illustration, the simulation procedure of encryption key distribution has been artificially stretched in time. It was assumed that it would be implemented between 60 and 118 sims. The whole experiment took 180 sims.

The basis for calculating the energy consumption during the encryption key distribution process was the speed of energy consumption of each node in the state of inactivity. For example, energy consumption of the ZigBee router and PAN Coordinator in the state of inactivity is shown in Fig. 7a.

Next, the energy consumption for each node was determined. For example, Fig. 7b shows the energy decrease curves for the coordinator in the inactive state and during encryption key distribution.

The curves of consumed energy and the difference in energy consumption in the PAN Coordinator's state of activity and inactivity are shown in Fig. 8.

The final results for each node are collected in Table 3. In the worst case (when has not the *pub_KDN* key) the total energy consumption is estimated at 0.464 joules, which is comparable with the results presented in [9] (1,5 joules for 10 active nodes).

At the current stage of research, important is that the model we used was simplified. In particular, the energy calculation procedure is very simple and does not take into account important parameters such as a distance between nodes, the temperature of nodes, time of information processing.

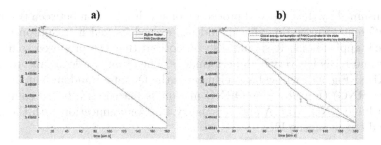

Fig. 7. Average remaining energy of router and PAN Coordinator (a) and global energy consumption of PAN Coordinator (b)

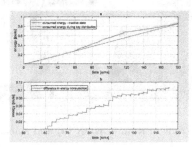

Fig. 8. Consumed energy (a) and the difference in energy consumption of PAN Coordinator (b)

Table 3. Collective results of simulation tests

	e_{14}	HC_1	KDN	HC_2	e_{24}
Energy consumption in inactive state [joule/s]	0.0018	0.0047	0.0019	0.0019	0.0018
Speed of energy consumption [joule/s]		0.0026	0.0064	0.0041	0.0038	0.0032
Total energy consumption [joule]		0.1508	0.3712	0.2378	0.2204	0.1856
Difference in energy consumption [joule]		0.0464	0.0986	0.1276	0.1102	0.0812
Total energy consumption [joule]	0.464				

It is, therefore, necessary to write the procedure for counting the energy consumption and the proper procedure (with event control) of generating network traffic during the distribution of encryption keys. It is also required to conduct tests in the conditions of normal communication between other nodes.

5 Conclusions and Future Work

From obtained results, it can be stated that energy consumption during the generation and distribution of the encryption key is small and does not significantly affect the overall energy condition of the network nodes. We estimated that the time consumed by the proposed procedure of key distribution between two nodes from different security domains should not exceed 5 s and it is appropriate for the MIoT applications. It should also be noticed that the proposed solution is scalable and KDN has very high performance. The primary role in this procedure plays KDN (Fig. 9a), whose basic element is the Quantis Random Number Generator (QRNG) (Fig. 9b). The QRNG module is placed on a board compatible with Arduino UNO/MEGA and Raspberry Pi. Communication with the board is carried out using the I2C.

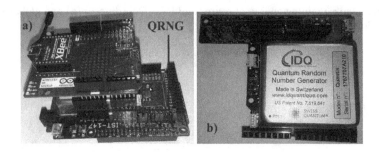

Fig. 9. An example configuration of KDN (a) and shield with QRNG (b)

Based on the results obtained in this work, some directions for follow-on research present themselves:

1. To develop fully integrated with the TPM v2.0 a LoRaWAN gateway with the QRNG device and embedded part of the proposed procedure to provide long-distance communication between domains and next to assess performance and scalability of this solution;
2. To conduct some comparison experimentations with the use of NB-IoT, Sigfox and LoRa WAN communications technologies to assess the applicability of these technologies in the proposed solution.

References

1. Zheng, D., Carter, W.A.: Leveraging the Internet of Things for a More Efficient and Effective Military. Rowman & Littlefield (2015)
2. Suri, N., et al.: Analyzing the applicability of internet of things to the battlefield environment. In: 2016 International Conference on Military Communications and Information Systems (ICMCIS) (2016)
3. Furtak, J., Zieliński, Z., Chudzikiewicz, J.: Security techniques for the WSN link layer within military IoT. In: 2016 IEEE 3rd World Forum on Internet of Things. WF-IoT. vol. 2016, pp. 233–238 (2016)
4. Suri, N., et al.: Leveraging LoRaWAN to support IoBT in urban environments. In: IEEE WF-IoT (2019)

5. Furtak, J., Zieliński, Z., Chudzikiewicz, J.: Secured domain of sensor nodes - a new concept. In: Hodoň, M., et al. (eds.) Innovations for Community Services. I4CS 2018, CCIS, vol. 863, pp. 1–11. Springer, Cham (2018). https://doi.org/10.1007/978-3-319-93408-2_15

6. Kodali, R.K., Chougule, S., Agarwal, A.: Key management technique for heterogeneous wireless sensor networks. In: IEEE 2013 Tencon - Spring, pp. 183–187 (2013)

7. Doyle, B., Bell, S., Smeaton, A., Mccusker, K., O'Connor, N.: Security considerations and key negotiation techniques for power constrained sensor networks. Comput. J. **49**(4), 443–453 (2006)

8. Oliveira, L., Scott, M., Lopez, J., Dahab, R.: TinyPBC: pairings for authenticated identity-based non-interactive key distribution in sensor networks. In: 5th International Conference on Networked Sensing Systems, 2008. INSS 2008, pp. 173–180 (2008)

9. Kulkarni, S., Ghosh, U., Pasupuleti, H.: Considering security for ZigBee protocol using message authentication code. In: IEEE INDICON, vol. 2015, pp. 1–6 (2015). https://doi.org/10.1109/INDICON.2015.7443625

10. Sun, M., Qian, Y.: Study and application of security based on ZigBee standard. In: 2011 Third International Conference on Multimedia Information Networking and Security (MINES), pp. 508–511 (2011)

11. Hoceini, O., Afifi, H., Aoudjit, R.: A new key management and authentication architecture for ZigBee networks (KAAZ). In: Int'l Conference on Security and Management, SAM 2017, CSREA Press (2017). ISBN: 1-60132-467-7

12. Corici, A., Emmelmann, M., Luo, J., Shrestha, R., Corici, M., Magedanz, T.: IoT inter-security domain trust transfer and service dispatch solution. In: WF-IoT, vol. 2016, pp. 694–699 (2016)

13. Chudzikiewicz, J., Zieliński, Z.: On some resources placement schemes in the 4-dimensional soft degradable hypercube processors network. In: Zamojski, W., Mazurkiewicz, J., Sugier, J., Walkowiak, T., Kacprzyk, J. (eds.) Proceedings of the Ninth International Conference on Dependability and Complex Systems DepCoS-RELCOMEX. June 30 – July 4, 2014, Brunów, Poland. AISC, vol. 286, pp. 133–143. Springer, Cham (2014). https://doi.org/10.1007/978-3-319-07013-1_13

14. Chudzikiewicz, J., Malinowski, T., Zieliński, Z.: The method for optimal server placement in the hypercube networks. In: Proceedings of the 2015 Federated Conference on Computer Science and Information Systems, ACSIS, vol. 2, pp. 947–954 (2015) https://doi.org/10.15439/2014F159

15. IDQ: Quantis AIS31 Brochure. https://marketing.idquantique.com/acton/attachment/11868/f-0220/1/-/-/-/-/Quantis

16. Jurcik, P., Koubaa, A.: The IEEE 802.15.4 OPNET Simulation Model: Reference Guide v2.0. Polytechnic Institute of Porto (ISEP-IPP) (2007)

17. Cunha, A., Alves, M., Koubaa, A.: Technical Report. Implementation of the ZigBee Network Layer with Cluster-tree Support. Polytechnic Institute of Porto (ISEP-IPP) (2007)

18. Sethi, S., Hnatyshin, V.Y.: The Practical OPNET User Guide for Computer Network Simulation. Chapman and Hall/CRC (2012)

Modeling of Computer Networks Using SAP HANA Smart Data Streaming

Monika Nycz[(✉)]

Institute of Informatics, Silesian University of Technology,
ul. Akademicka 16, 44-100 Gliwice, Poland
monika.nycz@polsl.pl

Abstract. This article presents an algorithm for transforming a mathematical model into a form accepted by the SAP HANA Smart Data Streaming (SDS). The implementation is intended to enable quick and easy processing of numerical data, generated from the model to evaluate the state of selected scenarios and network topologies. The mathematical model used is a fluid-flow approximation algorithm, describing changes in the dynamics of data transmission and changes in the queue length of packets waiting in network nodes over time. The adopted methodology assumes the use of an input stream, which transports the tuples with timestamps, as a trigger for determining the next steps of the model. Parameter values are stored in the column tables in the SAP HANA Platform database. In addition, the article proposes the extension of the model by the possibility of automatic and manual switching of parameters in the network. The purpose of the extension is to dynamically increase or reduce hardware resources for each of the network nodes, depending on the load of the given node. The modification introduces four additional thresholds, which are responsible for controlling the switching between operating modes: normal, high performance and energy saving. The introduction of modifications provides model's parameters reloads during numerical calculations. It also allows optimized resources utilization for network operations, with a small reduction in average flow throughput and a slight increase in the router's buffer occupancy.

Keywords: Fluid-flow approximation · Network topology ·
Power-saving · Energy-saving · SAP HANA · Smart data Streaming ·
SDS · Streaming · Data processing · Data mining ·
Computer networks · Dynamic parameters' switching

1 Introduction

Modern computer networks are constantly subjected to transient queue analysis, [3,6,8,11]. The main aim of modelling time-dependent flows and the dynamics of router queues changes is to have a possibility to predict QoS factors, such as packet loss probability and queuing delays. To achieve that we need efficient

© Springer Nature Switzerland AG 2019
P. Gaj et al. (Eds.): CN 2019, CCIS 1039, pp. 48–61, 2019.
https://doi.org/10.1007/978-3-030-21952-9_4

modelling tools. The ideal program should be able to generate, process and store large amounts of data, that are the results of the numerical calculations, in particular in real-time. However, the analysis of the changes in vast computer networks, like the Internet, assumes iterative, step-based calculations on large structures that depend on each other, so parallelization capabilities are limited, [10]. Therefore, the research has aimed to explore the possibility of transferring the modelling logic into a streaming processing tool, that is intended to operate on fast-changing, time-dependent data series. The focus is put on the use of streaming engine to carry out numerical modeling of Internet transmission dynamics with the use of well-known approximation method.

Making decisions based on the analysis of the infinite streams with variable events is one of the main goals of the Internet of Things, [4,12,13]. The possibility of combining a numerical analysis with real events occurring in the networks, would allow for an attempt to adapt the network traffic to the changing conditions, while maintaining QoS. In the era of growing demand for smart objects it is also necessary to enable the development of not only Internet of Things algorithms, but also the decision-making models through predictions of future events, especially that capturing data from an increasing number of devices can result in the transmission of millions of events per second. Storing such number of events and, what is more important, their subsequent analysis using classic database processing will be a time-consuming and impractical approach, because it assumes first collecting and storing all the data, and then running query on them. Moreover, with every data change, the query must be resend. Thus, it will be the analysis of the past rather than the present or the predition of the future. Even with the use of in-memory databases the query must be rerun, although in this case there is no problem with data refreshing. Stream processing, on the other hand, assumes the implementation of a continuous query. The input data is gathered from data sources and adapted to the necessary format so that the streaming engine can act on them (detect the information and react to it). The query (or multiple queries) is performed all the time. It is possible due to the fact that the streaming databases perform operations on a continuous flow of data (stream), represented in the form of ordered sequences of data tuples. This property has been used in this article to allow the evolution of the model in time. The proposed solution uses a Smart Data Streaming tool of SAP HANA Platform, [2], which is the improved and widely-used version of one of the original CEP (Complex Event Processing) platform, [5,14]. The main scripting language of SAP HANA SDS is CCL (Continuous Computation Language), which is based on SQL (Structured Query Language) and adapted for stream processing. It supports sophisticated data selection, calculation capabilities and manipulation of data during real-time continuous processing [1]. The basic contribution of the paper was to adapt the classic fluid-flow algorithm into a mixture of SQL language of database layer and CCL language of streaming processing engine. It is run using both engines. The main improvement is an extension of the model, which allows the dynamic change of the network parameters (which in classic model are constant) depending on the network nodes load. This in turn enables the energy-saving capabilities in the routers, without significant network throughput degradation.

The article presents in the following three sections a reference model of fluid-flow approximation, its implantation in CCL language and the extended model with energy-saving option. The last two sections present the numerical calculation performed on exemplary network scenario and the final conclusions.

2   Classic Fluid-Flow Approximation Model

Fluid-flow approximation is a well-known method for modeling transient states of computer networks, that considers only mean values of traffic intensity and service times. It was proposed in [7,9]. Mathematical model of fluid-flow approximation is based on first order differential equations, Eqs. (1) and (2), which are solved using step-wise numerical methods. Due to its simplicity, fluid-flow approximation has been used in the analysis of Internet broadcasts. The model used in the research was adapted to the analysis of TCP congestion window mechanisms and the RED (Random Early Detection) algorithm in routers.

A single network node (v) in a fluid-flow approximation model is defined as a service station with service intensity (C) along with a buffer for packets waiting for service (q) with finite occupancy (B). The model describes changes in the dynamics of data transmission and changes in the occupation of actual queues that buffer packets in network nodes. The changes are determined by a system of differential equations. Changes of the queue over time in the node, Eq. (1), are defined as the intensity of the input stream (defined as the sum of the throughput of K flows incoming to the node), and reduced by the intensity of the output stream from the node:

$$\frac{dq_v(t)}{dt} = \left(\sum_{i=1}^{K} \frac{W_i(t)}{R_i(q(t))} \right) \cdot (1 - p_v(t)) - \mathbf{1}(q_v(t) > 0) \cdot C_v \qquad (1)$$

A single TCP connection (i) is modelled in turn by the amount of data the source (client) can send (W) in a unit of time and the time delay of information flow in the network (R). Changes in the size of the congestion window, Eq. (2), are defined as an increment by 1 if there is no loss in the flow path at every RTT time, reduced manipulatively if the loss occurred:

$$\frac{dW_i(t)}{dt} = \frac{1}{R_i(q_i(t))} - \frac{W_i(t)}{2} \cdot \lambda_i(t), \qquad (2)$$

The supplementation of the model is the value of the loss ratio Eq. (3), which is determined as the current flows throughput multiplied by the probability of loss of a packet at any node (j) within the flow path (K):

$$\lambda_i(t) = \frac{W_i(t - \tau)}{R_i(t - \tau)} \cdot \left(1 - \prod_{j \in K}(1 - P_{ij}(t - \tau)) \right), \qquad (3)$$

The flow parameter specifying the RTT, Eq. (4), i.e. the round-trip time of the network information from the sender to the recipient and back, is directly dependent on the total current queue delay in all nodes on the route and on the total value of the propagation delay (Lp) on the links. The value of the RTT parameter affects the delay time (τ) of notifying the sender of the network status - the sender, only after the RTT time from the moment of sending the data packet, gets information whether the loss of the packet occurred and must either reduce the size of the congestion window or keep it unchanged.

$$R_i(\boldsymbol{q}(t)) = \sum_{j=1}^{K} \frac{q_j(t)}{C_j} + \sum_{j=1}^{K-1} Lp_j . \tag{4}$$

In addition, each router has a mechanism for active queue management, which is characterized by the probability of rejection of the packet in the node (p). This probability is determined by the moving average queue length of the router (x). The adopted mechanism in the classic model is RED, in which, depending on the value of the moving average queue length, Eq. (5), and two threshold values, a decision is made whether to launch the preventive packet rejection procedure. As a result the probability of packet loss is determined, Eq. (6).

$$x_{v_n}(t) = \alpha_v \cdot q_{v_n}(t) + (1 - \alpha_v) \cdot x_{v_{n-1}}(t). \tag{5}$$

$$p_v(x_v) = \begin{cases} 0, & 0 \leqslant x_v < t_{min_v} \\ \frac{x_v - t_{min_v}}{t_{max_v} - t_{min_v}} p_{max_v}, & t_{min_v} \leqslant x_v \leqslant t_{max_v} \\ 1, & t_{max_v} < x_v \leqslant B_v \end{cases} \tag{6}$$

3   Implementation of Classical Model Using Streaming Engine

The initial modelling algorithm with the use of a streaming engine assumed that two types of streams are defined in the system: stream of router parameters and stream of flow parameters. A single stream tuple consists of a timestamp and parameter values required to calculate new parameter values, depending on the type of stream: node or flow. Streams flow through three blocks of numerical calculations:

- determination of values of parameters of routers
- determination of values of parameters of flows
- determination of loss rates at network nodes

The tuples with the values of all nodes and flows for all time steps are completely stored in the windows - the full history of parameter changes is retained. In windows, it would not be possible to store only the values from the current step, due to the existing feedback of values in the fluid-flow approximation. Unfortunately, the analysis of data stream processing capabilities using a Smart Data Streaming engine has shown the limitations of built-in blocks. The most

serious of them turned out to be the inability to configure and/or program window retention policies according to custom needs. Both available options (keeping
a certain number of tuples or a specific time interval) do not meet the assumption of scheduling tuples at custom discretion. And so, the option of storing
tuples from the last time interval does not allow for inclusion of a timestamp
other than physical time. In turn, the option of keeping only a certain number
of tuples results in the loss of information necessary to correctly calculate the
model when next tuples arrive (older tuples according to physical time are automatically deleted). The use of a physical timestamp prevents the coherence of
the modelling algorithm (it assumes the necessity of having all the results from
the previous calculation step in order to determine values in the current step),
which leads to the potential occurrence of the race phenomenon. Resignation
from the built-in window mechanism caused another problem - the necessity of
introducing a blocking mechanism in individual blocks in anticipation of completing calculations related to the previous modelling step. Due to the disclosed
restrictions, the modelling algorithm has been modified in the CCL language in
such a way that currently a single tuple consists only of a timestamp consistent
with the current modelling step and acts as a trigger for calculations in individual blocks. Parameter values, on the other hand, are stored in dedicated tables,
Fig. 1, in the database:

- ROUTERS_SETTINGS - fixed and initial parameters of nodes,
- FLOWS_SETTINGS - fixed and initial parameters of flows,
- PATHS_SETTINGS - connections between routers based on flows flowing
  through them,
- LOSSES_HISTORY - history of loss rates for each flow
- ROUTERS_HISTORY - history of changes of parameters of routers
- FLOWS_HISTORY - history of changes of parameters of flows

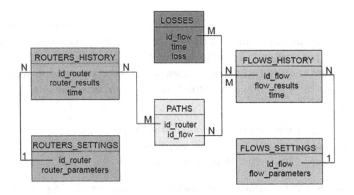

Fig. 1. Data scheme dedicated to the fluid-flow approximation model

Transferring the load of data storage to the database engine resulted in the
need to implement numerical calculations on the database side, for efficiency

reasons, as well as the limited mathematical capabilities of the environment itself. The resulting fluid-flow approximation algorithm, implemented in the SDS engine, Fig. 2, consisted of a tuple input stream with its own timestamp sent through the socket and published to the stream processing engine and passed between the parameter processing blocks. All processing blocks have been built using a dedicated Flex operator that enables software in the CCL language to process the event stream. The sequence of operations carried out in successive operators assumes in the first place the calculation of the parameter values for the zero step and, subsequently, the parameter values in the next steps. At the beginning of each non-zero step, a semaphore is placed waiting for the end of the previous block or the previous step (for the first Flex block). After the waiting is finished, a procedure is invoked that converts the model parameters adequate for the given block. At the end, a marker is set that allows to release the semaphore in the next block. The impossibility of storing model parameters on the side of the stream engine itself (due to the previously mentioned retention policies), results in the necessity of data storage and calculation of the model step on the database side. The procedure for determining parametric values is a procedure that uses only INSERT instructions (insert-only procedure). The individual processing blocks differ mainly with the called model procedures.

4   Model Extension with Automatic Modification of Parameters of Network Nodes

While creating a solution to the research problem, the concept of expanding the algorithm with the possibility of automatic modification of parameters of network nodes depending on the network load appeared, Fig. 3. The modification introduces four additional thresholds for switching parameter values, Eq. (7). It is assumed that in the zero step all nodes are in "Normal" mode (N), which in the real network may correspond to two service modules (working units). If the percentage of moving average queue occupancy of the node drops below the T_A threshold and the node is in the Normal mode, the node switches to "Energy-Saving" mode (ES) by reducing twofold: (1) the RED mechanism thresholds, (2) the service intensity and (3) the maximum buffer size; which may correspond to switching off one of the service modules available in Normal mode. If, on the other hand, the percentage of the average router's occupancy exceeds T_D threshold, while still in Normal mode, the node will switch to "High Performance" mode (HP) by increasing its parameters' values by half (this may correspond to turning on one additional working unit in the real device). If the node is in Energy-Saving mode and its percentage of x queue occupancy exceeds T_B threshold, the node will switch back to Normal mode. Similarly, the switch back to Normal mode occurs for the High Performance mode, when the percentage of moving average queue drops below T_C threshold. There is also the possibility of "manual" influence on parameter values by setting a special "reset" tag, transmitted along within a time tuple in the stream. This option is available only if the model is implemented in the stream processing engine.

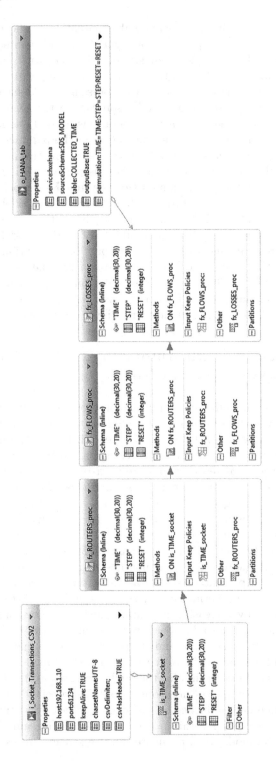

Fig. 2. Implementation of the model in the CCL query language

In addition, the "manual" modification option is independent of the duration of the modelling.

$$
mode_v = \begin{cases} ES, & \frac{100 \cdot x_v}{B_v} < T_{A_v} \quad \wedge N \; mode \\ N, & \frac{100 \cdot x_v}{B_v} > T_{B_v} \quad \wedge ES \; mode \\ N, & \frac{100 \cdot x_v}{B_v} < T_{C_v} \quad \wedge HP \; mode \\ HP, & \frac{100 \cdot x_v}{B_v} > T_{D_v} \quad \wedge N \; mode \end{cases} \tag{7}
$$

The algorithm allows for the formal elaboration of the extension of the fluid-flow approximation model. The definition of fixed parameters has changed compared to the classical model: the minimum, Eq. (8), and maximum threshold, Eq. (9), for the RED algorithm, the maximum buffer size, Eq. (10), and the node service time, Eq. (11). Note that the RED mechanism is still responsible for the packet dropping, while the extension enabled the packet rejection boundary to be shifted up, by providing the possibility to increase the hardware resources of network nodes when needed.

$$
t_{min_v} = \begin{cases} \frac{t_{min_v}}{2}, & ES \; mode \\ t_{min_v}, & N \; mode \\ \frac{3 t_{min_v}}{2}, & HP \; mode \end{cases} \tag{8}
$$

$$
t_{max_v} = \begin{cases} \frac{t_{max_v}}{2}, & ES \; mode \\ t_{max_v}, & N \; mode \\ \frac{3 t_{max_v}}{2}, & HP \; mode \end{cases} \tag{9}
$$

$$
B_v = \begin{cases} \frac{B_v}{2}, & ES \; mode \\ B_v, & N \; mode \\ \frac{3 B_v}{2}, & HP \; mode \end{cases} \tag{10}
$$

$$
C_v = \begin{cases} \frac{C_v}{2}, & ES \; mode \\ C_v, & N \; mode \\ \frac{3 C_v}{2}, & HP \; mode \end{cases} \tag{11}
$$

5   Numerical Example

The scenario assumed a comparative analysis of the classical model with the extended model with automatic reconfiguration of the network consisting of 494 nodes and 50 flows in the time interval of 100 s. Initial settings assumed that:

- queues in nodes are empty,
- the initial window value is one,
- the maximum threshold of the RED mechanism equals 75% of the buffer size,
- the minimum threshold of the RED mechanism equals 25% of the buffer size.

With such settings a number of modelling calculations were performed to determine the best settings for Extended model threshold values ($T_A, T_B, T_C, T_D,$

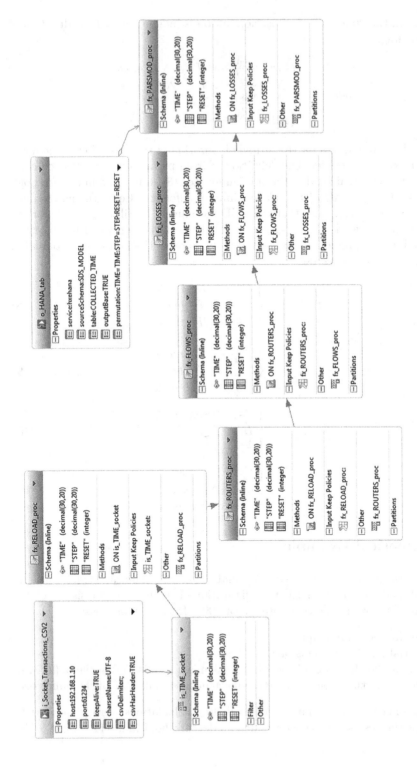

Fig. 3. Implementation of the extended model with the automatic reconfiguration of the network in the CCL query language

Eqs. (8) to (11)). This is really important because the choice of switching thresholds values has a significant impact on: (1) the performance of the whole switch algorithm; (2) the occurrence of value oscillations around the switching thresholds; (3) the cost minimization of the network infrastructure. As threshold values per node, that induce the best results, were chosen:

- $T_{A_v} = 25\%$ of the minimum RED threshold t_{min} of the current mode (N)
- $T_{B_v} = 25\%$ of the maximum RED threshold t_{max} of the current mode (ES)
- $T_{C_v} = 25\%$ of the minimum RED threshold t_{min} of the current mode (HP)
- $T_{D_v} = 25\%$ of the maximum RED threshold t_{max} of the current mode (N)

The case analysis was focused on the comparison of the average values of the entire network parameters, changing over time. The first one shows the average occupation of queues of network nodes, expressed as a percentage value over time, Fig. 4. The introduction of modifications increased the load on network nodes by an average of 1%. However, the average actual queue occupancy is still around 8% in both cases, so both models behave at a similar level. Moreover, in few moments the queue occupancy in extended model was smaller than in classic one.

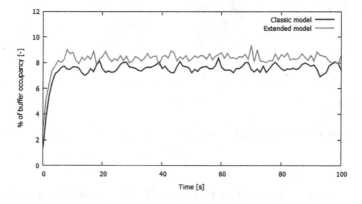

Fig. 4. Average queue occupancy of nodes in the network over time

The increase in the load of the nodes in the network generally results in a decrease of throughput in such network, Fig. 5. In this case, reducing the sizes of the windows in relation to the RTT times (throughput) turned out to be insignificant, on average at most 3 pac/s. In addition, with the increase of time, the difference between the throughput of classic and extended model was getting smaller.

The change in throughput in the Extended model has its source in increasing (by about 0.5%) the number of routers, in which the use of the average queue (x) exceeded 75% of the whole buffer size, Fig. 6. The number of the remaining nodes (with average queue less than 75%) in both models remains at a similar level.

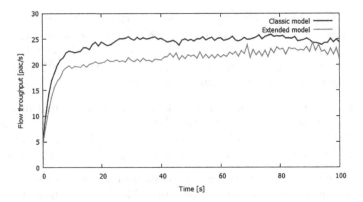

Fig. 5. Average flows throughput in the network over time

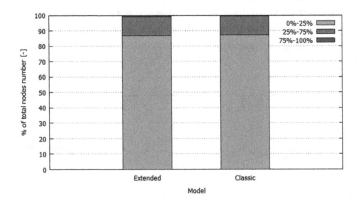

Fig. 6. Average percentage number of nodes with queue occupancy in three defined ranges

Overloading these few routers more (0.5%) in the extended model has not negatively affect the estimated total number of dropped packets in the network, Fig. 7. Even more, at certain times, the number of drops is smaller then in classic model.

The final step, in assessing the usefulness of modifying the model, was to compare the number of nodes working in one of three defined modes (Energy-Saving, Normal, High Performance), Fig. 8. The generated results of Extended model indicate a significant number of nodes working in the Energy-Saving mode (83.4%) and a small number of nodes working in Normal (8.5%) and High Performance mode (8.1%). While in Classic model all nodes may be allowed to work only in Normal mode.

As a result, the energy consumption of the working units in Classic model is much higher then it may be in Extended model, assuming the preservation of a similar network load and flows efficiency. Moreover, the Classic model option needs to sustain a larger number of working units than it is practically required

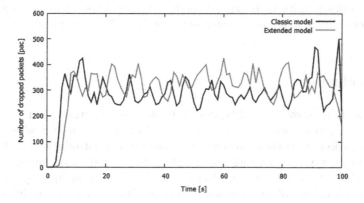

Fig. 7. Total estimated number of dropped packets in the network over time

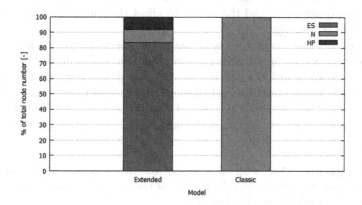

Fig. 8. Average number of routers working in each mode in each model

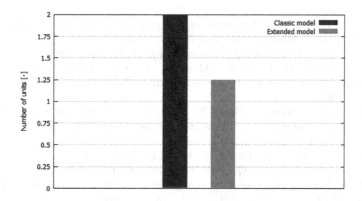

Fig. 9. Average number of working units in the network

(1.25 of working units in Extended model), Fig. 9. This results in significant increase of maintaining costs of such network by 37.5% compared to Extended model.

6 Conclusions

This research has shown that it is possible to use a streaming engine in numerical network calculations, its easy modification and creation of new models, including modification of network parameters on-the-fly. The introduction of dynamic switching of the parameters allows to reduce the cost of maintaining the network by 37.5% with simultaneous decrease in the total throughput by only 12%. Undoubted drawback of this solution is the rise of the time of determining of the values of the model in comparison to the classical solutions based on a dedicated structure. However, the unquestionable advantages are the ability to quickly, easily and unlimitedly extract the knowledge about the network status in a given time range of observation and the ability to affect the values of the model while the algorithm is running.

The tool, created because of work on the implementation with the use of the stream processing engine, allows for very fast configuration and reconfiguration operations of the model parameters. It gives the possibility of performing analyses even during numerical calculations and is also fully scalable. A large disadvantage of the solution is the lack of out-of-the-box support in the tool in case of a non-standard problem, that the stepwise solving of fluid-flow differential equations is.

Acknowledgments. This work was supported by Statutory funds for young researchers (grant no. BKM-509/RAU2/2017) of the Institute of Informatics, Silesian University of Technology, Gliwice, Poland.

References

1. CCL Language. https://help.sap.com/doc/saphelp_esp_51sp09_sug/5.1.9/en-US/e7/931cee6f0f101486068ade550250ad/frameset.htm
2. SAP HANA Platform. https://www.sap.com/products/hana.html
3. Abate, A., Chen, M., Wang, Y., Zakhor, A., Sastry, S.: Design and analysis of a flow control scheme over wireless networks. Int. J. Robust Nonlinear Control **23**(2), 208–228 (2013)
4. Armstrong, M.P., Wang, S., Zhang, Z.: The internet of things and fast data streams: prospects for geospatial data science in emerging information ecosystems. Cartography Geogr. Inf. Sci. **46**(1), 39–56 (2019)
5. Gualtieri, M., Curran, R., Kisker, H., Miller, E., Izzi, M.: The Forrester Wave: Big Data Streaming Analytics, Q1 2016. Technical report, Forrester (2016)
6. Johari, R., Tan, D.K.H.: End-to-end congestion control for the internet: delays and stability. IEEE/ACM Trans. Netw. **9**(6), 818–832 (2001)
7. Liu, Y., Presti, F.L., Misra, V., Towsley, D., Gu, Y.: Fluid models and solutions for large-scale IP networks. In: ACM/SigMetrics (2003)

8. Lv, H., Lin, J., Wang, H., Feng, G., Zhou, M.: Analyzing the service availability of mobile cloud computing systems by fluid-flow approximation. Front. Inf. Technol. Electron. Eng. **16**(7), 553–567 (2015)

9. Misra, V., Gong, W.-B., Towsley, D.: A fluid-based analysis of a network of AQM routers supporting TCP flows with an application to red. In: Proceedings of the Conference on Applications, Technologies, Architectures and Protocols for Computer Communication (SIGCOMM 2000), pp. 151–160 (2000)

10. Nycz, M., Nycz, T., Czachórski, T.: Modelling dynamics of TCP flows in very large network topologies. In: Abdelrahman, O.H., Gelenbe, E., Gorbil, G., Lent, R. (eds.) Information Sciences and Systems 2015. LNEE, vol. 363, pp. 251–259. Springer, Cham (2016). https://doi.org/10.1007/978-3-319-22635-4_23

11. Nycz, M., Nycz, T., Czachórski, T.: Performance modelling of transmissions in very large network topologies. In: Vishnevskiy, V.M., Samouylov, K.E., Kozyrev, D.V. (eds.) DCCN 2017. CCIS, vol. 700, pp. 49–62. Springer, Cham (2017). https://doi.org/10.1007/978-3-319-66836-9_5

12. Palattella, M.R., Dohler, M., Grieco, A., Rizzo, G., Torsner, J., Ladid, T.E.L.: Internet of things in the 5G era: Enablers, architecture, and business models. IEEE J. Sel. Areas Commun. **34**(3), 510–527 (2016)

13. Sivarajah, U., Kamal, M.M., Irani, Z., Weerakkody, V.: Critical analysis of big data challenges and analytical methods. J. Bus. Res. **70**, 263–286 (2017)

14. Woodie, A.: Streaming Analytics Ready for Prime Time, Forrester Says. https://www.datanami.com/2014/07/22/streaming-analytics-ready-prime-time-forrester-says/

An Impact of Jamming Signal
on the Energy Efficiency of ZigBee
Network Elements

Dariusz Czerwinski[1]([✉]), Jaroslaw Nowak[2], and Slawomir Przylucki[1]

[1] Lublin University of Technology, 38A Nadbystrzycka Str, 20-618 Lublin, Poland
{d.czerwinski,s.przylucki}@pollub.pl
[2] Polish Air Force Academy, Dywizjonu 303 no. 35 Str, 08-521 Deblin, Poland
j.nowak@wsosp.pl

Abstract. The paper presents evaluation of the energy efficiency of ZigBee network elements in the presence of disturbances caused by a jammer. The measurements were performed with the use of two Digi XBee-PRO S1 802.15.4 devices and Arduino boards configured for echo tests purposes. The CRJ4000 cell phone and WiFi ISM band handheld jammer was used as a jamming source. The influence of the radio power level of Xbee module during jamming conditions on the communication quality and the energy consumption was measured. Authors proposed the policy of switching the radio power level of Xbee modules in the presence of jamming signal.

Keywords: XBee · Energy efficiency · Jamming conditions ·
Power switching policy

1 Introduction

Wireless Sensor Networks (WSN) are commonly used nowadays and technologies, used to built these networks, become more and more advanced. Thanks to this, they can be expanded into many different areas [1,2]. A very important issue related to the WSN is power consumption and the ways of its reduction. This can be achieved in different ways: data reduction, duty cycling or control the transmitter power. Recently it can be observed, that there are many works devoted to energy efficiency in wireless sensor networks [3–6]. Horvat et al. in the paper [7] analyse the impact of several parameters on power consumption of Digi's XBee PRO modules. Based on the measurement results, the authors built a mathematical model of the node's power consumption. Some optimization of power consumption was made for various parameters using desirability functions.

In the earlier study of these authors [8] another approach in power consumption reduction i.e. the Adaptive Transmission Power Control (ATPC), where the end nodes reduce their transmit power in steady conditions, was proposed.

© Springer Nature Switzerland AG 2019
P. Gaj et al. (Eds.): CN 2019, CCIS 1039, pp. 62–75, 2019.
https://doi.org/10.1007/978-3-030-21952-9_5

In this study the authors presented the analysis of indoor propagation measurements in order to determine the minimum value of received signal strength for high reliability of data transfer.

Fedor et al. in the paper [9] explore the influence of the routing on the energy demand of WSNs. The authors showed when multi-hop routing is more energy efficient compared to direct transmission to the sink. They also specified the conditions, for which the two-hop strategy is optimal. Performed tests showed that the advantage of the multi-hop scheme depends on the source-node distance and packet collection cost.

One of the most important concerns shown in [10] is power optimization focused on the minimum of transmission power in order to extend as much as possible the WSNs lifetime. The authors proposed the transmission power self-optimization (TPSO) technique consisting of an algorithm able to ensure the connectivity and high quality of service (QoS). They demonstrate the TPSO technique effectiveness through the set of measurements with different type of noise and electromagnetic interference (EMI). Results of measurements showed that other radio signals (AM/FM and WiFi) have great impact on WSN performance which drops from 93% to 37%–67% depending on power mode used in the XBee PRO devices [10].

In [11] the authors presented the Waspmode device which great advantage is low power consumption. This energy solution is based on programmed algorithm supposed to pause the program operations until specific events appear. There are four operational modes of the device: normal mode with a 9 mA current consumption, sleep mode with a time interval between 32 ms and 8 s, deep sleep with a time interval from 8 s to minutes, hours, days, and hibernate mode. Time triggered power saving solution for WSN composed of XBee RF modules has been presented in [12]. The results showed that the proposed solution can help to achieve a great reduction in power consumption. An energy depletion attack on ZigBee wireless network was the research subject of Cao et al. in [13]. In this type of attack the bogus messages to lure a node are constructed, thus this leads to unnecessary security-related computations which intentionally deplete node's energy. Authors in [13] proposed several recommendations on how to proceed during the ghost energy depletion attack in ZigBee networks.

Our contribution, presented in this work consists of the measurements of energy consumption of ZigBee network elements in the presence of disturbances caused by the jammer. Based on the measurement results in can be noticed that during jamming, energy consumption of XBee PRO S1 devices increases significantly. We also propose the way of jamming conditions recognition with the use of software spectrum analyser and policy of energy saving with the use of CSMA/CA method. The remainder of the paper is organised as follows: Sect. 2 describes the test stand configuration for measurements of energy consumption. Section 3 describes the obtained results. Energy saving model in jamming conditions is presented and discussed in Sect. 4. Conclusions and some further research ideas are presented in Sect. 5.

2 Test Stand for Measurement the Influence of Jamming Signal

To measure the influence of the jamming signal on the energy efficiency of XBee network elements the test stand was set up. It consists of the following elements:

- two modules XBee-PRO S1 802.15.4 (transmitter and receiver) [14],
- PC notebook for data collection,
- CRJ4000 cell phone and WiFi handheld jammer [15],
- AirMagnet Spectrum XT Fluke analyser for data measurements and RF analysis [16].
- milli amperometer UT70A multimeter,
- Arduino Uno board for transmitter implementation.

The measurement test stand used in research is shown in Fig. 1. Measurements are based on recording send packet while increasing the jammer signal power. The XBee radio transmitter module was controlled by the Arduino Uno board with a program simulating the echo type behaviour, i.e. the transmitter (Fig. 2a) sends a packet of information counted in the `counter_pt` variable, then switches to receive mode and receives the packet sent by the twin device (receiver - Fig. 2b) located 100 m from the transmitter. The program checks if the packet received is identical to the packet sent. If packets are the same, than the program stores the number of packages in the `counter_pr` variable.

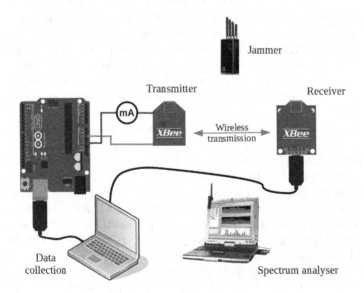

Fig. 1. The measurement test stand

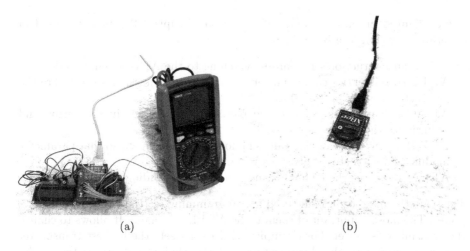

(a)   (b)

Fig. 2. Test stand main elements: (a) transmitter with multimeter, (b) receiver

The accuracy of the UT70A multimeter used in the experiment was as follows: range 200 mA, resolution 100 μA, accuracy $\pm(1.5\% + 1)$ [17].

Measurements presented in this article were performed in open area in the grassy terrain during early winter season in LoS (Line of Sight) conditions. The first scenario (scenario 1) of tests assumed measurements in the absence of interference (jammer was turned off) and the second scenario (scenario 2) assumed the communication while the jammer was turned on. Jammer was placed at the line of sight of both Xbee modules, at a distance 5 m from the receiver and 95 m from the transmitter. In second scenario the power signal level recorded by the spectrum analyser placed in place of receiver was equal to −60 dBm. In that circumstances the transmission is completely jammed. The summary of the jammer parameters was presented in Table 1.

In order to perform measurements the test stand should has an ability to change the transmission parameters of XBee-PRO S1 modules programmatically. It was done with the set of AT commands. This way, the XBee module

Table 1. Summary of jammer parameters

Model	CRJ4000
Output power	2.5 W
Shielding radius	up to 20 m at −75 dBm
Frequencies	2400–2500 MHz
Antennas	omnidirectional
Antennas gain	2.8 dBi (2.4 GHz)
Operating temperature	15 °C to 60 °C
Power supply	AC 110–240 V (50–60 Hz), DC 12 V, 1800 mAh Li-Ion

was configured in various modes of operation. Sample AT commands used in experiments were given below:

– +++ - an instruction that causes switching to the AT command mode,
– AT PLparameter/r - command for setting the transmission level, where the parameter are values from 0 to 4,
– AT AC/r - apply changes to the parameter values set in the queue and re-initialize the device,
– AT WR/r - saves parameter values in non-volatile memory, so that parameter modifications are stored after resetting,
– AT CN/r - leaves the command mode and apply pending changes.

These are the main commands used for programming (setting) the transmitter's power. The most important of them is the AT PL command. It allows to change the transmitter's power. For example: AT PL4/r - sets the highest transmitter power. According to the documentation of the Digi manufacturer the modes shown in Table 2 define the proper radio power of XBee Pro S1 international version used in experiment [18].

Table 2. Power mode and corresponding transmitter power levels

Mode	Power level
0	$-3\,$dBm
1	$-3\,$dBm
2	$2\,$dBm
3	$8\,$dBm
4	$10\,$dBm

The purpose of the experiments was to assess the of jamming signal on the energy efficiency of XBee network elements during communication. For this reason, different power levels of radio modules of XBee devices were taken into consideration, as well as different baud rates. Measurements were performed in the LoS conditions. Outside temperature was in range of 5–8 °C. Both XBee devices were placed on the wooden stands, 62 cm above the ground. Before turning on the jammer (measurement scenario 1) the background noise was measured with the Fluke XT analyser. Obtained value of the noise was in range $-110\,$dBm to $-105\,$dBm. In the measurement scenario 2 the jammer was placed 5 m from the transmitter and it was turned in order to jam the XBee modules communication. In this case 100% of packet loss in communication between transmitter and receiver was recorded.

Both XBee modules have been set as end devices and worked in Peer-to-Peer network topology. Firmware written to the radio modules was XBP24-15-4 for radio hardware XBee PRO 802.15.4.

3   Results of the Measurements

During the measurements the values of current flowing through the XBee module and number of packets sent and returned to the transmitter were recorded. The 10 independent registrations were carried out in each measurement and then the average value for given measurement was calculated. Current was measured with a 60 s time step. The experiment No. 1 was performed with standard baud rate set to 9600 bit/s and the successful round trip transmission (transmitter-receiver-transmitter) was recorded. Results for scenario 1 (no jammer) and scenario 2 (jammed signal) are presented in Table 3.

Table 3. Experiment No 1 results for baud rate 9600 bit/s

Power mode set by AT PL command	Maximum current - scenario 1 (mA)	Maximum current - scenario 2(mA)
0	62.5 ± 0.948	72.8 ± 1.102
1	63.5 ± 0.963	77.3 ± 1.170
2	64.7 ± 0.981	80.5 ± 1.218
3	65.8 ± 0.997	80.9 ± 1.224
4	64.6 ± 0.979	81.1 ± 1.227

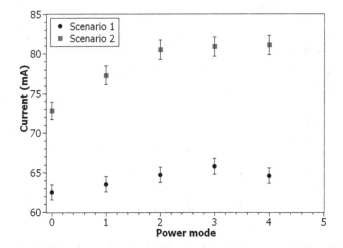

Fig. 3. Measurement results for baud rate 9600 bit/s and scenario 1 and 2

It can be noticed that there is difference in energy consumption comparing scenario 1 and scenario 2. To observe the trend of changes the chart with error bars has been prepared and shown in Fig. 3. What is interesting is that power modes higher or equal 2 have similar energy consumption in one scenario. In case of scenario 1 electrical power values, for supplying voltage 3.3 V, are in range from 213.1 MW to 217.1 MW and in case of scenario 2 in range 265.6 MW to 267.6 MW. The second interesting behaviour is that according to the Table 2, given by the manufacturer, modes 0 and 1 should use the same amount of power. This was not confirmed in measurements, especially in scenario 2 (Fig. 3) where difference between these two modes in energy consumption was about 15 MW (4.5 mA). The experiment No. 2 relied on setting the transmitter to transmitting only mode (no echo type behaviour) and increase the baud rate to 115200 bit/s. Measurement results were shown in Table 4 and in Fig. 4.

Table 4. Experiment No 2 results for baud rate 115 200 bit/s

Power mode set by AT PL command	Maximum current - scenario 1 (mA)	Maximum current - scenario 2 (mA)
0	92.1 ± 1.392	109.2 ± 1.648
1	92.5 ± 1.398	108.9 ± 1.644
2	95.8 ± 1.447	108.5 ± 1.638
3	92.8 ± 1.402	106.7 ± 1.611
4	103.5 ± 1.563	111.6 ± 1.684

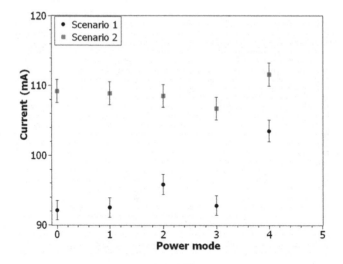

Fig. 4. Measurement results for baud rate 115 200 bit/s and scenario 1 and 2

Results of experiment No. 2 differ comparing to the previous one. One can noticed that energy consumption in modes 0–3 is very similar in each scenario. Only in mode 4 in both scenarios the current increases significantly (see Fig. 4). In case of mode 4 electrical power values are equal to 341.55 MW for scenario 1 and 368.28 MW for scenario 2. The average power consumption in modes 0–3 are equal to 307.89 MW in scenario 1 and 357.47 MW in scenario 2.

According to the manufacturer specification [18] the maximum transmit current for XBee-PRO S1 international variant (RPSMA module only) is equal to 150 mA while supplying voltage is equal to 3.3 V (electrical power consumption is equal to 495 MW). In experiment No. 2 the maximum current was equal to 112 mA, therefore we decided to do an additional experiment No. 3 in which the measurements were in line with scenario 2 (jammer was turned on) but the baud rate was set to 250 000 bit/s.

Table 5. Experiment No 3 results for baud rate 250 000 bit/s

Power mode set by AT PL command	Maximum current - scenario 2 (mA)	Electrical power - scenario 2 (MW)
0	115.1 ± 1.737	379.83 ± 5.730
1	128.7 ± 1.941	424.71 ± 6.404
2	130.8 ± 1.972	431.64 ± 6.508
3	131.5 ± 1.983	433.95 ± 6.542
4	139.5 ± 2.103	460.35 ± 6.938

The results of experiment No. 3 were shown in Table 5 and Fig. 5. It can be noticed that value of the current (about 140 mA) in power mode 4 is close to that given by the manufacturer i.e. 150 mA. The difference in electric power consumption comparing minimum and maximum value is much higher than in case of power mode 0 and 4. The difference between the maximum and minimum current it is equal to 24.4 mA and in case of electrical power is equal 80.52 MW.

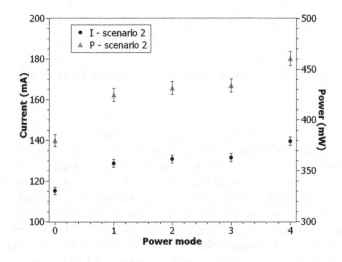

Fig. 5. Measurement results for baud rate 250 000 bit/s and scenario 2

Still, it is noticeable that according to the Table 2 modes 0 and 1 should use the same amount of power. This was not reflected in measurements (Fig. 5), where difference between these two modes in energy consumption was about 45 MW (13.6 mA). This situation is similar to that shown in experiment No. 1.

4   Energy Saving Model in Jamming Conditions

An XBee 802.15.4 PRO S1 modules have the implementation of carrier sense multiple access with collision avoidance protocol abbreviated as CSMA/CA. It can be disabled/enabled using AT commands and "RN" parameter [19].

In this article we propose to use CSMA/CA algorithm to stop the transmission in case of jammer detection. The XBee devices have some ability to detect the presence of jamming. This can be done with the use of ATCA command and setting the desired CCA (Clear Channel Assessment) threshold [20]. This allows to run simple spectrum analyser which is implemented in XCTU tool, or implement spectrum analysis in own code. The results of XCTU tool measurements, showing the difference in spectrum during normal and jammed conditions, were shown in Fig. 6.

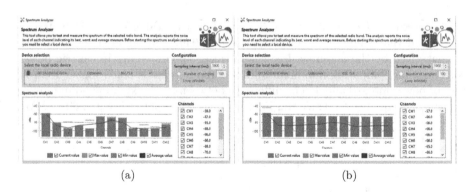

(a)   (b)

Fig. 6. The XCTU XBee spectrum analyser results: (a) without jamming, (b) with jamming

It can be noticed that in Fig. 6b all channels in spectrum are filled with jammer transmission, which maximum values are about −66 dBm. It can be assumed that detection of jamming can be done by checking the values of the signal in all channels with the use of spectrum analyser. In such case, to save the XBee devices energy, the policy of stopping the transmission can be applied. Similar assumptions are also analysed by other researches. In [24] the authors considered N nodes communicating directly with the sink using CSMA/CA access mechanism. Each node was equipped with PVC energy harvesting panel and energy prediction based on energy management algorithm (EPEM) was proposed. In the other work [25] the two dimensional Markov model is presented to study the

MAC layer behaviour of a single 802.15.4 node following slotted CSMA/CA for energy consumption optimization.

In this article, basing on the results presented in previous chapter, the theoretical worst case energy usage was calculated. To model the energy consumption during jamming conditions it is necessary to calculate the time of sending the frames and worst case backing off time. Basing on the XBee specification [20] and information in other articles [21–23], dealing with energy consumption and CSMA/CA topic, an initial values of the variables were set accordingly: $macMaxFrameRetries = 7$, $macMaxCSMABackoff = 5$, $macMaxBE = 8$, $macBeaconOrder = 15$, $macMaxCSMABackoffs = 4$, $macMinBE = 3$, $macSuperframeOrder = 15$. Next step is the calculation of the total worst case time according to Eq. 1.

$$T_{total} = macMaxFrameRetries \cdot (T_{backoff} + T_{frame} + T_{frack} + T_{ack} + T_{LIFS}) \quad (1)$$

where: $T_{backoff}$ is the worst case back off period expressed in s, T_{frame} is the time for a frame with maximum payload length, T_{frack} is a time between frame and corresponding ACK, T_{ack} is a transmission time for ACK, T_{LIFS} is a inframe spacing time, LIFS (Long InFrame Spacing) time is used when the MPDU is greater than 18 bytes.

The time $T_{backoff}$ is calculated in accordance with CSMA/CA algorithm, however the worst case values are calculated for maximum number of back off slots equal to $2^{macMaxBE} - 1 = 255$. Electrical power is calculated for supplying voltage equal 3.3 V and current measured in experiments (Tables 3 and 5) according to expression $P = U \cdot I$. Used energy can be express as $E = P \cdot t$, where t is the time of transmitting. The results of the power consumption for baud rate 9600 bit/s and scenario 2 (current values are shown in Table 3) were presented in Figs. 7 and 8. Calculated worst case total time after which there is no retransmission is equal to 75.48 s. Figure 7 shows the results of energy usage while there is no back off mechanism implemented. Total amount of cumulated power used for

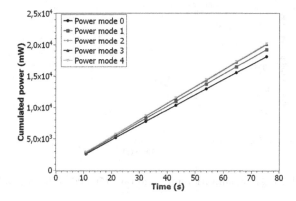

Fig. 7. Modelled energy consumption for baud rate 9600 bit/s and no backoff mechanism

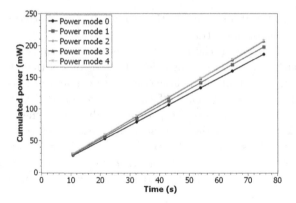

Fig. 8. Modelled energy consumption for baud rate 9600 bit/s and CSMA/CA implemented

data transmission is in range from 18 W to 20 W, depending on the power mode. The other situation can be observed in Fig. 8 where the results of calculations with implemented the unslotted CSMA/CA algorithm were presented. It can be noticed that energy usage is much lower and reaches the maximum value 0.2 W.

Calculations were also performed for baud rate 250 000 bit/s and current values presented in Table 5. Figures 9 and 10 presents the results of calculations. Calculated worst total time was equal to 2.83 s. Total amount of energy used for data transmission in this experiment was in range from 1 W to 1.3 W, depending on the power mode (Fig. 9). Values in case of implemented back off algorithm were shown in Fig. 10. In this case the total amount of energy used is equal to 13.39 MW for power mode 4.

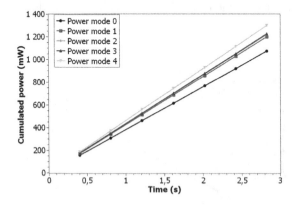

Fig. 9. Modelled energy consumption for baud rate 250 000 bit/s and no backoff mechanism

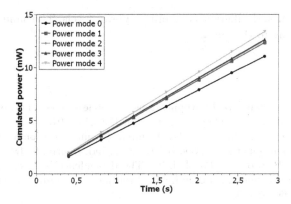

Fig. 10. Modelled energy consumption for baud rate 250 000 bit/s and CSMA/CA implemented

5   Conclusions and Future Work

The article presents the results of measurements of energy consumption of ZigBee network elements during jamming conditions. Setting the higher power modes causes the increase in power consumption which is shown in Tables 3, 4, 5. An interesting situation can be observed when power modes 0 and 1 are set. Accordingly to the manufacturer data radio power levels in these modes are the same (Table 2), however taking the measurements into consideration the radio module current consumption in these modes are different for two baud rates (Tables 3 and 5). Similar behaviour can be observed in [7] where switching between the power modes 0 and 1 in XBee Pro S1 device also increases the electrical current consumption.

Experimental results showed that jammed transmission causes the increase in current consumption (Figs. 3, 4, 5). This leads to the drop down in energy efficiency of ZigBee network elements. In the case of baud rate equal to 9600 bits/s power consumption increases in range 16.5–25.5% (average 22.22%). Lower values were obtained in case of baud rate equal to 115 200 bits/s. Power consumption increases in range 7.8–18.6% (average 14.5%).

To decrease the energy consumption the authors in this article propose the energy saving model. This model is based on the ability of XBee PRO S1 device software spectrum analyser to detect the jammer presence and use CSMA/CA algorithm to stop the transmission. Obtained values of worst case time T_{total} are accordant to values given in other authors work dealing with CSMA/CA method i.e: [26–28]. Cumulative power consumption in case of implemented CSMA/CA model is much lower (about 97 times) comparing to case with no backoff mechanism (see Figs. 7, 8, 9, 10).

In the future work the authors plan to extend measurements for checking how the CSMA/CA algorithm influence on power consumption during transmission howled down by the jammer. This can be measured for different values of jamming signal and different transmitter power levels. The simulation results are very promising in this area.

References

1. Rashid, B., Rehmani, M.H.: Applications of wireless sensor networks for urban areas: a survey. J. Netw. Comput. Appl. **60**, 192–219 (2016)
2. Gaj, P., Jasperneite, J., Felser, M.: Computer communication within industrial distributed environment-a survey. IEEE Trans. Ind. Inform. **9**, 182–189 (2013)
3. Mosorov, V., Biedron, S., Panskyi, T.: The dependence between the number of rounds and implemented nodes in LEACH routing protocol-based sensor networks. IAP-GOS **7**, 60–63 (2017)
4. Feltrin, G., Popovic, N., Flouri, K., Pietrzak, P.: A wireless sensor network with enhanced power efficiency and embedded strain cycle identification for fatigue monitoring of railway bridges. J. Sens. (2016)
5. Czerwinski, D., Milosz, M.: An inexpensive environmental monitoring system with IoT agents. In: ITM Web of Conferences, vol. 15, p. 01001. EDP Sciences (2017)
6. Fan, X., Wei, W., Wozniak, M., Li, Y.: Low energy consumption and data redundancy approach of wireless sensor networks with Bigdata. Inf. Technol. Control. **47**(3), 406–418 (2018)
7. Horvat, G., Sostaric, D., Zagar, D.: Response surface methodology based power consumption and RF propagation analysis and optimization on XBee WSN module. Telecommun. Syst. **59**(4), 437–452 (2015)
8. Horvat, G., Sostaric, D., Zagar, D.: Power consumption and RF propagation analysis on ZigBee XBee modules for ATPC. In: TSP, pp. 222–226 (2012)
9. Fedor, S., Collier, M.: On the problem of energy efficiency of multi-hop vs one-hop routing in wireless sensor networks. In: 21st International Conference on IEEE Advanced Information Networking and Applications Workshops, AINAW 2007, vol. 2, pp. 380–385 (2007)
10. Lavratti, F., et al.: A transmission power self-optimization technique for wireless sensor networks. ISRN Commun. Netw. (2012)
11. Luculescu, M.C., Zamfira, S.C., Cristea, L.: WiSeIn: wireless sensor network used for data acquisition from indoor locations. In: Visa, I. (ed.) The 11th IFToMM International Symposium on Science of Mechanisms and Machines. Mechanisms and Machine Science, vol. 18, pp. 391–399. Springer, Cham (2014). https://doi.org/10.1007/978-3-319-01845-4_39
12. Aurasopon, A.: Power saving approach based on time triggered architecture for wireless accelerometer system. CIGR J. Agric. Eng. Int. **20**(3), 245–252 (2018)
13. Cao, X., Shila, D.M., Cheng, Y., Yang, Z., Zhou, Y., Chen, J.: Ghost-in-ZigBee: energy depletion attack on ZigBee-based wireless networks. IEEE Internet Things J. **3**(5), 816–829 (2016)
14. Digi, XBee/XBee-PRO S1 802.15.4 Modules, December 2018. https://www.digi.com/support/productdetail?pid=3257
15. Ecer - cell mobile phone jammer company, Cell phone and Wifi jammer [CRJ4000] parameters, July 2016. http://www.ecer.com/products/cell_phone_and_wifi_jammer_crj4000-mpz234a62a-z1fcb0dd/showimage.html
16. Fluke Networks, Air Magnet Spectrum XT analyser datasheet. http://airmagnet.flukenetworks.com/assets/datasheets/AirMagnet_SpectrumXT_Datasheet.pdf. Accessed Dec 2017

17. TEM, UNI-T UT70A digital multimeter. https://www.tme.eu/gb/details/ut70a/portable-digital-multimeters/uni-t/. Accessed Dec 2018
18. Digi, XBee PRO S1 documentation, January 2019. https://www.digi.com/resources/documentation/digidocs/pdfs/90000982.pdf
19. Digi, XBee PRO S1 CSMA/CA documentation, January 2019. https://www.digi.com/resources/documentation/Digidocs/90001500/Reference/r_cmd_RN_xtend.htm
20. Digi International, XBee-PRO PKG-R RS-232 RF Modem User Guide, January 2019. https://www.digi.com/resources/documentation/digidocs/pdfs/90000829.pdf
21. Bouazzi, I., Bhar, J., Atri, M.: New CSMA/CA prioritisation based on fuzzy control mechanism. Int. J. Intell. Eng. Inform. **5**(3), 253–266 (2017)
22. Al-Anbagi, I., Erol-Kantarci, M., Mouftah, H.T.: Priority-and delay-aware medium access for wireless sensor networks in the smart grid. IEEE Syst. J. **8**(2), 608–618 (2014)
23. Dariz, L., Malaguti, G., Ruggeri, M.: Performance analysis of IEEE 802.15. 4 real-time enhancement. In: 2014 IEEE 23rd International Symposium on Industrial Electronics (ISIE), pp. 1475–1480 (2014)
24. Amjad, M., Qureshi, H.K., Lestas, M., Mumtaz, S., Rodrigues, J.J.: Energy prediction based MAC layer optimization for harvesting enabled WSNs in smart cities. In: 2018 IEEE 87th Vehicular Technology Conference (VTC Spring), pp. 1–6. IEEE, June 2018
25. Biswas, S., Roy, S.D., Chandra, A.: Cross-layer energy model for beacon-enabled 802.15.4 networks. J. Ambient. Intell. Hum. Comput., 1–16 (2018)
26. Chaari, L., Kamoun, L.: Performance analysis of IEEE 802.15. 4/Zigbee standard under real time constraints. Int. J. Comput. Netw. Commun. **3**(5), 235 (2011)
27. Lee, H.R., Park, J.H., Suh, Y.J.: Bursty-contention distribution for energy efficiency in large scale IEEE 802.15. 4 wireless sensor networks. Wirel. Pers. Commun. **84**(3), 1663–1687 (2015)
28. Musa, A., Baba, M.D., Mansor, H.M.A.H.: The design and implementation of IEEE 802.15.7 module with ns-2 simulator. In: 2014 International Conference on Computer, Communications, and Control Technology (I4CT), pp. 111–115 (2014)

Quantum Switch Realization
by the Quantum Lyapunov Control

Marek Sawerwain[1(✉)] and Joanna Wiśniewska[2]

[1] Institute of Control and Computation Engineering, University of Zielona Góra,
Licealna 9, 65-417 Zielona Góra, Poland
M.Sawerwain@issi.uz.zgora.pl
[2] Institute of Information Systems, Faculty of Cybernetics,
Military University of Technology, Gen. W. Urbanowicza 2, 00-908 Warsaw, Poland
jwisniewska@wat.edu.pl

Abstract. The quantum Lyapunov control is one of the methods
presently used in the quantum state engineering. The mentioned approach allows not only generating specified quantum states but also controlling the dynamics of the operators defining the computational process
in a quantum register. In this work, we present the quantum Lyapunov
control scheme which implements a quantum switch (it can be used as
a part of future quantum networks). However, this implementation is not
based on a quantum gates circuit. It directly concerns the spins of the
qubits which realize the quantum switch operation. We analyze also the
whole process of the quantum switching in terms of the accuracy of the
quantum Lyapunov control method for the switch's implementation.

Keywords: Quantum switch · Quantum Lyapunov control

1 Introduction

The dynamic development of the quantum computing [7,16] involves many fields
of the modern computer science, e.g. machine learning [4,11,23] quantum communication protocols and networks [17], quantum cryptography [1]. Solutions
referring to the quantum hardware [12] are also widely discussed. Next to the
mentioned fields, the quantum states engineering evolves quickly. In the quantum
states engineering [6,21] the usage of the quantum control methods is very popular – one of these methods is the Quantum Lyapunov Control (QLC) [2,20,22]
utilized to control [5,15] a process of the quantum states creation [9,14,24] and
a behavior of the unitary operator [10]. This second employment of the QLC
allows us to implement unitary gates in a noisy environment, too [8].

For the construction of future quantum networks, a quantum switch [18]
seems to be a significant element because it allows switching the information
between the system's input and output. The operation of switching may be
expressed as a unitary operator which may be defined as a circuit consisting
of three quantum gates, i.e. two CNOT gates and one Toffoli gate. However,

P. Gaj et al. (Eds.): CN 2019, CCIS 1039, pp. 76–85, 2019.
https://doi.org/10.1007/978-3-030-21952-9_6

the state-of-the-art does not offer any simple method of implementation for this circuit. Hence we would like to propose the other approach to solve the problem of implementation different than a circuit of quantum gates. The mentioned unitary operator may be considered as one, in this case three-qubit, gate. Using the QLC method simplifies the implementation because it allows seeing the switch as one system instead of three systems (one per each gate).

The paper is organized as follows: in Sect. 2 we present a preliminary information covering the notion of the quantum switch and the description of the QLC method. In Sect. 3 some examples of the quantum switch implementation with the use of the QLC method are shown. The examples are constructed for different steering models in which the control fields are defined by Hamiltonians. Particularly, Sect. 3.1 contains the numerical examples with values of the controlling function and the assessment of its quality with the use of the Fidelity measure. Conclusions are presented in Sect. 4. The chapter is ended with Acknowledgements and References to related works.

2 Preliminaries

This subsection contains the preliminary information referring to the notion of the quantum switch. Here, we describe how the switch works, and how it can be implemented as a quantum circuit. We define also some essential operators used in this work, i.e. the description of the QLC including the dynamics of the unitary operator U and the Hamiltonians playing different roles in the construction of the control field. Next, we present the Lyapunov function V and its derivative $\dot{V}$ which is also used to generate the control field.

2.1 Quantum Switch

The quantum switch is a three-qubit system. The qubits are denoted as: A, B, C. The states of A and B are arbitrary, but the qubit C may be only in one of the basis states:

$$|A\rangle = \alpha_A|0\rangle + \beta_A|1\rangle, \quad |B\rangle = \alpha_B|0\rangle + \beta_B|1\rangle, \quad |C\rangle = |0\rangle \text{ or } |1\rangle, \qquad (1)$$

where $\alpha_A, \beta_A, \alpha_B, \beta_B \in \mathbb{C}$, and $|0\rangle, |1\rangle$ represent the orthonormal vectors of the space $\mathbb{C}^2$. The vectors are expressed with the use of the Dirac notation what means that:

$$|0\rangle = \begin{bmatrix} 1 \\ 0 \end{bmatrix}, \quad |1\rangle = \begin{bmatrix} 0 \\ 1 \end{bmatrix}. \qquad (2)$$

Usually, the column vector $|0\rangle$ relates to zero and $|1\rangle$ to one, if we refer to the values of the classical bits. It should be also mentioned that the row vectors, i.e. in this work the complex conjugate of the column vectors, we denote as $\langle 0|$ and $\langle 1|$. The computational basis presented in Eq. 2 is termed as the standard basis.

The Dirac notation may be also utilized to define the projection operators P_0 and P_1:

$$P_0 = |0\rangle\langle 0|, \quad P_1 = |1\rangle\langle 1|, \qquad (3)$$

and the identity operator may be expressed as $\mathbb{I} = |0\rangle\langle 0| + |1\rangle\langle 1|$. Naturally, we can build the projection operators for the states of more than one qubit, e.g. $P = |011\rangle\langle 101| + |101\rangle\langle 011|$.

The main task, realized by the quantum switch, is switching the states of qubits $|A\rangle$ and $|B\rangle$ if the state of $|C\rangle$ equals $|1\rangle$. Otherwise, if $|C\rangle = |0\rangle$, the interchange of states between $|A\rangle$ and $|B\rangle$ is not performed.

$$|AB0\rangle \rightarrow |AB0\rangle, \quad |AB1\rangle \rightarrow |BA1\rangle. \tag{4}$$

The quantum switch may be implemented as a quantum circuit with the use of three quantum gates: two CNOT gates and one Toffoli gate. The exemplary circuit, switching $|A\rangle$ and $|B\rangle$ when $|C\rangle = |1\rangle$ and performing no action when $|C\rangle = |0\rangle$, is shown in Fig. 1. The figure contains also the matrix representation of the unitary operator U which describes how the switch works.

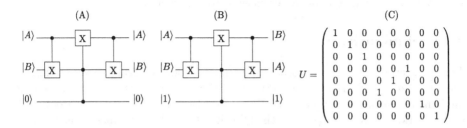

Fig. 1. The circuit of the quantum gates implementing the quantum switch. The case (a) refers to the situation when $|C\rangle = |0\rangle$, and the case (b) to the situation when $|C\rangle = |1\rangle$. Only when $|C\rangle = |1\rangle$ the states of qubits $|A\rangle$ and $|B\rangle$ are switched

However, in this work we want to describe the implementation of the switch not as the quantum circuit but as a Lyapunov control system. To achieve this goal, the control field must be defined. The control field influences the work of the switch and it is generated by the appropriate Lyapunov function and the Pauli operators. The Pauli operators X, Y, Z may be expressed as:

$$X = \begin{pmatrix} 0 & 1 \\ 1 & 0 \end{pmatrix}, \ Y = \begin{pmatrix} 0 & -i \\ i & 0 \end{pmatrix}, \ Z = \begin{pmatrix} 1 & 0 \\ 0 & -1 \end{pmatrix}. \tag{5}$$

The Lyapunov function is constructed with the use of the Fidelity measure's value which describes the difference between two unitary operators:

$$F = \frac{\left| \mathrm{Tr}\left(U_1^\dagger U_2 \right) \right|}{N}, \tag{6}$$

where N represents the dimension of the system, which may be expressed also as $N = \mathrm{Tr}\left(U_1^\dagger U_1 \right)$. The values of Fidelity satisfy $0 \leq F \leq 1$. If $F = 1$ then the operators are equal according to the global phase, which has no influence on the operator's form.

2.2   Quantum Lyapunov Control

In [10] the fundamental information concerning the QLC was presented. The main assumption of the QLC is manipulating the control field/fields to obtain a proper behavior of the quantum system. This proper behavior is realizing a given operation, e.g. a known quantum gate – in this work we want a quantum system to work as the switch. The following equation describes the time evolution of the operator U:

$$\frac{dU}{dt} = -i \left(H_0 + \sum_n f_n(t) H_n \right), \qquad (7)$$

where H_0 represents so-called free Hamiltonian and H_n, $n = 1, 2, \ldots$, are the control Hamiltonians with the control functions $f_n(t)$ describing the way the field is manipulated for each H_n. We assume that $U(0) = \mathbb{I}$ where $\mathbb{I}$ is the identity operator. This means that there is no field manipulation affecting the initial state of the system.

The main goal of the field manipulation is to evolve the operator U to the form O. We assume that the operators U and O are finite. If there is no control field, then the evolution $U(t)$ may be described as $U(t) = e^{-iH_0 t} O$. However, in the course of time the form of U will be significantly different than the final (eligible) form of O – we cannot obtain the proper form of O asymptotically. A solution to this problem is presenting the evolution of U as an orbit. Then, the operator U becomes O periodically. According to [10], we can denote this phenomenon as: $U \to \tilde{O}(t)$. The controlling field should steer the operator $\tilde{O}(t) = e^{-iH_0 t} O$ towards the right orbit. The orbit $\mathfrak{S}$ for the evolution $\tilde{O}(t)$ may be expressed as the following set of operators:

$$\mathfrak{S} = \left\{ U : U = e^{-iH_0 t_i} \tilde{O}(t), t_i \in \mathbb{R} \right\} = \left\{ U : U = e^{-iH_0 t_i} O, t_i \in \mathbb{R} \right\}. \qquad (8)$$

If there exists t_i for which $U = O$, then the orbit of the operator O is also the orbit of $\tilde{O}(t)$. The system's evolution converges to the final O and we need a function which points the difference between O and the current form of the operator U.

Utilizing the definition of the Fidelity as in Eq. (6), we can describe the Lyapunov function V as:

$$V = 1 - \frac{1}{N^2} \left| \mathrm{Tr} \left(\tilde{O}^\dagger(t) U \right) \right|^2, \qquad (9)$$

and the values of the Lyapunov function satisfy $0 \leq V \leq 1$. It is expected that $V = 0$ if $U = e^{ik} \tilde{O}(t)$ where $k \in \mathbb{R}$.

The function V may be utilized to manipulate the control field. To achieve this, first, we have to calculate the derivative of V with respect to time:

$$\dot{V} = -\frac{1}{N^2}\frac{d}{dt}\left(\mathrm{Tr}\left(O^\dagger e^{iH_0 t}U\right)\left(\mathrm{Tr}\left(O^\dagger e^{iH_0 t}U\right)\right)^\star\right)$$

$$= -\frac{1}{N^2}\left(\frac{d}{dt}\mathrm{Tr}\left(O^\dagger e^{iH_0 t}U\right)\left(\mathrm{Tr}\left(O^\dagger e^{iH_0 t}U\right)\right)^\star\right.$$

$$\left.+\mathrm{Tr}\left(O^\dagger e^{iH_0 t}U\right)\frac{d}{dt}\left(\mathrm{Tr}\left(O^\dagger e^{iH_0 t}U\right)\right)^\star\right)$$

$$= -\frac{2}{N^2}\sum_n f_n(t)\mathrm{Re}\left(\mathrm{Tr}\left(-iO^\dagger e^{iH_0 t}H_n U\right)\left(\mathrm{Tr}\left(O^\dagger e^{iH_0 t}U\right)\right)^\star\right), \quad (10)$$

where function $\mathrm{Re}(\cdot)$ returns the real part of the complex number.

Now, utilizing the last line from Eq. 10 and the operator definition given in Eq. 8, we can denote the function which controls the evolution of the quantum system:

$$f_n(t) = K\mathrm{Re}\left(\mathrm{Tr}\left(-i\tilde{O}^\dagger(t)H_n U\right)\left(\mathrm{Tr}\left(\tilde{O}^\dagger(t)U\right)\right)^\star\right), \quad (11)$$

where K is a real number describing the strength of the control field.

3 Quantum Switch Implementation by the Quantum Lyapunov Control

Let us analyze thoroughly the Lyapunov function V given in Eq. 9 to answer if the QLC method, with some properly defined Hamiltonians, allows obtaining the final form of the operator for the quantum switch.

In the previous section, we assumed that the symbol O denotes the operator which we want to estimate by the QLC method. In the case of the switch, the matrix form of the operator O is given in Fig. 1 as the matrix U.

Again, we define the operator $\tilde{O}^\dagger(t) = e^{-iH_0 t}O$ which describes the evolution of the free Hamiltonian to the form of the final operator O. It should be emphasized that at the beginning of the process the control function, obtained from the derivative of V, $f(t) = 0$. It means that the steering process was not launched. However, an additional constant τ might be used. This constant forces manipulating the control field at the beginning of the evolution of the operator U. The constant was introduced to the definition of the function $\tilde{O}^\dagger(t+\tau)$ and also to the definition of the function V:

$$V = 1 - \frac{1}{N^2}\left|\mathrm{Tr}\left(\tilde{O}^\dagger(t+\tau)U\right)\right|^2. \quad (12)$$

If we redo the calculation of $\dot{V}$, the strength of the control field is:

$$f_n(t) = K\mathrm{Re}\left(\mathrm{Tr}\left(-i\tilde{O}^\dagger(t+\tau)H_n U\right)\left(\mathrm{Tr}\left(\tilde{O}^\dagger(t+\tau)U\right)\right)^\star\right), \quad (13)$$

where H_n are the control Hamiltonians.

3.1   Numerical Examples

In the first example of the QLC method usage, we utilize the Ising model for the three-spin NMR system. The Hamiltonian H_0 is defined as follows:

$$H_0 = \frac{\omega_1}{2}\sigma_z^{(1)} + \frac{\omega_2}{2}\sigma_z^{(2)} + \frac{\omega_3}{2}\sigma_z^{(3)} + \frac{J}{4}\left(\sigma_z \otimes \sigma_z \otimes \sigma_z\right), \qquad (14)$$

where $\sigma_z^{(1)}$ denotes the Pauli gate Z utilized on the first qubit. The values ω_1, ω_2, ω_3 represent the precession frequencies for the adequate spins. The symbol J denotes the interaction between the spins.

Remark 1. One of the main goals in the field control process is utilizing only one Hamiltonian – H_1. This approach allows decreasing the number of needed parameters and operators which are used during the quantum system steering.

The control Hamiltonian is:

$$H_1 = J\left(|011\rangle\langle 101| + |101\rangle\langle 011|\right), \qquad (15)$$

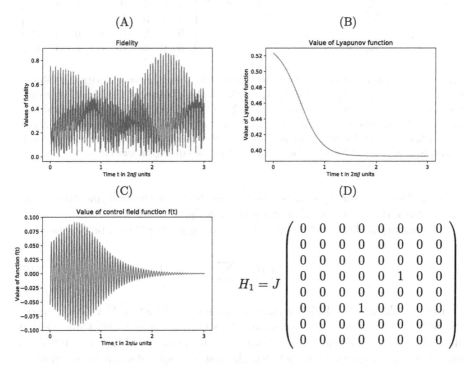

Fig. 2. An exemplary control process implementing the quantum switch with the usage of the Ising coupling. Figure shows the Fidelity measure values (A), the Lyapunov function values (B), and the strength of the control field (C). The model's parameters are: $\omega_1 = 5$, $\omega_2 = \omega_1$, $\omega_3 = \omega_1$, $J = 0.05\omega_1$, $K = 0.1\omega_1$, $\tau = 0.3/\omega_1$, and the Hamiltonian is given in (D). The model's configuration does not allow us to obtain the proper operator U for the quantum switch

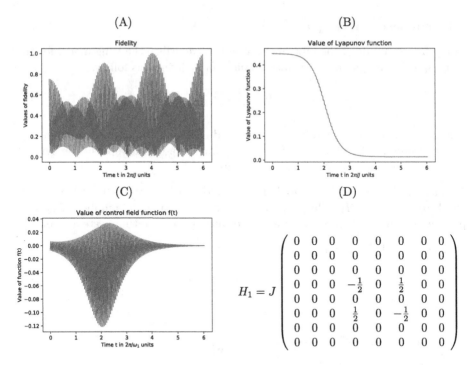

Fig. 3. The second control process implementing the quantum switch with the value of Fidelity measure reaching ≈ 0.997. Figure shows the changes of the Fidelity measure values (A), the Lyapunov function values (B), and the strength of the control field (C). The model's parameters are: $\omega_1 = 7$, $\omega_2 = \omega_1$, $\omega_3 = \omega_1$, $J = 0.05\omega_1$, $K = 0.05\omega_1$, $\tau = 0.1/\omega_1$, and the Hamiltonian is given in (D). The model's configuration allows us to obtain a very good form of the operator U for the quantum switch

and it is the operator σ_x spanned in a space between the first and the second qubit, where the exchange of information takes place.

The evolution of the system described by the operator U may be expressed as:

$$\frac{dU}{dt} = -\mathrm{i}\left(H_0 + \sum_n f_n(t)H_n\right). \tag{16}$$

Figure 2 depicts the Fidelity measure values, the function V values, and the strength of the control $f(t)$. Unfortunately, the chosen parameter values do not allow obtaining the efficient control: for $t = 2.45$ the best Fidelity, $F \approx 0.82$, was gained. We can ascertain the fact that the result is not sufficient to be called an effective approximation of the operator realizing the quantum switch.

Let us change the model's parameters and depict the Fidelity values, the function V value, and the strength $f(t)$ of the control field in Fig. 3. We can see that the proper values of the parameters allow gaining the efficient steering – for $t = 4.56$ the highest value of Fidelity was reached: $F \approx 0.997$. This result is

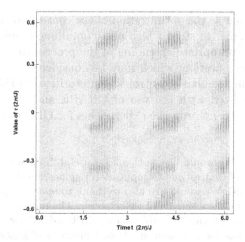

Fig. 4. The values of the Fidelity measure for the second example. The values change periodically thus we can point the values of τ and t for which the steering process is the most efficient. The black lines denote areas where values of the Fidelity are close to one

sufficient to start precise tuning of the system (to maximize the efficiency of the steering). Now, the control Hamiltonian is:

$$H_1 = J\left(-\frac{1}{2}\left(|011\rangle\langle101| + |101\rangle\langle011|\right) + \frac{1}{2}\left(|011\rangle\langle011| + |101\rangle\langle101|\right)\right), \quad (17)$$

and it is derived from the operators σ_x and σ_z which operate in the space (exchange the information) between the first and the second qubit.

   A very important issue, which should be taken into consideration during the field manipulation, is estimating the value of τ. This parameter influences the quality of the operator. A proper value of τ ensures faster calculation of U or allows us to decide when we want to obtain a sufficient form of the operator U. Figure 4 presents the values of Fidelity for the different values of t and τ during the evolution for both examples of the QLC usage.

4   Conclusions

As we could observe in the basic example, the presented approach does not allow us to obtain the proper results if the Hamiltonian directly applies to the qubits and it is realized by the change of the magnetic field with use of the operator σ_x. However, the appropriate control Hamiltonian improves on the whole process for the QLC method. The value of the constant τ is also a parameter which significantly influences the quality of the gained operator.

   The QLC allows easier implementing not only one- and two-qubit gates, like the CNOT gate [10], but also a three-qubit operator for the quantum switch, in which the control field was generated by quite complicated fields of interaction.

An interesting issue is the optimization of the control process, [3,19] i.e. the reduction of the number of steps which allow obtaining a quite high quality of the operation U or appropriate extending of the process time to gain the exact operator U. Another question, which may be considered as a task for further research, is comparing the basic control scheme (utilized in this work) with the optimal steering realized with the use of GRAPE and CRAB algorithms. It would be also interesting to run the experiments for QLC method with QuTIP package [13].

Acknowledgments. We would like to thank for useful discussions with the *Q-INFO* group at the Institute of Control and Computation Engineering (ISSI) of the University of Zielona Góra, Poland. We would like also to thank to anonymous referees for useful comments on the preliminary version of this chapter. The numerical results were done using the hardware and software available at the "GPU μ-Lab" located at the Institute of Control and Computation Engineering of the University of Zielona Góra, Poland.

References

1. Abubakar, M.Y., Jung, L.T., Foong, O.M.: Two channel quantum security modelling focusing on quantum key distribution technique. In: 5th International Conference on IT Convergence and Security (ICITCS), Kuala Lumpur, Malaysia, pp. 1–5 (2015). https://doi.org/10.1109/ICITCS.2015.7293032
2. Azodi, P.: Lyapunov Based Analysis of Continuously Observed Quantum Systems (2017). arXiv:1709.06801v1
3. Barreau, M., Seuret, A., Gouaisbaut F., Baudouin, L.: Lyapunov stability analysis of a string equation coupled with an ordinary differential system. arXiv:1706.09151v5 (2018)
4. Biamonte, J., Wittek, P., Pancotti, N., Rebentrost, P., Wiebe, N., Lloyd, S.: Quantum machine learning. Nature **549**, 195–202 (2017). https://doi.org/10.1038/nature23474
5. Chabir, K., Rhouma, T., Keller, J., Sauter, D.: State filtering for networked control systems subject to switching disturbances. Int. J. Appl. Math. Comput. Sci. **28**(3), 473–482 (2018). https://doi.org/10.2478/amcs-2018-0036
6. Emzir, M.F., Petersen, I.R., Woolley, M.J.: Lyapunov Stability Analysis for Invariant States of Quantum Systems. arXiv: 1707.07372v3 (2018)
7. Galindo, A., Martin-Delgado, M.A.: Information and computation: classical and quantum aspects. Rev. Mod. Phys. **74**(2), 347–423 (2002). https://doi.org/10.1103/RevModPhys.74.347
8. Ghaeminezhad, N., Cong, S.: Preparation of Hadamard gate for open quantum systems by the Lyapunov control method. IEEE/CAA J. Autom. Sin. **5**(3), 733–740 (2018). https://doi.org/10.1109/JAS.2018.7511084
9. Hou, S.C., Khan, M.A., Yi, X.X., Dong, D., Petersen, I.R.: Optimal Lyapunov-based quantum control for quantum systems. Phys. Rev. A **86**, 022321 (2012). https://doi.org/10.1103/PhysRevA.86.022321
10. Hou, S.C., Wang, L.C., Yi, X.X.: Realization of quantum gates by Lyapunov control. Phys. Lett. A **378**(9), 699–704 (2014). https://doi.org/10.1016/j.physleta.2014.01.008
11. Hou, S.C., Yi, X.X.: Quantum Lyapunov control with machine learning. arXiv:1808.02516v1 (2018)

12. IBM Q Homepage (2018). https://quantumexperience.ng.bluemix.net/. Accessed 17 Jan 2019
13. Johansson, J.R., Nation, P.D., Nori, F.: QuTiP 2: a Python framework for the dynamics of open quantum systems. Comput. Phys. Commun. **184**(4), 1234–1240 (2013). https://doi.org/10.1016/j.cpc.2012.11.019
14. Kuang, S., Dong, D., Petersen, I.R.: Rapid Lyapunov control of finite-dimensional quantum systems. Autom. Elsevier J. Int. Fed. Autom. Control **81**, 164–175 (2017). https://doi.org/10.1016/j.automatica.2017.02.041
15. Li, S., Wang, H., Aitouche, A., Tian, Y., Christov, N.: Active fault tolerance control of a wind turbine system using an unknown input observer with an actuator fault. Int. J. Appl. Math. Comput. Sci. **28**(1), 69–81 (2018). https://doi.org/10.2478/amcs-2018-0005
16. Nielsen, M.A., Chuang, I.L.: Quantum Computation and Quantum Information, 10th Anniversary Edition. Cambridge University Press, Cambridge (2010)
17. Nikolopoulos, G.M., Jex, I.: Quantum State Transfer and Network Engineering. Springer, Heidelberg (2014)
18. Ratan, R., Shukla, M.K., Oruc, A.Y.: Quantum switching networks with classical routing. In: 2007 41st Annual Conference on Information Sciences and Systems, Baltimore, MD, pp. 789–793 (2007). https://doi.org/10.1109/CISS.2007.4298416
19. Riaz, B., Cong, S., Qamar, S.: Lyapunov based control for one qubit quantum gates in coherence vector formulation. In: 37th Chinese Control Conference (CCC), Wuhan, China (2018). https://doi.org/10.23919/ChiCC.2018.8482596
20. Yusipov, I.I., Vershinina, O.S., Denisov, S.V., Kuznetsov, S.P., Ivanchenko, M.V.: Lyapunov exponents of quantum trajectories beyond continuous measurements. arXiv:1806.09295v1 (2018)
21. Wang, X., Schirmer, S.G.: Analysis of Lyapunov method for control of quantum states. IEEE Trans. Autom. Control **55**(10), 2259–2270 (2010). https://doi.org/10.1109/TAC.2010.2043292
22. Wang, L.C., Yi, X.X.: Lyapunov control on quantum systems. Int. J. Modern Phys. B **28**(30), 1430020 (2014). https://doi.org/10.1142/S0217979214300205
23. Wiebe, N., Kapoor, A., Svore, M.: Quantum algorithms for nearest-neighbor methods for supervised and unsupervised learning. Quantum Inf. Comput. **15**(3 & 4), 316–356 (2015). https://doi.org/10.26421/QIC15.3-4
24. Zhao, X.L., Shi, Z.C., Qin, M., Yi, X.X.: Edge state preparation in a one-dimensional lattice by quantum Lyapunov control. J. Phys. B Atomic Mol. Opt. Phys. **50**, 015301 (2017). https://doi.org/10.1088/1361-6455/50/1/015301

Simulation Analysis of Packet Delivery Probability in LoRa Networks

Rafał Marjasz[1]([✉])(iD), Krzysztof Grochla[1]([✉])(iD), Anna Strzoda[1]([✉])(iD),
and Zbigniew Laskarzewski[2]([✉])

[1] Institute of Theoretical and Applied Informatics, Polish Academy of Science,
Bałtycka 5, 44-100 Gliwice, Poland
{rmarjasz,kgrochla,astrzoda}@iitis.pl
[2] AIUT Sp. z o.o.,, Wyczółkowskiego 113, 44-109 Gliwice, Poland
zlaskarzewski@aiut.com

Abstract. The LoRa communication standard allows to transmit data in radio networks over distances of few kilometers. Although the LoRa WAN protocol assumes the use of ALOHA channel access, due to the use of multiple channels and multiple spreading factors in LoRa networks the evaluation of collision probability is challenging. We present the simulation model allowing to estimate the per node probability of successful packet delivery ratio. The model is evaluated for different network topologies, based on random distribution of nodes or based on the real location of meters in sample smart city deployments. The results show that the packet delivery ratio varies significantly depending on the location of the end nodes in the network. Due to the higher spreading factor used, the collision probability is very low for nodes located near to the access point. The use of proposed simulation model allows to identify the nodes which experience high collision probability and react e.g. by reconfiguration.

Keywords: LoRa · ALOHA · Loss ratio

1 Introduction and Motivation

The LoRa networks are becoming very popular communication solution for cheap devices in Smart City application. The LoRa uses unlicensed band (in EU it is typically 868 MHz, in US 902–928 MHz). The LoRa communication standard allows to transmit data in radio networks over distances of few kilometers thanks to the use of chirp spread spectrum (CSS) modulation, which is a spread spectrum wideband technique that uses linear frequency chirp pulses modulated to encode information.

The LoRa modulation can be used together with the LoRa WAN protocol, which defines the packet format and a star-of-stars topology in which gateways

Z. Laskarzewski—This research was partially funded by Polish National Center for Research and Development grant number POIR.04.01.04-00-0005/17.

© Springer Nature Switzerland AG 2019
P. Gaj et al. (Eds.): CN 2019, CCIS 1039, pp. 86–98, 2019.
https://doi.org/10.1007/978-3-030-21952-9_7

relay messages between end-devices and a central network server. All modes are capable of bi-directional communication. Although the LoRa WAN protocol assumes the use of ALOHA channel access, due to the use of multiple channels and multiple orthogonal spreading factors in LoRa networks the evaluation of collision probability is challenging. The spreading factor is a notion of the transmission coding rate and the gateway is capable of the simultaneous reception of multiple packets whenever they are transmitted using different spreading factors.

The rest of the paper is organized as follows: in Sect. 2 we present the literature review, next we describe the simulation model developed; in the fourth section we present the results of the analysis for random and real life topologies and we finish with a short conclusion in Sect. 5.

2   Literature Review

Owing to the fact that the majority of literature about LoRa network present theoretical simulation models, we propose the solution that could be applicable in authentic network topologies, based on the real location of nodes in a sample city.

LoRa network uses an ALOHA channel access method, which has been widely developed in the 1960's and 1970's i.e. [1] provides mathematical model of random access system for a slotted ALOHA or [2] that presents application of the slotted ALOHA random access method. LoRa specification is that it uses perfect orthogonality among different spreading factors depending on the distance or signal strength.

Furthermore, most current works have considered the subject of collision and packet loss in LoRa network. For instance, [4] provides a theoretical analysis of the collision and packet loss in LoRa when LoRaWAN protocol is taken into consideration. Theoretical expressions that have been presented are more exact to describe the collisions than the Poisson distributed process. The simulation is carried out on the small scale model consisting of one access point that can set connections with N nodes using the same spreading factors. Authors present closed-form expressions of collision and packet loss probabilities by using LoRaWAN properties.

Next example worth mentioning is [6] which presents LoRa throughput analysis with imperfect spreading factor orthogonality. Numerical results have shown the irrelevant affect of spreading factor's imperfect orthogonality, as well as the significant effects of spreading factor allocations on the general throughput.

Moreover, in [5] precisely have been described four cases of frame collisions. This work includes investigation under which conditions two colliding frames are both lost. Authors perform experiments on model composed of two nodes and a single-channel access point discreted 5 m apart whereas the simulation model we are considering is a network including more than 1700 nodes.

Both [7] and our work consider a LoRa network including nodes with various spreading factors and different parameters of bandwidth. Authors, also, calculate the average success probability for each configuration as a function of density taking into consideration both intra and inter-spreading factor collisions. Optimizing the amount of nodes having different spreading factor configurations is a technique designed to formulate and solve an optimization problem to maximize the node capacity for a given allocation area and frequency.

Referring to ALOHA system, in [8] have been investigated the throughput performance of the DS/SSMA unslotted ALOHA system with variable length data traffic. Assuming the geometrically distribution of number of packets in one message and constant packet length, authors model system as Markov chain. As a result of analytical and simulation researches it has been shown that throughput improvement can be gained by using an error-correcting code.

3   Model Description

In our research we are using OMNeT++ (Objective Modular Network Testbed in C++) which is a modular, component-based C++ simulation library and framework. We have created a simulation model in OMNeT++ to evaluate the packet loss probability in LoRa networks. While the Flora simulation is available to simulate the LoRa protocol [9], due to the performance issues with simulation of large number of devices and the simplifications of our model we have created a novel model representing the collision probability and based on the radio signal propagation models available in OMNeT++. The radio fading model is expressed by SUI (Stanford University Interim) propagation loss model, which is an extension of an early work carried out by AT&T Wireless and the further analysis done by Erceg et al. [3].

3.1   Packet Collisions in LoRa

In LoRa, like in any radio broadcast, collisions of frames occur, and are characterized by LoRaWAN protocol influence on frame collision conditions. A LoRa transmission is characterized by several key parameters:

- Bandwidth that defines the width of the radio frequency used for transmission of frames.
- Coding Rate that defines the Forward Error Correction rate used in LoRa modem.
- Spreading Factor (SF) that represents the ratio between chip rate and symbol rate. The higher the spreading factor is the greater the sensivity and range of transmission. This increases the power consumption and airtime. LoRaWAN uses spreading factors in range SF7 to SF12 which enables concurrent transmission receival on different spreading factors.
- Transmission Power that defines the power used for transmission. LoRa chipsets are supporting power transmissions from −137 dBm to 14 dBm.

According to results presented in [5] few cases of frame collision are distinguished. In all cases collision will occur when both interferring frames are transmitted on the same frequency posess the same spreading factor. The fact that spreading factors are not perfectly orthogonal among themselves, and thus the effect of inter-spreading factor interferences could cause a collision (especially for high spreading factors for which frames have a greater time on air), will be ignored for the sake of simplicity of the model we present. The differences in the types of collisions are following:

- Case 1: Both frames are lost when a frame with stronger signal arrives during the transmission of a frame with weaker signal. The difference between strong and weak signal must be equal to or higher than 8 dBm.
- Case 2: The stronger signal frame survives the collision with a weaker signal frame, and thus only this weaker frame is lost. The difference between strong and weak signal must be equal to or higher than 8 dBm.
- Case 3: Both frames are lost when their difference between strong and weak signal is less than 8 dBm.

3.2   Discrete Event Simulation Model

Our simulation model is composed of two types of modules:

- **End Point (EP)** - simulating a LoRa end device i.e. a sensor,
- **Access Point (AP)** - simulating a LoRa gateway serving as a router.

The communication between those two types of modules is represented by in and out OMNeT++ gate connections. Every EP in signal range of AP has a proper gate connection between them. If a non-chosen AP receives a frame it is treated as a noise and will not be received. Sending LoRa packets is simulated by sending two messages (a pair of messages connected with each other):

- **jobStart (jS)** - message with a timestamp for the moment of frame receival,
- **jobEnd (jE)** - message with a timestamp for the moment of frame transmission end.

Aforementioned packet collision and packet receival is simulated by five possible states of AP implemented system:

- **Listen** - AP is waiting in standby mode for jS arrival on all of 8 possible transmission channels (simulating the radio frequencies used by LoRa) ,
- **Reception** - AP is waiting in listening mode for jE arrival signaling the end of transmission on busy channel,
- **Collision** - AP received another jS from different EP on busy channel. The difference in transmission power between both simulated packets is either less than 8 dBm or the newly received jS has greater transmission power, and both packets spreading factor has the same value,
- **Reception with noise** - AP received another jS from different EP on busy channel. The difference in transmission power between both simulated packets is greater than 8 dB and the newly received jS has the lower transmission power, and both packets spreading factor has the same value,

- **Collision without reception** - AP is waiting for the arrival of the rest of jE on busy channel, or for the arrival of jS with power transmission value greater by 8 dBm than the noise transmission power values.

The simulation program code logic is defined by those states and executed by proper state changes according to their relationships presented in Fig. 1.

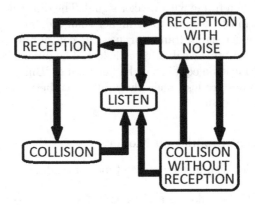

Fig. 1. The relationships between access point five possible states

3.3   Validation of the Model

The model has been validated by an analysis of a simplistic scenario, with small number of nodes. At the beginning scenario with a single access point and 2 clients using the same spreading factor was executed and the results were compared with the outcome of equations of ALOHA channel access effectiveness provided in [10]. Next the network size was increased and simultaneous transmission on different spreading factors was added. The results of channel usage and packet loss probability were compared between the simulation and the analytical model.

4   Results

In this section we present the results of simulations executed on random and real smart city existing topologies. In the network topologies, base stations have been selected in a way providing to cover the entire network. Every single node transmits to exactly one geographically closest base station, and is considered as a source of radio noise for the remaining access points. To compute the number of all edges and average node degree we take into account the edges resulting from the assignment of each node to one base station as well as those edges that are a result of creating radio noise. For both real and random type of simulation we assume two scenarios:

- All end points have a transmission window of 30 min to carry out their transmission to access points.
- All end points have a transmission window of 10 min to carry out their transmission to access points.

The moments of transmission are calculated randomly within window time according to a uniform distribution. The size of the packet is constant and it is 51 bytes. The single turn transmissions between all nodes is performed repeatedly more than 45000 times (this simulates a 123 year time of daily transmission between each LoRa end device and its corresponding gateways). One simulation covers this 123 year time period. We perform 30 independent simulations for each scenario and each topology, differing in RNG seed from each other.

The graphs show Experimental Probability Density Function that represents the percentage of nodes that correspond to probability of successful packet delivery. Experimental PDF value is calculated from the average of 30 independent simulation results. Confidence intervals are computed with confidence level of 90 percent and on the basis of standard deviation.

4.1 Random Topology

Random topology consists of 1794 end points and 6 access points. The number of edges is equal to 6819 and the average node degree is around 3.8. The location coordinates of all points are drawn according to the uniform distribution. Figure 2 presents the arrangement of points. Figure 3 presents the results for both scenarios. The increase of the size of transmissions window results in higher number of successful transmissions and a lower collision probability for all the nodes.

4.2 Smart City Topologies

We have created two models based on implementation of smart city topologies for two sample city-wide installations of smart meters. For the sake of anonymity we call these topologies City A and City B. City A topology consists of 1752 end points and 4 access points. The number of edges is equal to 3504 and the average node degree is 2. City B topology consists of 1713 end points and 6 access points. Number of edges is equal to 6455 and the average node degree is around 3.77. Figures 4, 5 present the arrangement of points in those cities. Figure 6 present the results of both scenarios for City A topology. Figure 7 present the results of both scenarios for City B simulation.

Due to the fact that topology of the City A model is divided in two separate areas, end points do not create radio noise for access points from different area. We can observe relatively low probability of collision for 10 min window transmission. There is an improvement for extended 30 min window and large part of

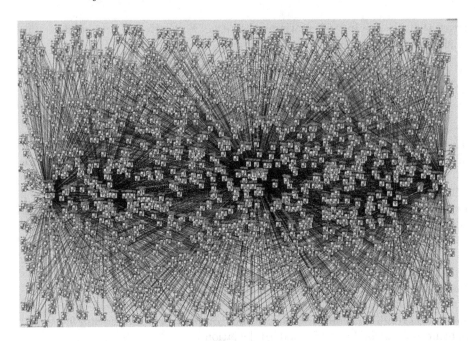

Fig. 2. The random topology model

the nodes have the packet collision probability between 1% and 2% showing that individual collisions are still present. Three times longer window transmission time gives relative packet delivery probability improvement of only 4%. It shows that either the positions of access points are not chosen well or the number of access points is slightly too small to handle all the incoming traffic.

City B topology is much more dense but with higher number of access points (6, comparing to 4 in City A) and is not divided into some separate areas. It causes higher level of interferences and lower probability of transmission success for shorter (10 min) time window, for which approximately 60% of the nodes have the packet collision probability of at least 22%. The 30 min time window presents improvement, as the number of nodes with the lowest packet collision probability of 14% is approximately equal to 7% and the number of nodes with the highest probability of transmission success is doubled. Nevertheless, we have observed that there are still some areas of the network with increased collision probability. This is analyzed in the next subsection.

4.3 Spatial Distribution of Collision Probability

The nodes which experience increased collisions probability are not distributed evenly in the network. During the deployment of the network and the selection of optimal location of the access point it is crucial to identify the locations in which the collision are occurring more often. This allows to react e.g. by relocating the access point or by reconfiguration of the nodes to use different access points.

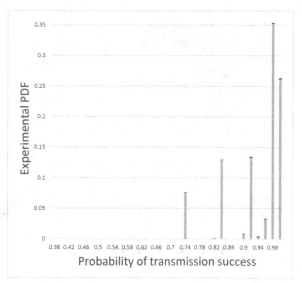

(a) 30 minutes window to carry out the transmission

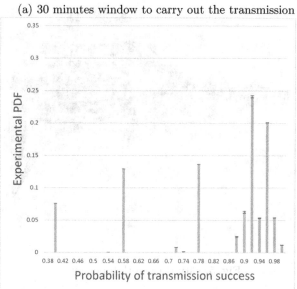

(b) 10 minutes window to carry out the transmission

Fig. 3. Comparison of dependencies between transmission window length and probability of transmission success in random model simulation.

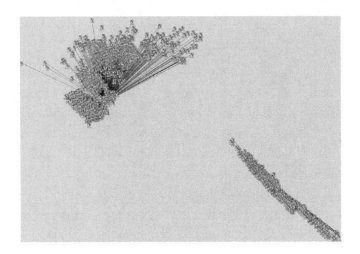

Fig. 4. City A network topology

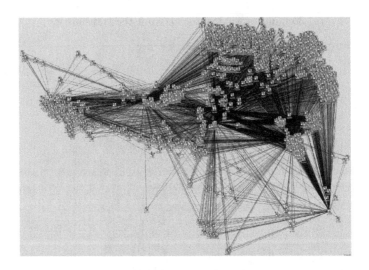

Fig. 5. City B network topology

On Fig. 8 we show how the packet delivery probability changes in the sample topology of City B in 30 min window scenario. The areas where the collisions are more probable can be identified and as can be seen, they include only a small part of the network on the border of the range of the access point.

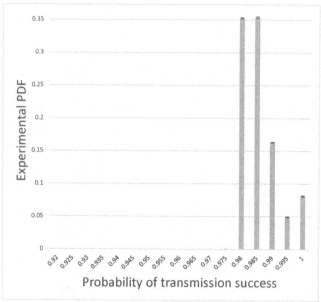

(a) 30 minutes window to carry out the transmission

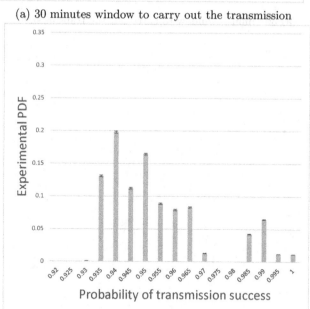

(b) 10 minutes window to carry out the transmission

Fig. 6. Comparison of dependencies between transmission window length and probability of transmission success in City A.

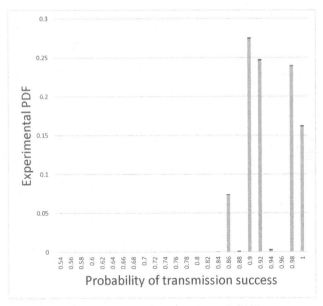

(a) 30 minutes window to carry out the transmission

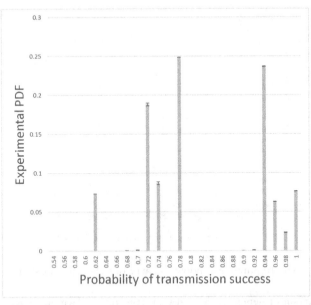

(b) 10 minutes window to carry out the transmission

Fig. 7. Comparison of dependencies between transmission window length and probability of transmission success in City B topology.

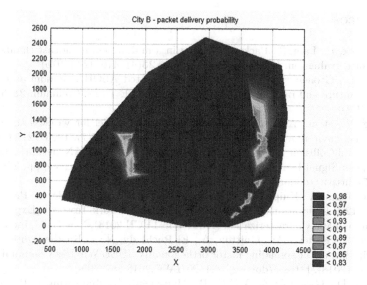

Fig. 8. Spatial distribution of packet delivery probability

5 Conclusions

The LoRa communication standards are a very cheap and attractive solution to provide wireless connectivity for IoT devices. Nevertheless, due to the operation in unlicensed band and low amount of bandwidth available in the 868 MHz/902 MHz band it is subject to collisions. The use of ALOHA channel access and the lack of coordination between the transmissions from different nodes increases the probability of the collision's occurrence. We present a simulation model allowing to evaluate the collision probability for different network topologies. Our evaluation of the packet delivery ratio for different sizes of the network and both for the random topologies and the real topologies of two sample cities shows that the collision probability if heavily influenced by the size of the time window in which transmission occurs.

The presented spatial distribution of the packet loss probability for one of the sample cities shows that the packet delivery chance changes significantly between different locations in the network and is few times lower in the densely populated areas on the border of the access point range. The experimental probability density function of the collision probability shows that in realistic topologies there are some areas in which the collision probability is unacceptably high (above 10%) and more access points are needed to achieve effective communication. In our future work we plan to investigate this further to propose a method for selection of locations for additional access points that can decrease the collision probability beyond a given threshold.

References

1. Kleinrock, L., Lam, S.: Packet switching in a multiaccess broadcast channel: performance evaluation. IEEE Trans. Commun. **23**(4), 410–423 (1975)
2. Davis, D., Gronemeyer, S.: Performance of slotted ALOHA random access with delay capture and randomized time of arrival. IEEE Trans. Commun. **28**(5), 703–710 (1980)
3. Erceg, V., et al.: An empirically-based path loss model for wireless channels in suburban environments. IEEE J. Sel. Areas Commun. **17**(7), 1205–1211 (1999)
4. Ferre, G.: Collision and packet loss analysis in a LoRaWAN network. In: 25th European Signal Processing Conference (EUSIPCO), Kos, Greece, pp. 2586–2590 (2017). https://doi.org/10.23919/EUSIPCO.2017.8081678
5. Rahmadhani, A., Kuipers, F.A.: When LoRaWAN frames collide. In: Proceedings of the 12th International Workshop on Wireless Network Testbeds, Experimental Evaluation and Characterization (ACM WiNTECH 2018), New Delhi, India (2018)
6. Waret, A., Kaneko, M., Guitton, A., El Rachkidy, N.: LoRa throughput analysis with imperfect spreading factor orthogonality. IEEE Wireless Communications Letters (2018). https://doi.org/10.1109/LWC.2018.2873705
7. Zorbas, D., Georgios, Z., Maillé, P., Montavont, N.: Improving LoRa network capacity using multiple spreading factor configurations. In: 25th International Conference on Telecommunications (ICT), pp. 516–520 (2018)
8. So, J., Han, I., Shin, B., Cho, D.: Performance analysis of DS/SSMA unslotted ALOHA system with variable length data traffic. IEEE J. Sel. Areas Commun. **19**, 2215–2224 (2001). https://doi.org/10.1109/49.963807
9. Słabicki, M., Premsankar, G., Di Francesco, M.: Adaptive configuration of LoRa networks for dense IoT deployments. In: Proceedings of IEEE NOMS 2018, Taipei (2018)
10. Wieselthier, J.E., Ephremides, A., Michaels, L.A.: An exact analysis and performance evaluation of framed ALOHA with capture. IEEE Trans. Commun. **37**(2), 125–137 (1989)

Comparative Study Between Reactive and Proactive Protocols of (MANET) in Terms of Power Consumption and Quality of Service

Hemin Akram Muhammad$^{(\boxtimes)}$, Tara Ali Yahiya$^{(\boxtimes)}$, and Nawzad Al-Salihi$^{(\boxtimes)}$

Department of Computer Science and Engineering, University of Kurdistan
Hawlêr (UKH), Erbil, Kurdistan Region, Iraq
{h.muhammad,t.ibrahim1,n.al-salihi}@ukh.edu.krd

Abstract. Recently, the lack of independent wireless networks deployment was the main reason not only loosing victims but also diminution of the performance of security forces and first aids services during natural disasters and wars. Mobile Ad Hoc Networks (MANET) is the technology of choice used in such critical situations where the infrastructure of wireless networks fails to work. MANET relies on its nodes to forward and route packets that gives it a characteristic of an independent network. The independence here means that the network relies on its battery power to achieve its routing. In this research work, we study two types of MANET protocols belonging to different kinds of routing protocol categories; namely reactive protocol and proactive protocol in terms of power consumption and quality of service. More specifically, we are interested on the Destination-Sequenced Distance-Vector (DSDV), Optimized Link State Routing (OLSR), Ad Hoc On-Demand Distance-Vector (AODV) and Dynamic Source Routing (DSR) protocols to investigate QoS and power they consume at different layers, operation modes, routing overhead and MAC load change. In order to achieve the goal of performance study, we choose some scenarios that can be adapted to different MANET contexts; such scenarios carried out when network area size, network density, pause time and mobile node speed are changing. Extensive simulations and results for these scenarios obtained by NS3 simulation software.

Keywords: MANET · AODV · DSR · OLSR · DSDV · NS3

1 Introduction

In today's era, there are many network systems, some of them are fixed (wired) and others are wireless. Ad hoc network is one of significant types of networks which use for connecting devices without depending on a Base Station (BS). As well as, it is used for particular purpose that is why it is called (ad-hoc) which is a Latin word that means (for this). In ad hoc network, any node can act as

© Springer Nature Switzerland AG 2019
P. Gaj et al. (Eds.): CN 2019, CCIS 1039, pp. 99–111, 2019.
https://doi.org/10.1007/978-3-030-21952-9_8

endpoint or as intermediate router. Connecting non-fixed nodes (mobile nodes) is a challenging issue in ad hoc network therefore it requires a technology to guarantee communication and error handling which is known as Mobile Ad hoc Network (MANET). It is an independent system for wireless connecting mobile nodes temporarily. There is no infrastructure or centralized administration for mobile nodes to be deployed in MANET system. Nowadays, many fields depend on this network for communication purpose such as military field, emergency and transportation field. Hidden terminal problem, packet loss due to transmission errors, mobility induced route changes, battery constrains, and limited wireless transmission range are the most challenges, which are experienced, in ad hoc network [1]. On the other hand, an ad hoc network can be created spontaneously with an adaptation to any eventual change in the topology and it can be set up shortly and easily deployed. In MANET, each node can work as a workstation and router in the same time, and can move freely, thus the topology of the network changes rapidly.

2   MANET Routing Protocols Categories

2.1   Reactive Routing Protocols

Ad-Hoc On-Demand Distance Vector (AODV): was created as a combination from on-demand and distance vector protocol [1]. AODV totally depends on hop by hop routing method. Its procedure starts by was created as a combination from ondemand and distance vector protocol [1]. AODV totally depends on hop-by-hop routing method. Its procedure starts by creating RoutRequest (RREQ) packet then sending it to all neighbors. The neighbors forward the RREQ to others or satisfy it by sending back Route Reply (RREP) packet to the source device. As well as, when RREQ sent from source device to destination device, it creates reverse path from all intermediate nodes to the source device. According to [4], the common problem in AODV is removing destination device or any intermediate device from the network. For solving this problem, RouteError (RERR) packet is used [6]. Based on receiving RERR packet, source device perceives that an intermediate device or destination device is removed. At that time, it has the chance to discover a new path or stop broadcasting [2].

Dynamic Source Routing (DSR): In this protocol nodes can discover source route to any destination device in the network. The main characteristic of DSR is that the header of data packet holds the entire nodes address that the packet must pass in order to reach the destination device [3]. Two mechanisms together are used in this protocol; the first one is Routing Discovery Mechanism. This is like the discovery mechanism in AODV. In this mechanism, the node checks its route cache to know whether the requesting route is already available in its routing cache, if not, then it sends (RREQ) packet to its neighbors. The second mechanism is Maintenance Route Mechanism. It is utilized when a disconnection is detected during transferring packet from source device to destination device.

These situations happen once network topology changed by the movement of an intermediate device or destination device [9].

2.2 Proactive Routing Protocols

Optimized Link State Routing (OLSR): is one of the proactive protocols that totally depend on routing table. Unlike reactive protocols, routing information table is continuously maintained. This protocol is an advanced version of traditional Link State protocol. It has controlling capability on broadcasting by any node in the network. Thus, the broadcasting flooding that causes overhead in the network will be diminished. For this purpose, OLSR fully depends on Multipoint Relay (MPR) technique. Each node in the network selects one of its neighbors as (MPR) for re-transmitting its messages. The nodes that are not selected as (MPR) are capable to receive messages but they do not have the same ability as (MPR) possesses. (MPR) utilizes (Hello) message for finding information about neighbor nodes and (Topology Control) message for sending information to the neighbors of (MPR) Selector [7]. However, normally time delay in OLSR is not long but when a link is broken, it requires more time for re-discovering [4,5].

Destination-Sequence Distance Vector (DSDV): is an advanced version of classical shortest-path routing protocol (Bellman-Ford algorithm) [8]. It has special technique in enhancing grouting performance which depends on the sequence number that is created by the destination device. In this protocol each node has a routing table which contains (Destination address, Next hop address, Metric of the routing to the destination device, Destination sequence number). The main benefit of utilizing sequence number is to make differentiation between old and new route and preventing from creating route loops. Routing table in each node is updated according to time change or event occurrence. Indeed, there are two different ways of updating routing table. The first one is called (full-dump) that updates all tables' entries (all records) and the second one updates only these entries that witness changes [1].

3   Simulation Methodology

3.1   Simulation Parameters

In this research network simulator NS-3 is chosen because it is more accurate and dependable simulator for research work and can provide simulation for wired and wireless network. As well as, it has full support for our selected protocols. Nodes are usually spending energy during their operations in the network, so calculating energy consumption for each element in the network should be considered by simulators. NS-3 has Energy Model which considers energy consumption by each entity of the network. Energy model is responsible for indicating initial energy for nodes at the beginning of simulation. Its components depend on the network type. Energy model with the following components are used as initial power for each node. (idlePower = 1.1 (Watt), rxPower = 1.0 (Watt), txPower = 1.2 (Watt), sleepPower = 0.001 (Watt), transitionPower = 0.2 (Watt), initialEnergy = 1000 (Joules) idlePower; indicates power consumption in idle state, rxPower; indicates power consumption in receiving process, txPower; indicates power consumption in transmitting process, sleepPower; indicates power consumption in sleep state. Radio propagation model must be taken into consideration because this model is responsible for expecting the received signal power of any nodes in the network. There are many propagation models in NS3. We used Two- Ray Ground Reflection Model because it is the most used in evaluating MANET protocols performance. In Two-Ray Ground Reflection Model, the received energy can be calculated by the sum of the direct line path (sender to receiver) and reflected path on the ground between the sender and receiver. Since we are working on a mobile environment, mobility model should be considered for the chosen protocols. In this research work, Random Waypoint Mobility Model (RWP) is used for simulating a movement which has stop time between movements besides to the random movement. RWP is chosen because it is used as a (benchmark) for assessing routing protocols of MANET in terms of mobility model and the most important features of this model refer to its wide obtain-ability and easiness. Table 1 shows the simulation parameters used in the scenarios

Table 1. Simulation parameters

Parameters	Values
Simulation time	250 s
Density	(14,20,30,40,50,60) nodes
Pause time	(0,50,100,150,200,250) s
Speed	(1,10,20,30,40,50) m/s
Area	(250 × 250, 500 × 500, 750 × 750, 1000 × 1000, 1250 × 1250,1500 × 1500) m
Traffic type	Constant bit rate "CBR"
Packet size	512 bytes
Channel type	Wireless channel
MAC type	IEEE 802.11
Protocols	DSDV, OLSR, AODV, DSR
Number of sources	10
Antenna model	Omni antenna
Mobility model	Random waypoint
Propagation model	Two ray ground
Packet sending rate	2 packets per second
Energy model	Energy model
Queue type	DropTail/PriQueue, CMUPriQueue

3.2   Power Consumption Metrics

In MANET, operation mode categories can be divided into four modes namely; transmit mode, receive mode, idle mode and sleep mode. Transmit and receive mode are called (Active mode) because they are used in transmitting and receiving packets respectively. The third mode is used when the mobile node is waiting for transmitting packets while in sleep mode, node does nothing (no sending, no receiving) that is why a minimum power is consumed in this mode. Network Interface Card (NIC) plays a great role in power consumption. Indeed, there are various types of NIC such as (Aironet PC, Aironet 350 PCI, Lucent Bronze, Lucent WaveLAN and Cabletron Roamabout). Power consumption in each mode varies according to the type of the NIC. Usually, in all events occurring in different network layer (Application layer, Network Layer, MAC Layer), the node will consume an amount of power. According to the following formula, total amount of power consumption can be calculated by summation of all power consumed of the nodes in the network.

$$Total\ Power\ =\ \sum_{i=0}^{n} Total\ power\ Consumed\ by\ Node(i) \qquad (1)$$

Where n = Total Number of Nodes and i = Node Number

In order to find the amount of power consumed in the layers (Application, Routing and MAC layer), power consumed in each layer should be divided over the total power consumption. The following formula shows power consumption at application layer.

$$Application\ Layer\ Power\ Consumption\ (\%) = \frac{Power\ Consumed\ at\ Application\ Layer}{Total\ Power\ Consumption}$$
$$(2)$$

On the other hand, power consumption in idle mode, receiving mode and transmitting mode are more considered and the amount of consumption is not equal in all modes. For understanding power consumption in each mode, the amount of consumption at each node should be divided over the total amount of power consumption. The following formula demonstrate the power consumption at Idle mode [10].

$$Idle\ Mode\ Power\ (\%) = \frac{Power\ Consumed\ at\ Idle\ Mode}{Total\ Power\ Consumption} \qquad (3)$$

Moreover, routing overhead is another factor of power consumption. The number of periodic rout messages sent by routing protocols is assumed as routing overhead. Power consumption due to routing overhead can be calculated by dividing value of power consumption due to routing overhead by the actual value of power consumption at routing layer as showed in the following formula [10].

$$Routing\ Overhead\ power\ (\%) = \frac{Power\ consumed\ by\ Routing\ Message}{Power\ Consumption\ at\ Routinglayer} \qquad (4)$$

Power consumption due to MAC load which is a power spent due to sending and forwarding short control messages (RTS, CTS and ACK) can be calculated by dividing value of power consumption due to MAC load by real value of power that spent by MAC layer.

$$MAC\ Load\ Power\ (\%) = \frac{Power\ Consumpted\ by\ Short\ Messages}{Total\ Power\ Consumption\ at\ MAC\ Layer} \quad (5)$$

3.3   Quality of Service Metrics in MANET

In this section, all equations that are needed for calculating QoS in terms of throughput, End-to-End delay and packet delivery ratio are shown. Throughput can be measured by dividing total number of successfully received data packet by simulation time as showed in below formula

$$Throughput = \frac{\sum_1^n Received\ Packet}{SimulationTime} \quad (6)$$

End-to-End delay can be calculated by measuring the average time which data packet needs for transferring from sender node to receiver node. Thus, average end to end delay can be obtained by dividing the difference in time between receive time and sent time by average number of received packet as showed in the following formula.

$$Avarage\ End-to-End\ Delay = \frac{\sum_1^n Received\ Time - Sent\ Time}{\sum_1^n Received\ Packet} \quad (7)$$

Packet Delivery Ratio (PDR) is a ratio of number of successfully received packet data by receiver node to the number of packet data sent by sender node. Below formula is used for calculating (PDR).

$$Packet\ Delivery\ Ratio\ (\%) = \frac{\sum_1^n Received\ Packet}{\sum_1^n Sent\ Packet} \times 100 \quad (8)$$

4   Performance Analysis and Simulation Results

4.1   Power Consumption

Total Power Consumption. Figure 1 shows total consumed power for all protocols corresponding to changing mobile node speed. Proactive protocols have steady behavior with increasing node speed, but reactive protocols increase harshly in the case of small motion (1 m/s and 10 m/s) then they increase very slowly. DSDV protocol consumes highest amount of power while DSR consumes lowest amount of power in changing mobile motion speed scenario.

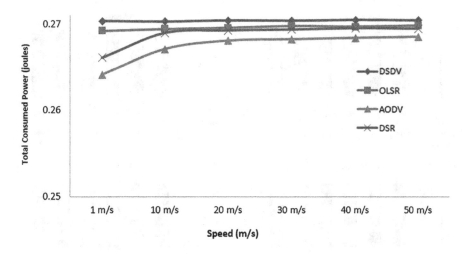

Fig. 1. Total consumed power percentage versus speed for all protocols

Power Consumption at Different Layers. Figure 2 shows how power consumption at different layers is affected by changing network density. It is noticed that both types of protocols have similar behavior in consuming power at MAC layer. They both consume higher power at MAC layer and increase by adding nodes. Regarding routing layer, proactive protocols consume higher power at routing protocol in comparison with reactive protocols and the rate of consuming increases by increasing density. Consuming power at application layer is very high when network density is small for all protocols, but it decreases by increasing density therefore the relationship between power consumption at application layer with density is indirect.

Power Consumption at Operation Modes. Figure 3 shows how changing network area size affecting power consumption at operation modes. Both types of protocols consume highest amount of energy at idle mode because when node senses the carrier it consumes a significant amount of power. According to the figure, consuming power at receive mode is higher than transfer mode by all protocols. Power consumption at idle mode is decreased by increasing area size for all protocols. In contrast, transfer mode and receive mode consume more power by increasing area size.

Power Consumption Due to Routing Overhead. The Fig. 4 shows the amount of consumed power for routing protocols overhead versus varying network area size. The figure illustrates similar behavior of proactive protocols (DSDV and OLSR). Both protocols spend approximately the same power in all area sizes. The percentage of consumed power due to routing overhead for both protocols (DSDV and OLSR) is near to (100%), which means that it is close to total power consumption at routing layer. While Reactive protocols (AODV

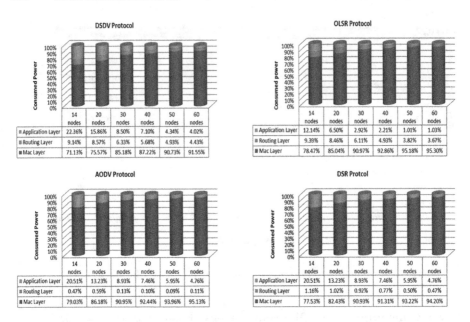

Fig. 2. Power consumption percentage across layers for all protocols corresponding to changing in network density

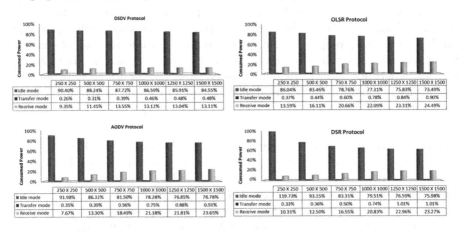

Fig. 3. Consumed power percentage at operation modes versus changing network area size

and DSR) do not behavior like proactive protocols because the amount of energy spent at routing layer is not all due to routing overhead. AODV has steady behavior with varying area size. It increases by increasing network area size but DSR has different behavior because it spends lowest power at area sizes (250×250 and 1500×1500). Whereas, for other area sizes (500×500 to 1250×1250), it consumes higher than AODV protocol.

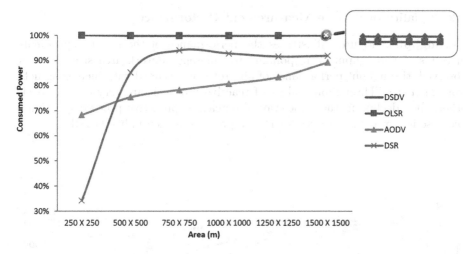

Fig. 4. Consumed power percentage due to routing overhead versus changing network area size for all protocols

Power Consumption Due to MAC Load. Figure 5 presents how changing network area size is affecting the power consumption due to MAC load. All protocols consume more power due to MAC load when area size is increasing. Proactive protocols (particularly DSDV protocol) consume less power in comparison to reactive protocols due to MAC load. At large network area size (1250 m × 1250 m and 1500 m × 1500 m) three protocols (DSDV, OLSR and AODV) have stability in consuming power while DSR protocol increase sharply at these area size values.

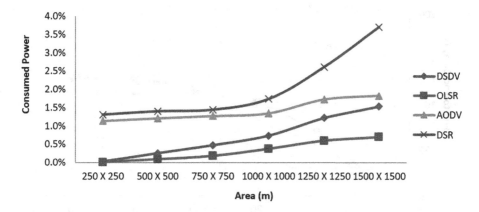

Fig. 5. Consumed power percentage due to MAC load versus changing network area size for all protocols

4.2   Quality of Service Measurement Performance

Throughput. Figure 6 illustrates the performance of the selected protocols in terms of throughput corresponding to different network area size. As it is observed, throughput performance of all protocols declined with increasing network area size. Throughput values of reactive protocols are very close to each other, they decline in the same way. Whereas, in proactive protocols, a sharp decrease in throughput values for DSDV protocol is seen unlike OLSR.

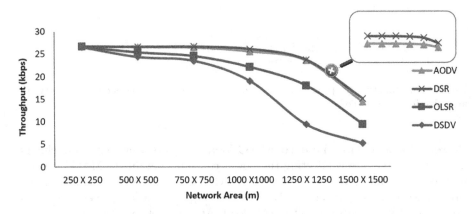

Fig. 6. Throughput versus changing area size for all protocols

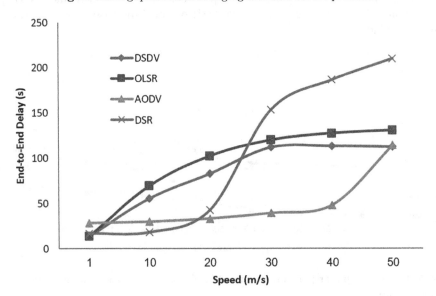

Fig. 7. Delay versus speed for all protocols

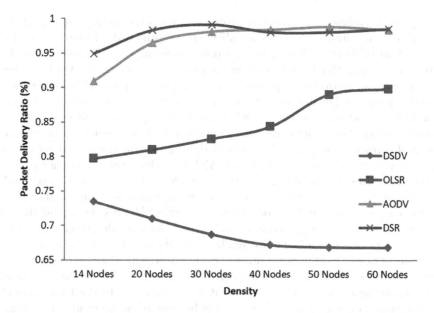

Fig. 8. Packet delivery ratio corresponding network density for all protocols

Delay. Figure 7 presents the relation between delay and node's speed for all protocols. Speed has obvious effect on end-to-end delay. Increasing speed of the nodes leads to increasing delay time but amount of that increasing is different. In proactive protocol, delay increases when increasing speed but at (40 m/s) and (50 m/s) they tend to be steadier. Reactive protocols have different behavior. Delay for DSR protocol increases slowly in the beginning then it increases sharply by increasing speed while AODV has stability but after (40 m/s) it starts to increase.

Packet Delivery Ratio (PDR). Figure 8 shows the PDR for all protocols with varying number of nodes. Reactive protocols (AODV and DSR) have well performance with increasing number of nodes but proactive protocols (OLSR and DSDV) do not reach the top value of PDR. OLSR has better performance than DSDV, but in overall reactive protocols have less packet loss than proactive protocols and the increase of the number of nodes is not affecting their performance.

5   Conclusion

Obtained simulation results for the different scenarios reveal a differentiation between routing protocols in consuming power. From the figures, we could understand that in total power consumption, proactive protocols consume more power than reactive protocols however, they have a steady behavior in all scenarios.

Besides, we perceived that proactive protocols have better performance in static network or a network with a less mobility; this is principally because of up-to-date routing table by nodes. Regarding power consumption at layers, generally MAC layer is an active layer in consuming power by all protocols. On the other hand, power dissipation at application layer by reactive protocol is higher than proactive protocols because reactive protocols spend more power on transferring data while proactive protocols spend most of their power by sending routing messages. Operation modes are another aspect that we took into consideration in this research work. As it is observed, all protocols consume most of the power at idle mode this is because in MANET, nodes sense the carrier almost all the time. About transfer and receiving modes, the results in this research work revealed that consuming power at receiving mode is always higher than consuming power at transfer mode by all protocols. The reason behind that is that the receiving mode is receiving neighbors' data besides receiving actual data while transfer mode is just for sending actual data. This research also paid attention to power consumption due to routing overhead. Based on results that achieved from simulating scenarios, consuming power due to routing overhead by proactive protocols is very high and it is almost equal to the total amount of consuming power at routing layer this is because of properly updating routing table. However, reactive protocols do not perform like proactive protocols because they do not spend most of the consumed power at routing layer due to routing overhead as they totally depend on discovery mechanism for finding routing and they create route when required. In analyzing power consuming due to MAC load, we see that reactive protocols consume more power than proactive protocols due to MAC load. This is because reactive protocols totally depend on hop-by-hop mechanism for finding a routing path. Increasing density or pause time has great effect on power dissipation.

According to the obtained results showed in Figs. 6, 7 and 8, Reactive protocols (AODV and DSR) present higher performance than proactive protocols (DSDV and OLSR) in term of throughput. Regarding end-to-end delay, behaviors of the protocols vary according to the scenario. For example, reactive protocols have lower end-to-end delay value than proactive protocols in changing speed scenario. As well as, reactive protocols have excellent performance in packet delivery ratio in all scenarios because their performance is a kind of steady and approximately equal to one, but packet delivery ratio performance of proactive protocols decline in all scenario and not in a good condition like reactive protocols. The main reason behind such behavior of reactive protocols is related to their operation mechanism as explained in chapter two section (2.2.1). AODV protocol does not update tables and fully depends on hop-to-hop methodology and it has a main problem with removing intermediate nodes. Regarding DSR, the main characteristic of DSR is that the header of data packet holds the entire nodes address that the packet must pass in order to reach the destination. On the other hand, proactive protocols (OLSR and DSDV) update routing table. DSDV has two techniques for updating routing table which are partially update and fully update, while OLSR has only fully update procedure and nodes in OLSR protocol totally depend on Multipoint Relay for transmitting their messages.

References

1. Sarkar, S.K., Basavaraju, T.G., Puttamadappa, C.: Ad Hoc Mobile Wireless Networks Principles Protocols and Applications. Auerbach Publications, New York (2007)
2. Gupta, A.K., Sadawarti, H., Verma, A.K.: Review of various routing protocols for MANETs. WSEAS Trans. Commun. **10** (2011)
3. Kumar, N.: Power aware routing protocols in mobile adhoc networks-survey. Int. J. Adv. Res. Comput. Sci. Softw. Eng. **2** (2012)
4. Kaur, D., Kumar, N.: Comparative analysis of AODV, OLSR, TORA, DSR and DSDV routing protocols in mobile Ad-Hoc networks. Comput. Netw. Inf. Secur. (2013)
5. Kumar, R., Singh, P.: Performance analysis of AODV, TORA, OLSR and DSDV routing protocols using NS2 simulation. Int. J. Innov. Res. Sci. Eng. Technol. **2**(8) (2013)
6. Kaushik, S.S., Deshmukh, P.R.: Comparison of Effectiveness of AODV, DSDV and DSR routing protocols in mobile ad hoc networks. Int. J. Inf. Technol. Knowl. Manag. **2**(2), 499–502 (2009)
7. Bakht, H.: Survey of routing protocols for mobile ad-hoc network. Int. J. Inf. Commun. Technol. Res. **1**(6) (2011)
8. Sochor, T., Gatek, T.: Ad-hoc routing protocols comparison using open-source simulation tool. In: Silhavy, R., Senkerik, R., Oplatkova, Z.K., Silhavy, P., Prokopova, Z. (eds.) Software Engineering Perspectives and Application in Intelligent Systems. AISC, vol. 465, pp. 425–433. Springer, Cham (2016). https://doi.org/10.1007/978-3-319-33622-0_38
9. Al-Dhief, F.T., Sabri, N., Salim, M.S., Fouad, S., Aljunid, S.A.: MANET routing protocols evaluation: AODV, DSR and DSDV Perspective, EDP Sciences, p. 06024 (2018)
10. Xiao, H., Ibrahim, D., Christianson, B.: Analysis of power consumption in ad hoc networks. In: IEEE Wireless Communications and Networking Conference (WCNC), pp. 2599–2604 (2014)

Extending Lifetime of Wireless Sensor Network in Application to Road Traffic Monitoring

Marcin Lewandowski[1], Marcin Bernas[2], Piotr Loska[3], Piotr Szymała[3],
and Bartłomiej Płaczek[1(✉)]

[1] Institute of Computer Science, University of Silesia, Sosnowiec, Poland
marcin.lewandowski@us.edu.pl, placzek.bartlomiej@gmail.com
[2] University of Bielsko-Biala, Bielsko-Biala, Poland
marcin.bernas@gmail.com
[3] Institute of Innovative Technologies EMAG, Katowice, Poland
{piotr.loska,piotr.szymala}@ibemag.pl

Abstract. The key issue in applications of wireless sensor networks is the limited lifetime of battery-powered sensor nodes. In this paper several methods were adapted and combined to extend the lifetime of the wireless sensor network. The considered network was designed for road traffic monitoring applications, such as detection of vehicles, bicycles, and pedestrians. Detailed experiments were conducted in order to determine lifetime of the sensor nodes in different real-world scenarios. The obtained results show that lifetime of the network for road traffic monitoring can be significantly extended by rotating the role of sensor nodes, suppressing unnecessary transmissions and putting the nodes to sleep mode according to a duty cycle.

Keywords: Vehicle detection · Pedestrian detection · Sensor fusion ·
Low-cost sensors · Intelligent transport systems

1 Introduction

Road traffic monitoring systems are necessary to provide intelligent services that enable smoother, safer, and environmentally friendly transportation [1]. An important task of the traffic monitoring system is detection of vehicles, bicycles, and pedestrians. The detection results are useful for adaptive traffic signals, variable speed limits, traveler information, and route guidance systems [2,3].

Recent trends in the development of traffic monitoring systems include applications of wireless sensor networks that are composed of small low-cost sensor nodes [4]. The sensor nodes can be quickly and effortlessly installed alongside traffic lanes, bicycle paths, and sidewalks [5]. The wireless sensor nodes are powered by batteries; hence lifetime of the wireless sensor network depends on energy consumption. Therefore, reducing the energy consumption of sensor nodes is considered as an important research issue [6].

© Springer Nature Switzerland AG 2019
P. Gaj et al. (Eds.): CN 2019, CCIS 1039, pp. 112–126, 2019.
https://doi.org/10.1007/978-3-030-21952-9_9

In this study a hybrid method was introduced to decrease energy consumption and extend lifetime of a wireless sensor network that detects the presence of traffic participants in predetermined detection zones. The presented method combines different approaches from the literature with data suppression. This method was experimentally evaluated using prototypes of sensor nodes.

The paper is organized as follows. Related works are surveyed in Sect. 2. Section 3 includes a presentation of the considered wireless sensor network. Details of the methods that enables extending lifetime of the sensor network are discussed in Sect. 4. Experiments and their results are described in Sect. 5. Finally, conclusions are given in Sect. 6.

2   Related Works

In the literature a number of methods have been proposed for extending lifetime of wireless sensor networks. A detailed survey of these methods can be found in [5]. This study is focused on the methods that are suitable for application in the traffic monitoring sensor network, which is presented in Sect. 3. The specific features of the considered network, that limit the choice of the lifetime extending methods include: low-cost hardware components, predetermined location of sensor nodes, permanent assignment of nodes to clusters, uniform resources distribution among sensor nodes, in-network decision making, and strict time limits for delivering detection results to sink. Based on literature review, the following applicable methods were selected for this research: rotating the role of cluster head among sensor nodes [9,10], suppressing unnecessary transmissions [11], putting the nodes to sleep mode according to a duty cycle [12], and energy harvesting [13].

Duty cycle is a critical feature, which extends lifetime of wireless of sensor network [12]. In the duty cycle approach, the sensor nodes periodically switch on and off their components. For the purpose of lifetime extension, sensor nodes have to be set to the sleep state as much as possible. However, the nodes cannot detect events and transmit data while in the sleep state. Detecting events and reporting them to the sink rapidly play a crucial role in the traffic monitoring applications. The dependencies between duty cycle, transmission delay, and network lifetime were discussed in [14].

For sensor networks with clusters, high energy expenditure is observed at cluster head nodes that have to remain active most of the time. The remaining sensor nodes are active only while they process and send their data. A balance of energy consumption is commonly restored by periodically reassigning (rotating) the cluster head role to different sensor nodes [9,10]. The research results presented in [15] show that a good balancing of sensor node lifetimes and optimized network lifetime can be achieved by the cluster head rotation with probabilistic sleep of ordinary nodes and opportunistic sleep of cluster head nodes.

In wireless sensor nodes, data collection operations consume relatively low energy whereas data transmission is energy expensive. Thus, the suppression schemes extends lifetime of sensor nodes by transmitting the data only if it is necessary [12]. In case of sensor nodes that make in-network decisions, large amounts of data often do not have to be transmitted as the decisions made with and without these data are the same. This fact was used in [16] to introduce a decision-aware data suppression scheme.

Recently sensor networks for road traffic monitoring were proposed that harvest energy from the surrounding by using solar panels [17], piezoelectric and thermoelectric technologies [18].

In this study, algorithms are presented that control operation of prototype sensor nodes in a wireless sensor network for traffic monitoring. The presented algorithms combine the above mentioned methods to extend lifetime of the sensor network.

3   Wireless Sensor Network for Road Traffic Monitoring

The wireless sensor network considered in this paper was designed for road traffic monitoring applications, such as detection of vehicles, bicycles, and pedestrians. This network consists of sensor nodes installed on side of a road or on pavement. The sensor nodes collect data readings from sensors and execute a detection algorithm to determine if an object (vehicle, bicycle or pedestrian) is present in its vicinity. The information delivered by single sensor node is a probability of object detection.

The sensor nodes are grouped into clusters. Each cluster corresponds to detection zone, i.e., a selected area on traffic lane or sidewalk, where the traffic participants have to be detected. One of the sensor nodes in a cluster takes the role of cluster head and collects detection probabilities from remaining members of the cluster. The cluster head aggregates collected probabilities and sends information to a sink. The information transmitted from cluster head to sink (e.g., to traffic signal controller) indicates if the detection zone is empty or occupied. When a detection zone is occupied, the type or class of the detected object (e.g., personal car, semi-truck, truck) can be also reported to the sink.

Architecture of the sensor node is composed of microcontroller, sensor modules (accelerometer, magnetometer, light sensor), and XBee communication module. The node is powered with battery, which optionally can be charged from a photo voltaic panel.

4   Methods for Extending Network Lifetime

All sensor nodes in the considered network perform data collection and transmission tasks periodically (in cycles). The sensor nodes are put in a sleep mode if their required operations at a given cycle are finished. During sleep mode the energy consumed by sensor node is reduced, thus the lifetime can be extended. An example of duty cycle with sleep times for a cluster of 4 sensor nodes is presented in Fig. 2.

At each cycle, the sensor nodes (that are not the cluster heads) have to collect data from their sensors, perform detection algorithm, and report the evaluated

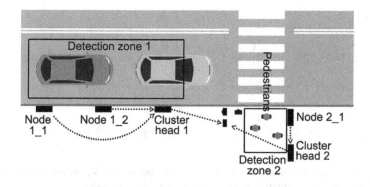

Fig. 1. Structure of wireless sensor network for road traffic monitoring.

detection probability to the cluster head. The cluster head (denoted as CH in Fig. 2) has to execute the same data collection and processing operations as the remaining nodes. Additionally, the cluster head receive transmissions from the cluster members, aggregate received detection probabilities, and send information about the detection result to sink. The operations executed by cluster head require more time and more energy than the operations performed by the non-cluster head nodes. Thus, battery of the cluster head is depleted faster than for the other nodes.

Node 1 (CH)	active		sleep
Node 2	active	sleep	
Node 3	sleep	active	sleep

time

Fig. 2. Duty cycle of sensor nodes operating in one cluster (Node 1 is cluster head).

In order to balance the energy consumption for all nodes in the cluster, the role of cluster head is rotated among the nodes as shown in Fig. 3(a). As a result, each node acts as the cluster head for a similar number of cycles. This method enables the operation of sensor network to be continued even if the cluster head fails due to energy depletion. Moreover, the dead node can be put back into operation after recharging its battery with a photo voltaic panel.

It should be noted here that shorter active time of sensor node leads to lower energy consumption and longer lifetime of the node. The active time of sensor nodes can be reduced by executing the data processing and transmission tasks in parallel. Thus, in this study, one of the analysed scenarios assumes that the Xbee node transmits object detection result obtained at previous cycle, while in the same time the microcontroller acquires new data from sensors and executes

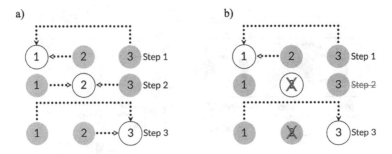

Fig. 3. Rotation of cluster head role: (a) normal operation, (b) failure of cluster head (white circles represent cluster heads, arrows show data transfers).

the detection algorithm. Disadvantage of this solution is a delay of one cycle in delivering the detection results to the Sink. However, such delay is acceptable for the road traffic monitoring applications.

The most energy consuming operation of sensor node is wireless data transmission. Therefore, an important approach to extend node lifetime is to suppress the transmissions of data that are not necessary to determine the detection result. In the proposed solution, the transmission from sensor node to cluster head is suppressed, if the detection probability determined by the sensor node is low (below a predetermined threshold). If the cluster head does not receive transmission from a given sensor node then it assumes that this sensor node did not detect the target object at current cycle.

Algorithm 1. Initialization of sensor node

1: $step = 1$
2: $counter = 0$
3: **while** $SN[step]$ *is not active* **do**
4:     $step = step + 1$
5:     **if** $step > s_{max}$ **then**
6:         $step = 1$
7:     **end if**
8: **end while**
9: **if** $step = id$ **then**
10:     $role = CH$
11: **else**
12:     $role = CM$
13: **end if**

In this study a method was proposed, which combines the above-mentioned techniques for extending network lifetime. This method was implemented in sensor node prototypes, and verified in real-world experiments. The proposed method is summarized by pseudo codes of Algorithms 1–3. Algorithm 1 describes

Algorithm 2. Operation of sensor node with role of cluster member (CM)

```
 1: //role = CM
 2: wait(id · delay)
 3: wake up
 4: collect readings from sensors
 5: p = detection_probability(readings)
 6: if p > threshold then
 7:     send p to SN[step]
 8: end if
 9: counter = counter + 1
10: if counter = c_max then
11:     repeat
12:         step = step + 1
13:         if step > s_max then
14:             step = 1
15:         end if
16:     until SN[step] is active
17:     if step = id then
18:         role = CH
19:     end if
20:     counter = 0
21: end if
22: sleep
```

initialization of sensor node. This algorithm has to be performed by each sensor node at start of the sensor network operation. It was assumed that the sensor nodes have unique identifiers ($id = 1, 2, ...s_{max}$). The symbol $SN[identifier]$ in Algorithms 1–3 denote sensor node with a given $identifier$. The abbreviations CH and CM are used to describe the role of sensor node (cluster head or cluster member). The operations presented in Algorithm 2 are executed by all cluster members, at each cycle of the sensor network operation after initialization. Similarly, the operations performed by cluster head are shown in Algorithm 3. It should be noted that c_{max} determines the number of cycles between changes of the cluster head. The *step* variable corresponds to the step of the cluster head rotation procedure (see Fig. 3), and variable id denotes identifier of the considered sensor node.

5   Experiments

The wireless sensor network has been subjected to detailed tests in order to determine the current consumption of the nodes in different scenarios. The considered test scenarios correspond to implementation of the methods presented in Sect. 3. On this basis, lifetime of the sensor network was evaluated and compared for different combinations of the above-discussed methods.

During experiments, the sensor nodes were powered with 1350 mAh/3.7 V battery. Length of the sensor node operation cycle was 500 ms. For such settings,

Algorithm 3. Operation of sensor node with role of cluster head (CH)

1: //$role = CH$
2: *wake up*
3: *collect readings from sensors*
4: $p = detection\_probability(readings)$
5: *wait for messages from CMs*
6: *send detection result to sink*
7: *counter = counter* + 1
8: **if** *counter* = c_{max} **then**
9:     **repeat**
10:         step = step + 1
11:         **if** $step > s_{max}$ **then**
12:             step = 1
13:         **end if**
14:     **until** SN[step] is active
15:     **if** $step <> id$ **then**
16:         $role = CM$
17:     **end if**
18:     *counter* = 0
19: **end if**
20: *sleep*

---

detailed measurements were performed in order to determine current consumption and duration of particular operations executed by the sensor nodes. Results of these measurements are presented in Tables 1, 2, 3, 4 and 5, where particular rows describe successive operations of the sensor node modules during one cycle.

Table 1 shows the results obtained for the simplest test scenario, where just one basic method was implemented to reduce the energy consumption, i.e., the node was put in sleep mode after finishing its tasks. The sleep mode means that all node elements are in sleep state: sensors, microcontroller and XBee modules. It can be observed in Table 1 that the current consumption of sensor node in the sleep (idle) mode is significantly lower (see the first row in Table 1) when comparing with the active mode. There are several sleep levels available for the microcontroller. If deeper sleep level is used then lower power consumption can be achieved. However, the deeper sleep level requires longer wakeup procedure and involves lower timer precision. In this study the sleep level with RAM support was used. The highest current consumption was experienced when activating the wireless communication module (XBee).

The results shown in Table 2 correspond to node operation cycle with parallel sensor readings collection and data transmission. This modification has enabled reducing the active time of sensor node by 22 ms. As a result, the average current consumption was reduced by 97 μA. It should be also noted here that Tables 1 and 2 include the results of measurements that were conducted with universal asynchronous receiver-transmitter (UART) used to preview the state of the node. The utilization of UART interface has caused an additional current

Table 1. Node operation with sleep mode (data preview via UART).

Module	Task	Time[ms]	Current consumption [μA]
Microcontroller	Idle mode	422	530
Microcontroller	Wake up	4	
	Change system clock		
	Initialize internal modules		
Light sensor	Wake up	23	12000
	Read data		
	Sleep		
Accelerometer and magnetometer	Wake up	9	
	Read data		
	Sleep		
Microcontroller	Detection algorithm	38	34000
XBee	Wake up		
	Send data		
	Sleep		
Microcontroller	Sleep	4	12000
Average current consumption			**3.991 mA**

consumption. Without the preview via UART interface, the current consumption is significantly lower (see Table 3).

Tables 3 and 4 show the measurement results for two variants of the node operation. In the first case (Table 3), the sensors are put to sleep mode after finishing the node tasks at a given cycle. Thus, it is necessary to wake the sensors up before the next data reading is registered. This results in longer active time of the sensor node. In the second case (Table 4), the sensors (light sensor, accelerometer, and magnetometer) stay active during whole cycle. The current consumption in sleep mode is increased by 195 μA, but it is not necessary to wake up the sensors. The higher energy consumption is therefore compensated by the shorter active time of the node. As a consequence, comparable average energy consumption at the level of 2.5 mA was observed for both cases.

Table 5 shows the work cycle of the sensor node in case when the data transmission in a given cycle is suppressed due to low object detection probability. In this case the current consumption is 5 times lower.

The average current consumption for sensor nodes that have the role of ordinary cluster member (CM) and cluster head (CH) is compared in Fig. 4. Additionally, Fig. 4 compares the current consumption for sleeping and non-sleeping nodes as well as for nodes that suppress their transmissions.

Table 2. Node operation with parallel data collection and transmission (data preview via UART).

Module		Task		Time [ms]	Current consumption [μA]
Microcontroller		Idle mode		444	530
Microcontroller		Wake up		8	10000
		Change system clock			
		Initialize internal modules			
XBee		Wake up		48	34000
Sensors	XBee	Wake up	Send data		
		Read data	from previous		
		Sleep	cycle		
XBee		Sleep			
Microcontroller		Detection algorithm			
Average current consumption					**3.895 mA**

Table 3. Node operation with parallel data collection and transmission (no data preview via UART).

Module		Task		Time [ms]	Current consumption [μA]
Microcontroller		Idle mode		460	200
Microcontroller		Wake up		2	10000
		Change system clock			
		Initialize internal modules			
XBee		Wake up		38	30530
Sensors	XBee	Wake up	Send data		
		Read data	from previous		
		Sleep	cycle		
XBee		Sleep			
Microcontroller		Detection algorithm			
Average current consumption					**2.544 mA**

Expected lifetime of the wireless sensor network was evaluated on the basis of the experimental results presented in Tables 1, 2, 3, 4 and 5. In this evaluation two different battery capacities were taken into account (1350 mAh and 3500 mAh). It was also assumed that the network consists of 4 sensor nodes. The residual energy of each sensor node was calculated recurrently, using the

Table 4. Node operation with non-sleeping sensor modules (no data preview via UART).

Module	Task		Time [ms]	Current consumption [μA]
Microcontroller	Idle mode		467	395
Microcontroller	Wake up		3	10000
	Change system clock			
	Initialize internal modules			
XBee	Wake up		30	34000
Sensors XBee	Wake up	Send data		
	Read data	from previous		
	Sleep	cycle		
XBee	Sleep			
Microcontroller	Detection algorithm			
Average current consumption				**2.469 mA**

following formula, with a time step of one hour:

$$E(t) = E(t-1) - E_s(t) + E_f(t)[mAh], \qquad (1)$$

where: E(0) is energy capacity of the battery, is average usage of a sensor node energy within one hour, and denotes energy harvested from photovoltaic panel.

The usage of energy of sensor node s is calculated based on fraction of time when the sensor node is in one of three modes (active with transmission, active without transmission, and sleeping):

$$E_s(t) = 34\alpha + 16\beta + 0.2 \cdot (1 - \alpha - \beta)[mAh], \qquad (2)$$

Table 5. Node operation without data transmission (no data preview via UART).

Module	Task	Time [ms]	Current consumption [μA]
Microcontroller	Idle mode	478	395
Microcontroller	Wake up	3	10000
	Change system clcok		
	Initialize internal modules		
Sensors	Read data	19	1800
Microcontroller	Detection algorithm		
Average current consumption			**0.506 mA**

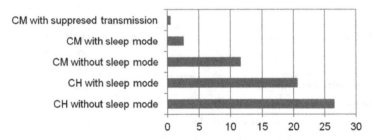

Fig. 4. Comparison of average current consumption for different role of sensor node.

where: α, β are the fractions of time when particular sensor node (s) is in active mode with and without communication, respectively. The values of α and β were determined during experiments, independently for each cycle of the sensor network operation. The supply voltage of sensor nodes was 3.7 V. The harvested energy was calculated as follows:

$$E_f(t) = \frac{\gamma}{24} e_g \, [mAh], \qquad (3)$$

where the energy gain from photovoltaic panel $e_g = 27\,mAh$ was estimated based on its parameters (U = 9 V, P = 1 W, $\eta = 10$) and number of sunny hours a day for a given month ($\gamma \in [2.8, 11.2]$).

The objective was to decrease α and β values, while retaining the functionality of sensor node. As a result the network lifetime is extended. The lifetime lt was determined as the maximum t value, for which all sensor nodes have non-zero residual energy (all sensor nodes are operational):

$$lt = max\{t : min_{s \in (1,2,...,s_{max})} E_s(t) > 0\} \qquad (4)$$

Figure 5 compares the network lifetime for three scenarios. According to the first scenario (fixed CH) one of the nodes takes the role of cluster head for the whole lifetime of the network. In second scenario (rotating CH) the cluster head role is rotated, i.e., each sensor node operates as the cluster head for approximately one quarter of the network lifetime. Finally, in the third scenario the cluster head role is rotated, and additionally the transmissions of unnecessary information are suppressed, i.e., the sensor nodes send messages to cluster head only if a new vehicle is detected. It should be noted here that in this case the energy consumption related to data transfers from sensor nodes to cluster head depends on the number of passing vehicles (traffic volume). The results shown in Fig. 5 for the third scenario correspond to traffic volume of 500 vehicles per hour. This value was selected as average traffic volume for roads in Poland. It can be observed in Fig. 5 that for the CH rotation method the network lifetime was about 2.8 times longer in comparison with the fixed CH scenario. In case of the third approach with data suppression, the network lifetime is 5 times

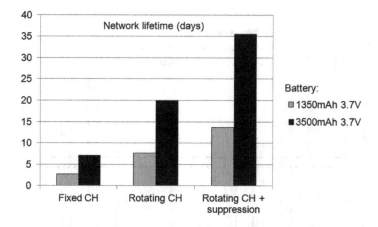

Fig. 5. Impact of cluster head (CH) rotation and data suppression on network lifetime.

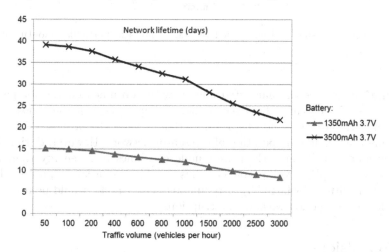

Fig. 6. Lifetime of sensor network for different traffic volumes.

longer. Dependency between traffic volume and network lifetime for the third scenario is illustrated in Fig. 6. These results show that the suppression method significantly extends the network lifetime for a wide range of traffic volumes.

Further analysis was devoted to the possibility of extending network lifetime by recharging battery of sensor node with use of photo voltaic panel. Typical parameters of a low-cost photo voltaic panel were taken into consideration, i.e., 9 V, 1 W power and 10% efficiency. The lifetime estimation was based on average number of sunny days per month and sunny hours per day for Warsaw in years 2000–2017 [19]. Based on the above assumptions it was estimated that average harvested energy equals 3.15 mAh for winter days and 13.06 mAh for summer days.

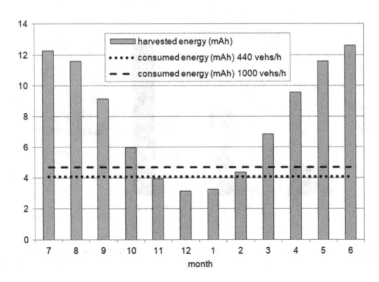

Fig. 7. Energy harvested and consumed by sensor node in particular months of the year.

Figure 7 compares the harvested energy with consumed energy for particular months of the year. These results show that battery of the sensor node can be fully recharged in months 3–10. In that months the proposed approach will ensure uninterrupted operation of the sensor network. However, during winter (November - February) the sensor node consumes more energy than it is harvested by the photo voltaic panel. Thus, in this case a larger photo voltaic panel, larger battery (3500 mAh) or redundant sensor nodes would be necessary to enable data collection without interruptions.

6   Conculsions

Main aim of the research reported in this paper is to extend the lifetime of a wireless sensor network that was designed for road traffic monitoring applications. During experiments detailed measurements were conducted for prototypes of battery-powered sensor nodes to determine the energy consumption in real network operation scenarios. Based on the measurement results, it was shown that lifetime of the considered sensor network can be extended 5 times when using the proposed method, which combines rotation of cluster head role with suppression of unnecessary data transfers. The experiments have also confirmed that the energy consumption in sensor network can be significantly reduced when the sensor nodes are periodically put to sleep. However, the experiments also show that sleep mode in case of short time periods could be energy inefficient. Additional saving of the battery power was achieved by executing data processing and transmission tasks in parallel at the level of sensor node. Long-term uninterrupted operation of the sensor network can be ensured by using the proposed approach for the sensor nodes that are equipped with low-cost battery and photo voltaic panel. The presented approach can be enhanced in the future by introducing redundant sensor nodes to improve robustness of the network to node failure, battery depletion, and low insolation periods.

Acknowledgements. This research was funded by the National Centre for Research and Development (NCBR) Grant No. LIDER/18/0064/L-7/15/NCBR/2016.

References

1. Zhuhadar, L., Thrasher, E., Marklin, S., de Pablos, P.O.: The next wave of innovation-Review of smart cities intelligent operation systems. Comput. Hum. Behav. **66**, 273–281 (2017)
2. Kamkar, S., Safabakhsh, R.: Vehicle detection, counting and classification in various conditions. IET Intel. Transport Syst. **10**(6), 406–413 (2016)
3. Płaczek, B.: A self-organizing system for urban traffic control based on predictive interval microscopic model. Eng. Appl. Artif. Intell. **34**, 75–84 (2014)
4. Anisi, M.H., Abdullah, A.H.: Efficient data reporting in intelligent transportation systems. Netw. Spat. Econ. **16**(2), 623–642 (2016)
5. Bernas, M., Płaczek, B., Korski, W., Loska, P., Smyła, J., Szymała, P.: A survey and comparison of low-cost sensing technologies for road traffic monitoring. Sensors **18**(10), 3243 (2018)
6. Du, R., Gkatzikis, L., Fischione, C., Xiao, M.: On maximizing sensor network lifetime by energy balancing. IEEE Trans. Control Netw. Syst. **5**(3), 1206–1218 (2018)
7. Yetgin, H., Cheung, K.T.K., El-Hajjar, M., Hanzo, L.H.: A survey of network lifetime maximization techniques in wireless sensor networks. IEEE Commun. Surv. Tutorials **19**(2), 828–854 (2017)
8. Curry, R.M., Smith, J.C.: A survey of optimization algorithms for wireless sensor network lifetime maximization. Comput. Ind. Eng. **101**, 145–166 (2016)
9. Gamwarige, S., Kulasekere, C.: Optimization of cluster head rotation in energy constrained wireless sensor networks. In: IFIP International Conference on Wireless and Optical Communications Networks. WOCN 2007, pp. 1–5. IEEE, July 2007
10. Nam, D.H., Min, H.K.: An energy-efficient clustering using a round-robin method in a wireless sensor network. In: 5th ACIS International Conference on Software Engineering Research, Management & Applications. SERA 2007, pp. 54–60. IEEE, August 2007
11. Kim, D.H., Lee, J., Kim, Y.: Information analysis of local suppression scheme based on a spatial-temporal model. J. Appl. Stat. **45**(16), 2929–2942 (2018)
12. Barboni, L., Valle, M.: Experimental analysis of wireless sensor nodes current consumption. In: Second International Conference on Sensor Technologies and Applications. SENSORCOMM 2008, pp. 401–406. IEEE, August 2008
13. Shaikh, F.K., Zeadally, S.: Energy harvesting in wireless sensor networks: a comprehensive review. Renew. Sustain. Energy Rev. **55**, 1041–1054 (2016)
14. Xie, R., Liu, A., Gao, J.: A residual energy aware schedule scheme for WSNs employing adjustable awake/sleep duty cycle. Wireless Pers. Commun. **90**(4), 1859–1887 (2016)
15. Tavakoli, H., Mišić, J., Mišić, V.B., Naderi, M.: Energy-efficient cluster-head rotation in beacon-enabled IEEE 802.15. 4 networks. IEEE Trans. Parallel Distrib. Syst. **26**(12), 3371–3380 (2015)

16. Płaczek, B.: Decision-aware data suppression in wireless sensor networks for target tracking applications. Front. Comput. Sci. **11**(6), 1050–1060 (2017)
17. Ali, Q.I.: Event driven duty cycling: an efficient power management scheme for a solar-energy harvested road side unit. IET Electr. Syst. Transp. **6**(3), 222–235 (2016)
18. Guo, L., Lu, Q.: Potentials of piezoelectric and thermoelectric technologies for harvesting energy from pavements. Renew. Sustain. Energy Rev. **72**, 761–773 (2017)
19. Weather. https://www.weatheronline.pl/. Accessed 15 Nov 2018

BotGRABBER: SVM-Based Self-Adaptive System for the Network Resilience Against the Botnets' Cyberattacks

Sergii Lysenko[✉], Kira Bobrovnikova, Oleg Savenko, and Andrii Kryshchuk

Department of Computer Engineering and System Programming,
Khmelnitsky National University, Instytutska, 11, Khmelnitsky, Ukraine
{sirogyk,savenko_oleg_st,rtandrey}@ukr.net, bobrovnikova.kira@gmail.com
http://ki.khnu.km.ua

Abstract. The paper presents a SVM-based self-adaptive system for the network resilience against the botnets cyberattacks named BotGRAB-BER. It is a novel framework, which combines the ability to reveal the cyberattacks executed by botnets, to detect the botnets that use the evasion techniques, to execute the self-adaptive appliance of the security scenarios in the situation of cyberattacks. As the inference engine for botnet detection the support vector machine is used. It provides the ability to recognize the cyberattacks, taking into account the hosts' network activity and the captured network traffic. In case of the attack, the BotGRABBER system is able to produce the security scenario concerning the attack in order to mitigate it, and to ensure the network's resilient functioning.

Keywords: Botnet · Cyberattack · Botnet detection ·
Network resilience · Self-adaptive systems · Resilience ·
Security scenario · Malware · DDoS attack

1 Introduction

Today cybercriminals find more ways to obtain the profit from the legitimate businesses and enterprises, which are the target of extortion and a lucrative source of income for organized crime groups because of the personally identifiable information stored and processed by these establishments. Botnets are one of the most powerful tools used by cybercriminals to commit such malicious acts [1].

Botnet infection is capable of spreading to any end device, including servers, routers, Network Attached Storage devices, digital video recorders, IP cameras and other smart devices. Bots use exploits to take over devices and enlist them with their command and control server [2].

The wide spread and development of Internet technologies have led to a significant expansion of botnet capabilities in launching denial of service attacks,

© Springer Nature Switzerland AG 2019
P. Gaj et al. (Eds.): CN 2019, CCIS 1039, pp. 127–143, 2019.
https://doi.org/10.1007/978-3-030-21952-9_10

infecting millions of computers with malicious code, theft of confidential data, large-scale spamming, blackmail and extortion. Botnets can use the combinations of multiple attacks targeting known and unknown end devices vulnerabilities [3].

Cybercriminals are developing new ways to evade modern botnet detection methods, therefore existing approaches are not able to counter the growing botnets' threat. Meanwhile the botnets' attacks are becoming increasingly dangerous and devastating.

The described situation actuates the development of new approaches able to detect, prevent, and mitigate the botnets' attacks against the networks. Furthermore, it is very important to ensure the stable network's functioning in the situation of attacks – its resilience – the ability of the network and its components to recover quickly and continue operating [4]. One of the way to solve this problem is to construct a security system which will have the self-adaptive nature and will be able to accumulate information for the purpose of assessing changes in external and/or internal conditions and to adapt to these changes by modification of their own behaviour [5]. The aim of such an adaptation is the self-reconfiguration of the network infrastructure, depending on the type of attack, type of the victim in the network, etc. Moreover, the development of a self-adaptive system for the network resilience against the botnets' cyberattacks requires the involvement of effective inference engines for attacks' recognition in its structure.

2   Related Work

2.1   Botnet Detection Approaches

In recent years, the great number of the botnet detection approaches, such as honeynets, DNS (Domain Name System)-based, network traffic analysis, as well as techniques for Internet of Things (IoT) have been developed.

Approaches [6,7] are focused on a detailed analysis of attackers against windows emulating honeypots in various types of networks, as well as the analysis of the behaviour of various attacks. Using honeynets, the attacks are analysed according to the threat type, session duration, AS (autonomous system), country and RIR (Regional Internet Registry) of the attack origin.

In [8] a mechanism for analysis of botnet activities in the IoT, based on machine learning techniques is presented. It is based on network flow identifiers that can track suspicious activities of botnets.

In [9,10] investigations of IoT botnets, IoT-specific network behaviours concerning the DDoS attacks are presented. Techniques deal with the IoT network traffic with a variety of machine learning algorithms. Approaches provide remedies and recommendations to mitigate IoT-related cyber risks and briefly illustrate the importance of cyber insurance in the networks.

In [11] a survey of the botnet detection via DNS traffic analysis is presented. It is focused on open issues within each category of a detection taxonomy presented earlier. The article provides evidences to many cases where the existing approaches still have some limitations related to the accuracy and deployed location.

In [12,13] the machine learning-based methods for malware domain names detection are presented. The approaches use a neural network Extreme Learning Machine to analyse alphanumeric characteristics of the domain names and to classify them taking into account the features extracted from multiple resources.

A variety of approaches devoted to botnet detection which use machine learning algorithms have been presented. In [14] a botnet detection approach uses the genetic algorithm and the decision tree algorithm C4.5 and operates with a set of features that are able to indicate botnets connections to C&C (command and control servers) in different phases of its functioning. In [15] a botnet detection approach uses the unsupervised machine learning and similarity analysis between benign traffic data and botnet's traffic data. In [16,17] methods for P2P botnet detection based on an adaptive multilayer feed-forward neural network in cooperation with decision trees are presented, where the classification and the regression tree are applied as a feature selection technique to choose the relevant features. In [18] a botnet detection method uses the high-speed network environment and proposes an efficient quasi-real-time intrusion detection system, which is based on a random forest model. In [19] an approach uses an active learning under the streaming data context, decouples the machine learning algorithm from the raw throughput of the stream and provides the opportunity to manipulate the distribution of data used for model building.

In [20] a correlation-based communication histogram analysis approach to detect HTTP botnets is proposed. It is based on the similarity and the correlation of botnets' group activity in the network.

In [21] an approach operates with the sequence and syntax of SMTP (Simple Mail Transfer Protocol) commands observed during the email delivery. Several improvements for detecting unsolicited email sources from different botnets (fingerprinting) that can be used during network forensic investigation are presented.

Furthermore, there are approaches based on the anomaly detection. [22] proposes a method for botnet detection which applies a sliding window to the network traffic and monitors anomalies in the network based on large deviation results, for flow and packet level data. In [23] a FlowIds framework for anomaly detection on SMTP traffic flows makes a decision about a botnet presence using the decision tree classification and deep learning algorithms.

The common weakness of the aforementioned approaches is the requirement of large amounts of computing resources and the fact that they are not able to respond adaptively to known and unknown attacks performed by botnets. Moreover, the mentioned techniques are unable to assure resilient functioning of the network under the botnet cyberattacks.

2.2 Self-Adaptive System for the Corporate Area Networks Resilience in the Presence of Botnet Cyberattacks

Due to the high importance of the botnet spreading problem a great amount of suggestions for the botnet detection approaches have been produced during the last years. Among them, an effective system for botnet detection – BotGRAB-BER – was presented. The very first idea for the efficient botnet detection was to use a multi-agent system. A conclusion about the botnet's presence in the network was made using the fuzzy logic and was based on the information gathered from agents' allocated to each host of the network [24].

The next generation of the BotGRABBER system obtained the possibility to detect botnets via DNS traffic analysis. It took into account the group activity of bots in the DNS-traffic, which appeared in a small period of time in the group DNS-queries of hosts during trying to access the C&C-servers, migrations, running commands, or downloading updates of the malware. Moreover, the method took into account abnormal behaviours of the hosts' group, which are similar to botnets [25].

As the cycling of IP mapping, "domain flux", "fast flux", DNS-tunneling are the evasion techniques often used by botnets, the further development of the BotGRABBER system was the inclusion of the possibility to execute the passive monitoring and active DNS probing of the network. It was based on a cluster analysis of features obtained from the payload of incoming DNS-messages. The method uses the semi-supervised fuzzy c-means clustering [26]. To increase the accuracy of the botnet detection, the mentioned system obtained the possibility to analyse the software's behaviour in the host, which may indicate the possible presence of the bot directly in the host [27].

The most important upgrade of BotGRABBER system was its transformation into the self-adaptive system for the corporate area networks' resilience in the presence of botnets' cyberattacks. Based on the gathered Internet traffic features inherent to cyberattacks, the BotGRABBER system was able to produce the security scenarios according to the cyberattacks performed by the botnets in order to mitigate the attacks and ensure the network's resilient functioning. The proposed approach used the semi-supervised fuzzy c-means clustering, where the objects of clustering were feature vectors which elements may indicate the appearance of cyber threats in the corporate area networks [28].

In spite of promising BotGRABBER's results it had several disadvantages. As it used the semi-supervised fuzzy c-means clustering as the inference engine, the minor situations could occur:

1. The BotGRABBER system could produce uncertain results of clustering when the observed object could belong to more than one cluster. In case when those clusters correspond to malicious traffic, the BotGRABBER system was able to conclude that the botnet was present in the network but at the same time it was unable to choose the only one needed security scenario, which was more relevant to the botnet's cyberattacks.
2. Another possible situation could occur when the observed object was referred to more than one cluster, when one cluster indicated the botnet's presence and another one – that the traffic was normal. In that case the efficiency of the botnet detection decreased greatly.
3. The lack of knowledge about multi-vector attacks also leads to the detection efficiency reduction and demands some improvements in this area.

In order to eliminate aforementioned drawbacks, it is required to change the instrument of classification.

3 Support Vector Machine

The support vector machines (SVMs) are the high-potential approach for the object classification. SVMs are the supervised learning models with associated learning algorithms [29,30]. They are able to produce accurate and robust classification results, even when input data is non-monotone and non-linearly separable, and to evaluate more relevant information in a convenient way, providing high accuracy of classification with small training sets [31].

Basically, SVM performs classification by finding the hyperplane that maximizes the margin between two classes. Vectors (cases) that define the hyperplane are the support vectors. To define an optimal hyperplane we need to maximize the width of the margin

$$w \cdot x + b = 0, \tag{1}$$

where x is a classification object lying on the hyperplane, w is normal to the hyperplane, b is the bias and $\frac{|b|}{\|w\|}$ is the perpendicular distance from the hyperplane to the origin, with $\|w\|$ the Euclidean norm of w [29,32].

Let φ be a mapping function which projects the training data into a Hilbert high dimensional space H, $\varphi : R^q \longrightarrow H$. The data point x is represented in space H as $\varphi(x)$.

In the context of SVM, the kernel function defines the hypothesis space, and is defined as

$$K(x, x_i) = (\varphi(x) \cdot \varphi(x_i)), \tag{2}$$

where x_i is a training data.

It leads to decision functions of the following form

$$f(x) = sgn\left(\sum_{i=1}^{r} \alpha_i a_i K(x, x_i) + b\right), \tag{3}$$

where $\alpha_i, i = 1, \ldots, r$ are Lagrange multipliers, the maximal magnitude of which is governed by C.

In order to compute the separating hyperplane without explicitly carrying out the mapping into the feature space, different kernel functions can be used [33,34]. In this approach, the classification process involved the kernels: linear, polynomial, Gaussian, exponential, and B-Spline.

The linear kernel is the simplest kernel function. It is given by the inner product (x, x_i) plus an optional constant c:

$$K(x, x_i) = x^T x_i + c, \ c \in R. \tag{4}$$

The polynomial kernel is a non-stationary kernel, where the adjustable parameters are the slope α, the constant term c and the polynomial degree p:

$$K(x, x_i) = \left(\alpha x^T x_i + c\right)^p, \ \alpha \in R, \ c \in R, \ p \in N. \tag{5}$$

The Gaussian kernel is an example of radial basis function kernel:

$$K(x, x_i) = e^{-\frac{1}{2\sigma^2}\|(x-x_i)\|^2}, \sigma > 0, \tag{6}$$

where σ is the parameter that controls the width of the Gaussian kernel.

The exponential kernel is closely related to the Gaussian kernel, with only the square of the norm left out. It is also a radial basis function kernel:

$$K(x, x_i) = e^{-\frac{1}{2\sigma^2}\|(x-x_i)\|}. \tag{7}$$

The B-Spline is a radial basis function kernel, and is defined on the interval $[-1, 1]$. It is given by the recursive formula [33]:

$$K(x, x_i) = B_{2p+1}(x - x_i), \ where \ p \in N \ with \ B_{i+1} := B_i \otimes B_0. \tag{8}$$

In order to perform the multi-class classification, the "one against all" and "one against one" SVM-based methods are used [31].

"One Against All" SVM Classifier. This method constructs k SVM models, where k is the number of classes. The mth SVM is trained with all of the samples in the mth class with positive labels, and all other samples with negative labels. The mth SVM solves the task of training data mapping to a higher dimensional space for the given l training data $(x_1, a_1), \ldots, (x_l, a_l)$, where $x_i \in R^n, i = 1, \ldots, l$, $a_i \in \{i = 1, \ldots, k\}$ is the class of training data x_i.

In order to perform the classification for x_i, we use k decision functions, where k is the number of classes

$$(w_i)^T \varphi(x_i) + b_i, where \ i = 1, \ldots, k, \tag{9}$$

where φ – is the mapping function, $\varphi : R^q \longrightarrow H$.

The data x_i then belongs to class a, for which the above decision function has the largest value

$$a \equiv argmax_{i=1,\ldots,k}((w_i)^T \varphi(x_i) + b_i). \tag{10}$$

"One Against One" SVM Classifier. The proposed technique uses the "one against one" SVM classification method as well. Here, $k(k - 1)/2$ classifiers are to be constructed for each pair of classes and the max-win strategy is to be followed. Specifically, if $sgn(w_{jl})^T \varphi(x_i) + b_{jl}$ evaluates x_i to be in jth class, then the vote for the jth class is incremented by one, else that for the lth class is increased by one. Finally, the training data vector x_i is predicted to belong to the class with maximum number of votes [31].

Taking into account the advantages of the SVM, it can be used as the inference engine in the structure of the BotGRABBER for making the decision about the presence of the cyberattacks performed by botnets in the network.

4 BotGRABBER: SVM-Based Self-Adaptive System for the Network Resilience Against the Botnet Cyberattacks

4.1 The Key Features of BotGRABBER

BotGRABBER presents the SVM-based system for botnet detection with the ability to produce security scenarios for assuring the corporate area networks' resilience in the presence of botnets' cyberattacks. The network's resilience is ensured by the network's dynamic adaptive reconfiguration.

BotGRABBER is a multi-vector protection system as it combines the analysis at both network and host activity. The combined information allows one not only to detect the botnet's cyberattacks but also to produce the needed security scenario of the network reconfiguration according to the type of cyberattack performed by the detected botnet.

The BotGRABBER system provides a novel botnet detection framework with the key features given below:

1. ability to detect the most known botnets' cyberattacks;
2. ability to detect botnets that use DNS evasion techniques (cycling of IP mapping, "domain flux", "fast flux" and DNS-tunneling);
3. ability to self-adaptive appliance of the security scenarios for the botnet cyberattacks mitigation;
4. assuring the corporate area networks' resilience in the presence of botnets' cyberattacks;
5. assurance of the multi-vector protection for corporate area networks.

4.2 Architecture of the BotGRABBER System

Let us consider and explain the main components of the framework (Fig. 1).

Knowledge base. A knowledge base provides the information storage concerning the cyberattacks performed by a botnet in the network and in the hosts. Here, each cyberattack is presented as a feature vector which consists of functional botnets' features.

To increase the efficiency of the botnet detection each stage of the possible botnet's life cycle functioning (infection, initial registration or connection to C&C server, performance of a malicious activity, maintenance, its functioning termination) is presented by own feature vector.

A list of detectable attacks and the attacks' features, combined into groups, which are analysed by the BotGRABBER system, are presented in Fig. 2. These features are the base of the set of feature vectors $X = \{x_i\}_{i=1}^{N_X}$, where each of the feature vector x_i describes a cyberattack or its absence, N_X – is the number of the feature vectors.

The updated version of the BotGRABBER system has involved a set of new attacks' features, which are very significant for the accurate botnet detection such as: the number of send/received null packets, the number of send/received

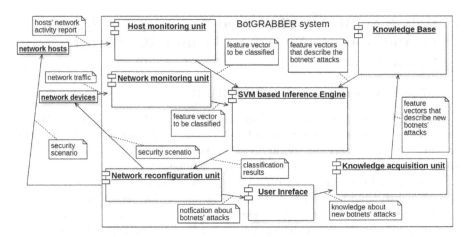

Fig. 1. BotGRABBER's architecture.

small packets, the number of send/received packets with the same length over the total number of packets, a ratio between the number of send/received packets over the total number of send/received packets, the length of the first packet, a ratio between the number of small packets over the total number of send/received packets.

Taking into account the aforementioned features at the present time Bot-GRABBER system operates with 58 features (a full list of features is given in [28]).

Knowledge acquisition unit. Taking into account the increasing number of new ways to perform cyberattacks, the proposed tool has an ability to update the knowledge about new botnets.

Network monitoring unit. This unit implements the network monitoring via gathering the inbound and outbound network traffic. The collected information is converted into the feature vectors and is sent to the SVM-based inference engine for further data processing.

Host monitoring unit. This unit implements gathering the information about the hosts' network activity and reports of the hosts' antiviruses. It also converts the collected information into the feature vectors and sends it to the SVM-based inference engine for further data processing.

SVM-based inference engine. The main task of the SVM-based inference engine is to assign feature vector x_i obtained from the network to class a_t, where $x_i \in X$, $a_t \in A$, $A = \{a_t\}_{t=1}^{N_A}$, N_A – is the number of classes, where each class corresponds to one specified type of attacks, performed by botnet. The SVM-based inference engine makes a conclusion about the presence or absence of a cyberattacks and detected possible type of the attack.

Depending on the detected type of attack a_t, the security scenario s_q is to be applied for the network reconfiguration, $S = \{s_q\}_{q=1}^{N_S}$, where S – is the set of all security scenarios, N_S – is the number of security scenarios.

Thus, function f, selecting a security scenario for network reconfiguration, is defined as: $f : d_u \times a_t \rightarrow s_m$, where $d_u \in D$, $D = \{d_u\}_{u=1}^{N_D}$, where d_u – is the network component attacked by botnet, N_D – is number of network components.

Network reconfiguration unit. If an attack is observed, the security scenario, produced by the SVM-based inference engine, is to be applied in order to mitigate the attack's consequences. This unit applies the security scenario. The aim of the security scenario is to reconfigure the network infrastructure, depending on the type of attack.

5   Experiments

The experimental part of the study involves the examination of the BotGRAB-BER's efficiency by solving the following tasks:

1. to investigate the applicability of SVM for making decisions about the presence of cyberattacks performed by botnets in the network;
2. to verify whether the SVM-based approach is able to increase the botnet's detection efficiency and to avoid the results' uncertainness when the analysed network activity object can belong to more than one class, and therefore the BotGRABBER system is not able to choose the needed security scenario;
3. to investigate the system's detection efficiency after adding new features to the knowledge base, and to study the mitigation potential of the system;
4. to compare results of BotGRABBER's efficiency with c-means clustering engine [28] and SVM-based approach.

5.1   Evaluation Setting

In order to evaluate the BotGRABBER's performance, a detection accuracy tests using real world network traffic were carried out. For this purpose, a botnet dataset [35] was employed. It combines generality, realism and representativeness. The dataset contains both malicious (e.g. traces of Storm, Zeus Neris, Rbot, Virut, NSIS, Menti, Sogou, and Murlo botnets) and non malicious traffic (gaming packets, HTTP traffic, and P2P applications, such as bittorrent). Furthermore, it contains generated real traffic that mimics users' behaviour (e.g. SSH, HTTP, and SMTP). The dataset is divided into training T and evaluation (test) E datasets that include botnets that perform attacks presented in Fig. 2. The training dataset includes 19755 samples, 49.56% of which are malicious and the reminder contains normal flows. The test dataset includes 18917 samples, 55.77% of which represent malicious flows.

To carry out experiments, the university local area network of hosts including 50 hosts (hosts with Microsoft Windows operating system), one dedicated

BotGRABBER

Botnets' attacks	Analyzed botnets' features types

Botnets' attacks

- ✓ DDoS
- ✓ ping flooding
- ✓ smurf attack;

- ✓ TCP SYN Flood
- ✓ Fragmented UDP Flood
- ✓ DNS Amplification
- ✓ TCP Reset

- ✓ ICMP Flood

- ✓ RUDY
- ✓ SIP INVITE Flood
- ✓ Encrypted SSL DDoS
- ✓ ping sweep attack
- ✓ SQL /PHP injection
- ✓ Cross-Site Scripting (XSS)
- ✓ DNS spoofing

- ✓ TCP scan
- ✓ UDP scan
- ✓ Phishing

Analyzed botnets' features types

- ✓ velocity of outbound/inbound traffic
- ✓ average payload length per connection
- ✓ number of bytes/packets transmitted per connection
- ✓ flow duration
- ✓ transmission protocol
- ✓ size for the session in bytes/packets
- ✓ self-similarity of the outbound/inbound packets in the session
- ✓ standard deviation of packet size within the session
- ✓ TCP flags
- ✓ ARP-requests
- ✓ network devices parameters
- ✓ amount of denied packets
- ✓ domain name characteristics
- ✓ TTL-periods
- ✓ A-records/IP-addresses corresponding to the domain name
- ✓ DNS-records
- ✓ size of the DNS-messages
- ✓ successful of DNS-query

Fig. 2. The list of the detectable attacks and the attacks' features types analysed by the BotGRABBER system.

server (Linux OpenSusE operating system with nginx HTTP server) and network devices (MikroTik CCR1009-8G-1S-1S+PC routers) was employed. Network traffic was captured by the means of tcpdump utility. All experiments were organized in real time and real networks, and lasted during from several seconds (e.g. phishing, TCP reset, SQL/PHP injection, XSS) to one hour (e.g. DDoS, ping flooding, RUDY, Fragmented UDP Flood, TCP SYN Flood etc.) depending on the type of the attack. Support vector machine as the inference engine in the BotGRABBER's structure was implemented using Matlab [36].

5.2   Performance Measures of the BotGRABBER System

The experimental results were estimated via standard sensitivity (SN), specificity (SP), and overall accuracy (Q) performance measures, taking into account the quantity measures of True Positives (TP), True Negatives (TN), False Positives (FP), False Negatives (FN):

$$SN = \frac{TP}{(TP+FN)}, \; SP = \frac{TN}{(TN+FP)}, \; Q = \frac{TP+TN}{TP+TN+FP+FN}, \quad (11)$$

Furthermore, the ability of BotGRABBER system to assure the corporate area networks' resilience in the presence of botnets' cyberattacks was assessed using the formula [37]:

$$GR = R \times \left[\frac{RAPI_{RP}}{RAPI_{DP}}\right] \times (TAPL)^{-1} \times RA, \qquad (12)$$

where R – robustness, which measures the network's performance between t_d and t_{ns} values, $R \in [0, 1]$, where 0 indicates a total loss of operation and 1 – normal network's functioning, t_d – time when the network is under a botnet attack, t_{ns} – time when the network was reconfigured according to security scenario produced by the BotGRABBER system; $RAPI_{DP}$ – the approximated rapidity value during the phase of the botnet attack; $RAPI_{RP}$ – the rapidity value during the network's reconfiguration phase; $TAPL$ – time-averaged network's performance loss value, which considers the time of appearance of a botnet attack up to network's reconfiguration; RA – the network's reconfiguration ability that describes the network's performance, reached after the produced security scenario was applied.

In order to obtain the number of the successful network's reconfiguration, we had to compute resilience measure GR. It is a dimensionless metric, that makes it possible to evaluate the resilience of various systems under different disruptive events. In our case, the resilience is the ability of a server, network, storage system, or an entire data centre, to recover quickly and continue operating. Thus, we consider, that the value of GR metric higher than specified threshold ($\gamma > 0.7$) means, that the stable network's functioning is ensured. The achievement of the needed value of the GR metric after the security scenario appliance means that the network's reconfiguration was successful.

5.3 Research of the SVM-Based Inference Engine Efficiency for Making the Decision About the Presence of the Cyberattacks

In order to investigate the SVM-based inference engine efficiency we used different SVM kernel functions. Examples of classification results using linear, polynomial, Gaussian, exponential, and B-Spline kernels are presented in Fig. 3.

The process of classification is divided into several iterations. In the first iteration, the classification objects are divided into two classes: malicious traffic and benign one. Then classifiers divide objects into other two classes, for instance: malicious traffic and spoofing traffic. Next iterations separate malicious traffic and other classes of attacks and so on until all of them are totally divided. Figure 3 presents the placement of the classification objects on the 2-D plane and the objects' separation into two classes.

Experimental results of different SVM classifiers elucidated that the linear and polynomial classifiers had provided the worst results (Table 1). They were characterized by longer execution times, and higher rates of the overall classification accuracy. Non-linear classifiers demonstrated better results, where B-Spline provided better results then others. Thus, for an experimental evaluation samples, the most effective classifier using the SVM was the B-spline since it provided

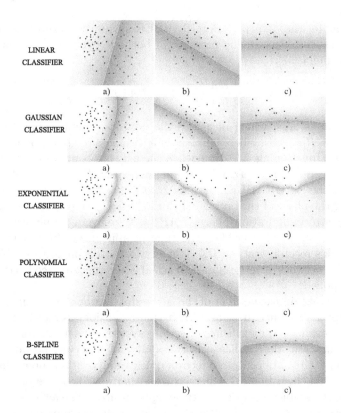

Fig. 3. Results of SVM classification using different kernel functions: (a) malicious traffic/benign traffic; (b) malicious traffic/spoofing traffic, performed by botnet; (c) malicious traffic/smurf attack's traffic.

Table 1. Experimental results for different SVM classifiers.

Classifier kernel	Classification accuracy (%)		Execution time (sec)	Distance between hyperplanes
	One against one SVM	One against all SVM		
Linear	91.59	96.50	0.5	0.17
Gaussian	95.48	97.39	0.4	0.21
B-spline	99.01	98.41	0.4	0.39
Polynomial	93.77	96.49	0.5	0.35
Exponential	97.02	97.83	0.5	0.31

the greatest distance between hyperplanes, the shortest time of evaluation, and the best accuracy of the classification; thus, it was employed as a basic kernel function in the SVM-based inference engine of the BotGRABBER system.

5.4   Results

This subsection presents overall results of BotGRABBER's efficiency, taking into account sensitivity, specificity, overall accuracy and the rate of system's resilience. Employing the dataset [35] this stage involved the evaluation of the SVM-based inference engine accuracy concerning each type of cyberattacks separately. The combined results are given in Table 2.

It presents the results of updated BotGRABBER's efficiency in comparison with its previous version [28], which was based on the usage of the c-means clustering. In addition, Table 2 demonstrates results of the successful network's reconfigurations in order to assure network's resilience in the presence of botnets' cyberattacks via the appliance of the security scenarios.

Table 2 shows that the overall accuracy of BotGRABBER is in the range from 90.40% to 98.42%. Moreover, sensitivity and specificity are in the range of 91.52–99.13% and 88.46–97.52%, respectively. Therefore, this approach indicates the capability of SVM for the botnets classification.

Table 2. Test results for different classes of the botnets attacks: number of training set samples T, number of evaluation samples E (including malicious and benign traffic samples), sensitivity (SN), specificity (SP), overall accuracy of the updated BotGRABBER (Q), overall accuracy of the previous version (Q_{old}) [28], true positives (TP), true negatives (TN), false positives (FP), false negatives (FN), the rate of successful reconfigurations (SR).

Attack, performed by botnet	T	E				Results				
		Malicious		Benign		SN, %	SP, %	Q, %	Q_{old}, %	SR, %
		TP	FN	TN	FP					
DDoS	574	661	16	489	14	97.64	97.22	97.46	93.23	73
Ping flooding	564	585	11	465	13	98.15	97.28	97.77	93.90	76
Smurf attack	364	568	15	429	21	97.43	95.33	96.52	92.28	76
TCP SYN flood	563	567	10	321	7	98.27	97.87	98.12	91.33	58
Fragmented UDP flood	554	384	33	323	29	92.09	91.76	91.94	88.98	77
DNS amplification	421	435	38	553	41	91.97	93.10	92.60	91.99	73
TCP reset	671	575	31	644	19	94.88	97.13	96.06	93.42	85
ICMP flood	764	541	23	565	7	95.92	98.78	97.36	93.23	54
RUDY	198	764	36	548	39	95.50	93.36	94.59	92.53	77
SIP INVITE flood	611	434	21	561	22	95.38	96.23	95.86	93.86	79
Encrypted SSL DDoS	571	554	41	464	35	93.11	92.99	93.05	90.62	77
Ping sweep attack	521	494	8	198	8	98.41	96.12	97.74	89.25	69
SQL/PHP injection	381	653	29	328	35	95.75	90.36	93.88	90.81	67
Cross-Site Scripting (XSS)	439	642	39	461	41	94.27	91.83	93.24	91.15	77
Phishing	555	457	4	354	9	99.13	97.52	98.42	92.53	70
DNS spoofing	571	453	42	253	33	91.52	88.46	90.40	86.60	76
TCP scan	345	451	21	326	12	95.55	96.45	95.93	90.36	67
UDP scan	231	432	12	326	12	97.30	96.45	96.93	91.46	73
Smurf	237	344	15	433	8	95.82	98.19	97.13	92.28	68
MAC flooding	655	556	13	326	11	97.72	96.74	97.35	91.57	52

Another aspect of the BotGRABBER functioning is the ability to apply the security scenarios for cyberattacks' by networks reconfiguration. In order to find out the ability of the network's functioning under the cyberattacks, different types of botnets' attacks against the network hosts, server, and available network devices were emulated. Table 2 demonstrates that the number of network's successful reconfigurations is in the range from 52.0% to 85%, with the average value at about 71.2%. Previous result was at about 70% [28].

6   Discussion

As BotGRABBER system uses the SVM-based engine there are several factors which may affect the prediction accuracy. One of them is the diversity of training samples. Most conspicuously, that not all possible feature vectors, that describe different cyberattacks, are adequately represented in the training set. Thus, system may be further improved by choosing more refined set of malicious traffic samples for each attack classes.

In order to increase the classification accuracy the SVM prediction may be further improved by using different classification kernels, as well as the SVM optimization procedure and feature vector selection may also be improved.

Results of the experiments demonstrated, that the BotGRABBER achieves the best results for detection of such attacks as DDoS, ping flooding, smurf, TCP SYN Flood, ping sweep, phishing etc. At the same time, the efficiency of the system concerning DNS Amplification, TCP Reset, RUDY, Encrypted SSL DDoS, XSS and DNS spoofing attacks is lower. This is because the traffic flow of some attacks is very similar to users' ones and some of botnets' features weren't taken into account for the detection process. On the other hand, the main disadvantage of the system is not very high rate of the resilience, which the system is able to assure for the attacked network. For this reason, the further work may be devoted to the development more relevant and effective security scenarios and addition to knowledge base the new features, relevant to the botnet's cyberattacks.

7   Conclusions

The paper presents an SVM-based self-adaptive system for the network resilience against the botnets cyberattacks, named BotGRABBER. The novel framework provides the ability to detect cyberattacks executed by botnets. As the inference engine for botnet detection the support vector machine was used. The detection process is performed by taking into account the hosts' network activity and the captured network traffic.

Experimental research showed that the SVMs are able to produce the accurate classification results. Implementation of the SVM-based inference engine into the BotGRABBER's structure as well as the addition of new botnets' attacks features allowed to increase its mean accuracy from 88.16% to 95.62%

in comparison with previous version. Experiments demonstrated, that system is able to detect different types of attack in the range from 90.40% to 98.42%.

Another important feature of the BotGRABBER system is the ability to produce a security scenario concerning the attack in order to mitigate it to ensure the network's resilient functioning. Experiments show that the average number of network's successful reconfigurations are at the rate of about 71.72%.

References

1. Trend Micro. https://www.trendmicro.com/vinfo/us/security/news/botnets. Accessed 10 Jan 2019
2. Virus Bulletin. https://www.virusbulletin.com/. Accessed 10 Jan 2019
3. Nexusguard. https://www.nexusguard.com/. Accessed 10 Jan 2019
4. Giudice, M., Wilkinson, C.: Crowe Horwath. Resilience Going Beyond Security to a New Level of Readiness (2016). https://www.crowehorwath.com/insights/asset/cyber-resilience-readiness-level
5. Macas-Escriv, F.D., Haber, R., Del Toro, R., Hernandez, V.: Self-adaptive systems: a survey of current approaches, research challenges and applications. Expert Syst. Appl. **40**(18), 7267–7279 (2013)
6. Zuzcak, M., Sochor, T.: Behavioral analysis of bot activity in infected systems using honeypots. In: Gaj, P., Kwiecień, A., Sawicki, M. (eds.) CN 2017. CCIS, vol. 718, pp. 118–133. Springer, Cham (2017). https://doi.org/10.1007/978-3-319-59767-6_10
7. Sochor, T., Zuzcak, M., Bujok, P.: Analysis of attackers against windows emulating honeypots in various types of networks and regions. In: 2016 Eighth International Conference on Ubiquitous and Future Networks (ICUFN), pp. 863–868. IEEE (2016)
8. Koroniotis, N., Moustafa, N., Sitnikova, E., Slay, J.: Towards developing network forensic mechanism for botnet activities in the IoT based on machine learning techniques. In: Hu, J., Khalil, I., Tari, Z., Wen, S. (eds.) MONAMI 2017. LNICST, vol. 235, pp. 30–44. Springer, Cham (2018). https://doi.org/10.1007/978-3-319-90775-8_3
9. Doshi, R., Apthorpe, N., Feamster, N.: Machine Learning DDoS Detection for Consumer Internet of Things Devices. arXiv preprint arXiv:1804.04159 (2018)
10. Angrishi, K.: Turning Internet of Things (IoT) into Internet of Vulnerabilities (IoV): Iot botnets. arXiv preprint arXiv:1702.03681 (2017)
11. Alieyan, K., ALmomani, A., Manasrah, A., Kadhum, M.M.: A survey of botnet detection based on DNS. Neural Comput. Appl. **28**(7), 1541–1558 (2017)
12. Shi, Y., Chen, G., Li, J.: Malicious domain name detection based on extreme machine learning. Neural Process. Lett. **48**(3), 1347–1357 (2018)
13. Baruch, M., David, G.: Domain generation algorithm detection using machine learning methods. In: Lehto, M., Neittaanmäki, P. (eds.) Cyber Security: Power and Technology. ISCASE, vol. 93, pp. 133–161. Springer, Cham (2018). https://doi.org/10.1007/978-3-319-75307-2_9
14. Alejandre, F.V., Cortés, N.C., Anaya, E.A.: Feature selection to detect botnets using machine learning algorithms. In: 2017 International Conference on Electronics, Communications and Computers (CONIELECOMP), pp. 1–7. IEEE (2017)
15. Wu, W., Alvarez, J., Liu, C., Sun, H.M.: Bot detection using unsupervised machine learning. Microsyst. Technol. **24**(1), 209–217 (2018)

16. Alauthaman, M., Aslam, N., Zhang, L., Alasem, R., Hossain, M.A.: A P2P Botnet detection scheme based on decision tree and adaptive multilayer neural networks. Neural Comput. Appl. **29**(11), 991–1004 (2018)

17. Ye, W., Cho, K.: P2P and P2P botnet traffic classification in two stages. Soft Comput. **21**(5), 1315–1326 (2017)

18. Chen, R., Niu, W., Zhang, X., Zhuo, Z., Lv, F.: An effective conversation-based botnet detection method. Math. Prob. Eng. **2017**, 9 pages (2017)

19. Khanchi, S., Vahdat, A., Heywood, M.I., Zincir-Heywood, A.N.: On botnet detection with genetic programming under streaming data label budgets and class imbalance. Swarm Evol. Comput. **39**, 123–140 (2018)

20. Eslahi, M., Abidin, W.Z., Naseri, M.V.: Correlation-based HTTP Botnet detection using network communication histogram analysis. In: 2017 IEEE Conference on Application, Information and Network Security (AINS), pp. 7–12. IEEE (2017)

21. Bazydło, P., Lasota, K., Kozakiewicz, A.: Botnet fingerprinting: anomaly detection in SMTP conversations. IEEE Secur. Priv. **15**(6), 25–32 (2017)

22. Wang, J., Paschalidis, I.C.: Botnet detection based on anomaly and community detection. IEEE Trans. Control Netw. Syst. **4**(2), 392–404 (2017)

23. Aziz, M.Z.A., Okamura, K.: Leveraging SDN for detection and mitigation SMTP flood attack through deep learning analysis techniques. Int. J. Comput. Sci. Netw. Secur. **17**(10), 166–172 (2017)

24. Savenko, O., Lysenko, S., Kryschuk, A.: Multi-agent based approach of botnet detection in computer systems. In: Kwiecień, A., Gaj, P., Stera, P. (eds.) CN 2012. CCIS, vol. 291, pp. 171–180. Springer, Heidelberg (2012). https://doi.org/10.1007/978-3-642-31217-5_19

25. Lysenko, S., Pomorova, O., Savenko, O., Kryshchuk, A., Bobrovnikova, K.: DNS-based anti-evasion technique for botnets detection. In: 2015 IEEE 8th International Conference on Intelligent Data Acquisition and Advanced Computing Systems: Technology and Applications (IDAACS), vol. 1, pp. 453–458. IEEE (2015)

26. Pomorova, O., Savenko, O., Lysenko, S., Kryshchuk, A., Bobrovnikova, K.: Anti-evasion technique for the botnets detection based on the passive DNS monitoring and active DNS probing. In: Gaj, P., Kwiecień, A., Stera, P. (eds.) CN 2016. CCIS, vol. 608, pp. 83–95. Springer, Cham (2016). https://doi.org/10.1007/978-3-319-39207-3_8

27. Lysenko, S., Savenko, O., Bobrovnikova, K., Kryshchuk, A., Savenko, B.: Information technology for botnets detection based on their behaviour in the corporate area network. In: Gaj, P., Kwiecień, A., Sawicki, M. (eds.) CN 2017. CCIS, vol. 718, pp. 166–181. Springer, Cham (2017). https://doi.org/10.1007/978-3-319-59767-6_14

28. Lysenko, S., Savenko, O., Bobrovnikova, K., Kryshchuk, A.: Self-adaptive system for the corporate area network resilience in the presence of botnet cyberattacks. In: Gaj, P., Sawicki, M., Suchacka, G., Kwiecień, A. (eds.) CN 2018. CCIS, vol. 860, pp. 385–401. Springer, Cham (2018). https://doi.org/10.1007/978-3-319-92459-5_31

29. Weston, J., Mukherjee, S., Chapelle, O., Pontil, M., Poggio, T. Vapnik, V.: Feature selection for SVMs. In: Advances in Neural Information Processing Systems, pp. 668–674 (2001)

30. Chapelle, O., Vapnik, V., Bousquet, O., Mukherjee, S.: Choosing multiple parameters for support vector machines. Machine Learn. **46**(1–3), 131–159 (2002)

31. Foody, G.M., Mathur, A.: A relative evaluation of multiclass image classification by support vector machines. IEEE Trans. Geosci. Remote Sens. **42**(6), 1335–1343 (2004)

32. Deng, N., Tian, Y., Zhang, C.: Support Vector Machines: Optimization Based Theory, Algorithms, and Extensions, 363 pages. Chapman and Hall/CRC. London (2012)
33. Hofmann, T., Schölkopf, B., Smola, A.J.: Kernel methods in machine learning. Ann. Stat., 1171–1220 (2008)
34. Larrañaga, P., Atienza, D., Diaz-Rozo, J., Ogbechie, A., Puerto-Santana, C.E., Bielza, C.: Industrial Applications of Machine Learning, 336 pages. CRC Press, Boca Raton (2018)
35. Canadian Institute for Cybersecurity. Botnet Dataset. https://www.unb.ca/cic/datasets/botnet.html. Accessed 10 Jan 2019
36. MathWorks. https://www.mathworks.com/. Accessed 10 Jan 2019
37. Linkov, I., Palma-Oliveira, J.M. (eds.): Resilience and Risk: Methods and Application in Environment, Cyber and Social Domains. NSPSSCES, 580 pages. Springer, Dordrecht (2017). https://doi.org/10.1007/978-94-024-1123-2

Cyber Attack Detection by Using Neural Network Approaches: Shallow Neural Network, Deep Neural Network and AutoEncoder

Serpil Ustebay[1], Zeynep Turgut[2(✉)], and M. Ali Aydin[3]

[1] Istanbul Medeniyet University, Istanbul, Turkey
serpil.ustebay@medeniyet.edu.tr
[2] Halic University, Istanbul, Turkey
zeynepturgut@halic.edu.tr
[3] Istanbul University-Cerrahpasa, Istanbul, Turkey
aydinali@istanbul.edu.tr

Abstract. As the accuracy rate of artificial intelligence based applications increased, they have started to be used in different areas. Artifical Neural Networks (ANN) can be very successful for extracting meaningful data from features by processing complex data. Well-trained models can solve difficult problems with high a high accuracy rate. In this study, 2 different ANN models have been developed to detect malicious users who want to access high-security servers. These models are tested from simple to complex: Shallow Neural Network (SNN), Deep Neural Network (DNN), and Auto Encoder are used to reduce features. All models are trained with CICIDS2017 dataset. Server connection requests are classified as normal or malicious (Brute Force, Web Attack, In ltration, Botnet or DDoS) with 98.45% accuracy rate.

Keywords: Shallow Neural Network · Auto Encoder ·
Deep Neural Network · IDS · Cyberattack

1 Introduction

In recent years, cyber security has become a topic that draws great interest. In today's world the new threats are now coming through the digital world. For this reason, cybersecurity has become very important for providing security against cyber attacks. Attacks which is conducted against institutions, states, and individuals to access information illegally are called cyber threats. Rendering servers unserviceable is one of the most common examples of cyber threats.

Cybersecurity is one of the most important subjects of commercial/public and military organizations. Possible threats can cause huge damages. For this reason, cybercrime expenditures are increasing significantly. Followings are the high-lights of the cybersecurity costs obtained from Ponemon Institute [1], Data Breach Study in 2017.

© Springer Nature Switzerland AG 2019
P. Gaj et al. (Eds.): CN 2019, CCIS 1039, pp. 144–155, 2019.
https://doi.org/10.1007/978-3-030-21952-9_11

- When costs of cybercrime caused in 2017 compared with the costs in 2016 it is seen that the cost rates have been increased 23% (11.7 $ million) and the average time for malware attacks is 50 days.
- Cybersecurity costs increased by 22.7 from 2016 to 2017.
- The average cost of data breaches is more than $ 6.3 million in 50.000 registered companies.
- It is estimated that the damage related to cybercrime will be 6 trillion dollars by 2021.

Firewalls can restrict packets on the network, but the number of attacks they can detect is low. Firewalls analyze source, destination IP addresses, protocols, and destination ports. Specifically, they cannot detect port forwarding attacks. As different from firewalls, IDS can detect port forwarding attack types and then access requests can be denied/dropped.

2 Related Works

IDS can be classified by different approaches [2–4]. Although there are lots of studies on the taxonomy of IDS, it is divided into three different groups in this paper as shown Fig. 1. System commands, system accounting information, system log data and system security log features are used for monitoring a network. Network IDS (NIDS) analyzes traffic between computers and devices in a network. NIDS listens the entire workstation. When an attack or abnormal behavior is detected, it activates the warning system. Host-Based IDS (HIDS) works on a single computer. It monitors the traffic which comes to the computer it is working on. It can't watch the entire workstation. It is often used to track possible attack attempts on critical servers. It is not preferred to oftenly use because of difficulties for maintaining in large networks with different operating systems and configurations. Attacker who seizes the entire system can easily disable the host-based IDS system.

In order to detect cyber attacks, anomaly based and signature based approaches are developed [6]. Signature-based IDS systems use attack patterns. Network traffic is analyzed by using a matching algorithm based on these patterns. When mapping occurs, the attacker information is transmitted to the system administrator or to the related alarm unit. Although registered patterns are detected with high accuracy, the system is vulnerable to different types of attacks. Anomaly-based IDS try to determine if there is a different data flow activity in the system. Although the accuracy rate of intrusion detection is high in self-learning anomaly based systems, false positive rate is also high. For example, when a large number of data is downloaded from the system, it can be classified as an attack and connection access requests can be rejected.

In this study, the performance of artificial neural networks in anomaly based approaches is investigated. In the literature:

[7] developed an IDS system with NSL-KDD dataset. NSL-KDD dataset contains about 126.000 samples for 22 attacks/intrusion activities. The model

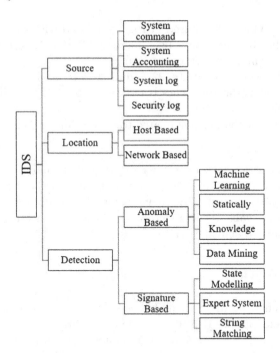

Fig. 1. Taxonomy of IDS.

consists of four hidden layers and each hidden layer includes 60 neural nodes. Adam optimizer is used with 32 batch size. Input Perturbation is used to reduce features. Before feature reduction (with 121 feature) model achieves 96.2% F1 score, after feature reduction (20 feature) model achieves 93.6% F1 score.

[8] KDD Cup 1999 dataset is used to train Deep Learning model to identify the users' behaviors as normal and attack activities. Two hidden layers and an output layer with softmax activation function is developed. 99% accuracy is obtained.

[9] IDS based system is developed on Deep Learning architecture: Recurrent Neural Network (RNN) with different variations: bi-directional RNN, Long Short Term Memory (LSTM) and bi-directional (LSTM). The highest precision (93%) is obtained by using RNN. The highest F1-score (93%) and recall (98%) are obtained with BLSTM.

[10] a hybrid model is used for classify connection requests. Data package considered as a picture and each byte of data are considered. Distributed embedding for byte representation is used for this transformation. The transformed data package is sent to RBM for feature extraction. The output of RBM is sent to RNN to classify data normal or attack. Darpha and ISCX datasets are used to test the hybrid model. 99.3% accuracy for Darpha and 99.73% accuracy for ISCX are obtained.

[11] is developed a DNN model with four hidden layers and 100 hidden units and it is employed for the intrusion detection. The study used the ReLU function as the activation function of the hidden layers and the adaptive moment (Adam) optimizer, a stochastic optimization method for DNN learning.

[12] a deep learning framework is developed by combining Sparse Autoencoder and Taguchi Method. Their aim is to extract more quantitative data.

By making experimental trials and testing their model with different data set. 99% accuracy is obtained for DDoS Detection.

[13] CICIDS2017 dataset is used to develop a hybrid Traffic-User behavior detection for the application layer. Network connection requests are considered as time window and the model identified DDoS attacks at two levels: Traffic level and IP level. User behaviors are identified as DDoS attack or normal request with the 99% accuracy rate.

[14] CICIDS2017 data set is used to train a Multi-Layer Perceptron (MLP). A MLP is developed which has 3 hidden layers and Linear activation function is used on each layer. For the output layer, the sigmoid activation function is used. Researches indicate that despite the 95% accuracy of model, the training time is a disadvantage.

Neural networks have an important place in machine learning techniques which are included in anomaly based IDS approaches. In the related works, all of the features which are in the data set are used. In order to obtain the highest accuracy with low time complexity the performances of the shallow neural network, deep neural network and autoencoder are tested with different number of features.

3   Method

In this section, technical details of Shallow Neural Network (SNN), Deep Neural Network (DNN), and Autoencoder (DAE) are explained. Data set which is used for training the models and attack types are presented.

3.1   Dataset

CICIDS2017 is provided by Canadian Institute of Cybersecurity and contains most up-to-date attack scenarios [15]. The user information which sent a request to access the server has been registered and tagged. Types of attacks in the related dataset and the days attacks were captured are listed in Table 1.

Dataset has 15 distinct classes, 83 features, and 2830540 instances. Though CICIDS2017 contains 288602 instances there are some class labels are missing and also some of 203 instances have missing information [16].

3.2   Shallow Neural Network (SNN) and Deep Neural Network (DNN)

Artificial Neural Networks are a model which can collect information from data, make generalizations, and make decisions using previous knowledge for data that they have never encountered before.

$$d = \{(x_i, y_i)\}_i^N \tag{1}$$

Table 1. Types of attacks in the CICIDS2017 and the days the attacks were captured.

Day	Attack type(s)
Monday	Benign
Tuesday	Benign, FTP-Patator, SSH-Patator
Wednesday	Benign, Dos GoldenEye, Dos Hulk, Dos Slowhttptest, Dos slow loris Heartbleed
Thursday	Benign, Web attack-Brute Force, Web Attack SQL Injection, Web Attack XSS, Infiltration
Friday	Benign, Bot, PortScan, DDos

Let be d as a training set and N is number of training examples. y_i and x_i represented as a vector. m refers to number of features and c refers to number of labels.

$$x_i = x_{1_i}, x_{2_i}, ..., x_{m_i} \tag{2}$$

$$y_i = y_{1_i}, y_{2_i}, ..., y_{c_i} \tag{3}$$

The smallest unit of ANN is Perceptron which transmits $f(x)$ function to an activation function after multiplying multiple input values by different weight coefficients. The activation function transmits a value output between 0 and 1 according to next perceptron as [17].

In neural networks, the layers contain more than one perceptron. Data that comes to the layer are processed simultaneously by each perceptron and results are transmitted as input to next layer. The input layer, which is also starting point of the network, is the layer from which the original value of incoming data is received. The output layer is the layer where the inferences from the data based on the type of application ends. The layers between input and output layers are called 'Hidden Layer'. SNN is an artificial neural network model which consists of a single hidden layer as shown at Fig. 2 x_i is a vector includes m features, and h is number of hidden layer perceptron size.

In this study, Rectified Linear Units (Relu) is used as the activation function of the hidden layer. The selection of activation functions is intuitive. Different functionalities have been tried and ReLu, which gives the best results, is preferred for training of the network. ReLu function (Eq. 4) gives an output if a is positive and 0 otherwise.

$$ReLu(x) = \begin{cases} 1 & x > 0 \\ 0 & x \leq 0 \end{cases} \tag{4}$$

For dealing with classification problems softmax activation function is recommended at output layer. The softmax function can be used to represent a categorical probability distribution. Mathematically the softmax function is Eq. 5.

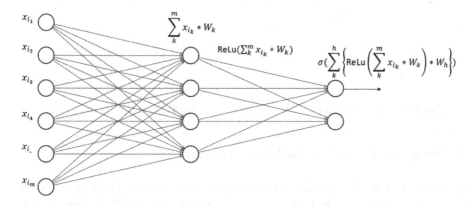

Fig. 2. Shallow Neural Network

z is a j-dimensional vector of the inputs to C dimensional output layer. Usually C is label number in dataset $k \in (1, 2, ..., C)$.

$$\sigma(z_j) = \frac{e^{z_j}}{\sum_k^C e^{z_k}} \tag{5}$$

Deep Neural Network is a network structure that contains more than one hidden layer. Each node layer runs on a different set of properties based on the output of the previous layer. As the depth increases, the properties in the previous layer are combined so the features that the layers can recognize are so complex.

3.3   AutoEncoder (AE)

AutoEncoder is an unsupervised neural network method which aims to decode data at hidden layer and then reconstruct them from compressed data. The simplest AE is called Vanilla AE and it is presented in Fig. 3. AutoEncoder [15] consists of two basic structures as encoder and decoder. Encode layers reduce the size by compressing incoming data. Decode layer tries to obtain the original data by decoding the encoded data. Usually number of nodes in input layer and number of nodes in the output layer are equal. Number of nodes in the output layer can be reduced and used for feature reduction.

In this study, an AE is designed with 3 decode layer and 3 encode layer. ReLu activation function in hidden layer and softmax activation function in output layer are used (Fig. 4).

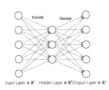

Fig. 3. Structure of Vanilla AE   **Fig. 4.** Structure of deep AE.

4 Proposed Models and Test Results

In this study, CICIDS2017 dataset is divided into two parts which are training data and testing data. Both parts of data are randomly selected. Training set contains 1979513 records and testing set contains 848363 records with six different labels. 'NaN' values are deleted from dataset. Distribution of each label group of the whole data set is given in Fig. 5. The most of the data are labeled data as 'Benign' and the least recorded label is 'Infiltration' (only 36 records is included).

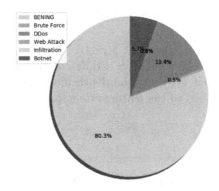

Fig. 5. Percentage distribution of labels in the data set.

SNN consists of a single hidden layer which has 78 nodes. For DNN, 4 hidden layer transmit data to the classification layer. The number of nodes at each layers are reduced like 60→50→30→20→. As the number of hidden layers used is increased, it is ensured that the relations between the features are learned better despite the increasing in time requirement. The number of hidden layers are chosen by experimenting in the best way to determine the relationship between the features and to avoid overfit. ReLu is used as activation function in each hidden layer. ReLU function only activates some of the neurons at one step and this cause sparsity which makes efficient computation. In the classification layer, the softmax function activates incoming data to 6 outputs. Each output represents the type of attack in the data set. Both of DNN and SNN models use Adaptive Moment Estimation (Adam) optimization at training phase. Adam is a combination of gradient descent with momentum and RMSprop algorithms [19]. One of the greatest advantage of Adam optimization function is that it needs less memory and calculation. Table 2 describes performance metrics to compare models. True Positive (TP) is number of intrusions correctly detected. True Negative (TN) is number of non-intrusions correctly detected. False Positive (FP) is number of non-intrusions incorrectly detected. False Negative (FN) is number of intrusions incorrectly detected.

Table 2. Performance metrics.

Metric	Formula
Accuracy	$\frac{TP+TN}{TP+FP+FN+TN}$
Precision	$\frac{TP}{TP+FP}$
Recall	$\frac{TP}{TP+FN}$
F1 score	$2 * \frac{Precision*Recall}{Precision+Recall}$

Table 3. Results obtained by using all features in the dataset.

	SNN			DNN		
	Accuracy = 98.05%			Accuracy = 98.40%		
	Precision	Recall	F1 score	Precision	Recall	F1 score
BENIGN	1.00	0.98	0.99	0.99	0.99	0.99
Brute Force	0.94	0.97	0.96	0.94	0.99	0.96
DDoS	0.96	0.99	0.98	0.97	0.99	0.98
Web Attack	0.91	0.03	0.07	0.87	0.07	0.13
Infiltration	0.00	0.00	0.00	0.00	0.00	0.00
Botnet	0.83	0.99	0.90	0.88	0.95	0.92
Micro avg	0.98	0.98	0.98	0.98	0.98	0.98

SNN and DNN models are trained using all features. Test results are given in Table 3. Both models have high accuracy, 98.40% for DNN and 98.05% for SNN. Although DNN model has a multi-layered deep learning process, it does not identify 'Infiltration' attacks as similar SNN.

AE has been used to reduce the size of the multi-dimensional data set. For this, a ReLu activation function which has 3 encode end 3 decode layers is used to train AE model. Firstly, 3 differert AE models are trained to reduce data set to 30, 20 and 10 features. Secondly, Reduced datasets are trained by using DNN and SNN models. All models (SNN, DNN and AE) are run with 20 epoch on 500 batch sized data. Results are obtained by using Python programming language and Keras library.

When 78 features are reduced into 10 significant features, accuracy rate is calculated as 91.08% for SNN and 94.72% for DNN. Brute Force, Infiltration and

Table 4. Test results obtained on reduced dataset (10 features).

	SNN			DNN		
	Accuracy = 91.08%			Accuracy = 94.72%		
	Precision	Recall	F1 score	Precision	Recall	F1 score
BENIGN	0.92	0.97	0.95	0.97	0.97	0.97
Brute Force	0	0	0	0.26	0.08	0.13
DDoS	0.90	0.65	0.76	0.95	0.86	0.90
Web Attack	0	0	0	0	0	0
Infiltration	0	0	0	0	0	0
Botnet	0.77	0.75	0.76	0.74	0.95	0.83
Micro avg	0.91	0.91	0.91	0.95	0.95	0.95

Web Attacks aren't classified as attack when SNN model is used. As different from SNN, DNN model is able to detect Brute Force attacks which is shown in Table 4.

Table 5 lists performance of the models that are trained on AE by using reduced dataset which contains 20 features. Table 6 lists performance results of the models that are trained on AE by using reduced dataset which contains 30 features. Tables 5 and 6 shows that performance results of both models are similar.

Although AE model reduces features to most significant size, it could not prevent decreasing accuracy rates. In both of models' accuracy are decreased when feature size is reduced according to its original size. Confusion matrices are examined to see difference. Figure 6 shows confusion matrix of DNN model which is trained by using reduced data on AE. Figure 7 shows confusion matrix of DNN model which is trained with original dataset. Model, that is trained with reduced data, has a tend to classify attacks with 'Benign' label. This causes FN

Table 5. Test results of reduced dataset with 20 features.

	SNN			DNN		
	Accuracy = 94.98%			Accuracy = 95.92%		
	Precision	Recall	F1 score	Precision	Recall	F1 score
BENIGN	0.97	0.97	0.97	0.98	0.97	0.97
Brute Force	0.86	0.28	0.42	0.95	0.49	0.65
DDoS	0.95	0.86	0.90	0.95	0.92	0.94
Web Attack	0	0	0	0	0	0
Infiltration	0	0	0	0	0	0
Botnet	0.74	0.98	0.85	0.73	0.99	0.84
Micro avg	0.95	0.95	0.95	0.96	0.96	0.96

Table 6. Test results of reduced dataset with 30 features.

	SNN			DNN		
	Accuracy = 95.12%			Accuracy = 96.71%		
	Precision	Recall	F1 score	Precision	Recall	F1 score
BENIGN	0.96	0.98	0.97	0.98	0.98	0.98
Brute Force	0.99	0.11	0.20	0.97	0.51	0.67
DDoS	0.97	0.83	0.89	0.97	0.90	0.93
Web Attack	0	0	0	0	0	0
Infiltration	0	0	0	0	0	0
Botnet	0.78	0.97	0.86	0.83	0.98	0.90
Micro avg	0.95	0.95	0.95	0.97	0.97	0.97

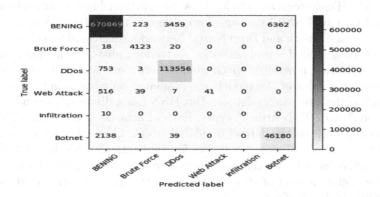

Fig. 6. Confusion matrix of DNN model trained by using all features.

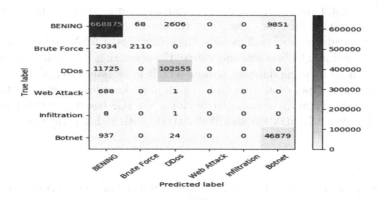

Fig. 7. Confusion matrix of DNN model trained by using reduced (30) features.

numbers increase. 'Infitration' attack type record numbers are too few according to others attack types. But in Fig. 7 an Infitration attack is predicted as DDoS. Despite the misclassification, it is able to identify the attack type.

CICIDS2017 is one of the most comprehensive data sets used in the literature. When studies using the same data set are examined; [10] can perform DDoS detection with 99% accuracy. In this study, 5 different attacks are determined by using artificial neural networks. [11] developed a NN model with 95% accuracy, in this study 98.45% accuracy rate is obtained which is higher. In the related study, it is shown that attacks with 98.45% accuracy can be detected with a single classifier and by using fewer data in the training dataset.

5   Conclusion

In this study, IDS systems are developed by using artificial neural networks which are tested on CICIDS2017 dataset. Two different artificial neural approaches: Shallow Neural Network and Deep Neural Networks are created and the data are trained by using related neural networks. At training phase, learning parametres (such as activation functions, epoch number, etc.) are sent to models. When the achievement rates of the models are evaluated only according to accuracy rate, DNN is the most successful one. But DNN has a disadvantage on learning "Infiltration" and "Web Attack" types. Both of attacks classified as BENIGN. Number of examples in the dataset of related attacks has made difficult to detect them. There are only 36 record examples for 'Infiltration attack' in CICIDS2017 dataset. ANNs learn by making inferences from the training data presented to them. The higher number of attack examples in the training data provide better learning and higher performance rates for ANNs.

Although ANN resembles biological neural networks, there are different learning parameters that have not yet been fully explored in biological neural networks. Different parameters has to be compared, such as the number of layers used, activation functions, etc., and the learning style of network must be changed for producing different models. Disadvantages of ANNs are also encountered at this point. The best learning rate of the network is based on the result of which is realized by using different parameters. It is necessary to update training data to find new types of attacks and to gain more knowledge current attacks.

In future study, we are planning to develop a new IDS based on network flow at byte level to detect Infiltration and 'Web Attack' with higher performance rates.

References

1. Ponemon Institute: 2017 Cost of Data Breach Study: Global Overview, NorthTraverse City, Michigan 49686 USA (2017)
2. Gyanchandani, M., Rana, J.L., Yadav, R.N.: Taxonomy of anomaly based intrusion detection system: a review. Int. J. Sci. Res. Publ. **2**(12), 1–13 (2012)
3. Shameli-Sendi, A., Ezzati-Jivan, N., Jabbarifar, M., Dagenais, M.: Intrusion response systems: survey and taxonomy. Int. J. Comput. Sci. Netw. Secur. **12**(1), 1–14 (2012)

4. Taylor, D.E.: Survey and taxonomy of packet classification techniques. ACM Comput. Surv. (CSUR) **37**(3), 238–275 (2005)
5. Axelsson, S.: Intrusion detection systems: a survey and taxonomy, vol. 99. Technical report (2000)
6. Sabahi, F., Movaghar, A.: Intrusion detection: a survey. In: Third International Conference on Systems and Networks Communications, pp. 23–26. IEEE, Sliema (2008)
7. Diep, N.N.: Intrusion detection using deep neural network. Southeast Asian J. Sci. **5**(2), 111–125 (2017)
8. Roy, S.S., Mallik, A., Gulati, R., Obaidat, M.S., Krishna, P.V.: A deep learning based artificial neural network approach for intrusion detection. In: Giri, D., Mohapatra, R.N., Begehr, H., Obaidat, M.S. (eds.) ICMC 2017. CCIS, vol. 655, pp. 44–53. Springer, Singapore (2017). https://doi.org/10.1007/978-981-10-4642-1_5
9. Elsherif, A.: Automatic intrusion detection system using deep recurrent neural network paradigm. J. Inf. Secur. Cybercrimes Res. (JISCR) **1**, 28–41 (2018). Naif Arab University for Security Sciences
10. Li, C., Wang, J., Ye, X.: Using a recurrent neural network and restricted Boltzmann machines for malicious traffic detection. NeuroQuantology **16**(5), 823–831 (2018). https://doi.org/10.14704/nq.2018.16.5.1391
11. Kim, J., Shin, N., Jo, S.Y., Kim, S.H.: Method of intrusion detection using deep neural network. In: IEEE International Conference on Big Data and Smart Computing, pp. 313–316 (2017). https://doi.org/10.1109/BIGCOMP.2017.7881684
12. Karim, A.M., Güzel, M.S., Tolun, M.R., Kaya, H., Çelebi, F.V.: A new generalized deep learning framework combining sparse autoencoder and Taguchi method for novel data classification and processing. In: Mathematical Problems in Engineering (2018). https://doi.org/10.1155/2018/3145947
13. Jiang, J., et al.: ALDD: a hybrid traffic-user behavior detection method for application layer DDoS. In: 2018 17th IEEE International Conference on Trust, Security and Privacy in Computing and Communications/12th IEEE International Conference on Big Data Science and Engineering (TrustCom/BigDataSE), pp. 1565–1569. IEEE (2018)
14. Watson, G.: A comparison of header and deep packet features when detecting network intrusions. Digital Repository at the University of Maryland (2018). https://doi.org/10.13016/M2G737680
15. Sharafaldin, I., Habibi Lashkari, A., Ghorbani, A.A.: Toward generating a new intrusion detection dataset and intrusion traffic characterization. In: Proceedings of the 4th International Conference on Information Systems Security and Privacy (Cic), pp. 108–116 (2018). https://doi.org/10.5220/0006639801080116
16. Panigrahi, R., Borah, S.: A detailed analysis of CIC2017 dataset for designing Intrusion Detection Systems. Int. J. Eng. Technol. **7**, 479–482 (2018)
17. Ibnkahla, M.: Applications of neural networks to digital communications-a survey. Signal Process. **80**(7), 1185–1215 (200)
18. Liu, W., Wang, Z., Liu, X., Zeng, N., Liu, Y., Alsaadi, F.E.: A survey of deep neural network architectures and their applications. Neurocomputing **234**, 11–26 (2017)
19. Kingma, D.P., Ba, J.: Adam: a method for stochastic optimization. arXiv preprint arXiv:1412.6980 (2014)

Temporal Characteristics of CodeSys Programmed Raspberry Pi and Beaglebone Black Embedded Devices

Jacek Stój[(✉)] and Ireneusz Smołka[(✉)]

Institute of Informatics, Silesian University of Technology,
Akademicka 16, 44-100 Gliwice, Poland
{jacek.stoj,ireneusz.smolka}@polsl.pl

Abstract. The area of application of embedded devices has grown rapidly in last years. From more than couple of years it is even possible to program them using software dedicated formerly only to industrial purposes, i.e. CODESYS development platform. As a result, embedded devices are a lot cheaper alternative to Programmable Logic Controller PLC. In that case however, their temporal characteristics should be taken into consideration. Shortly speaking, one should consider the time needed for the embedded device to react to external stimulus. Moreover, the response time is not constant and depends on other tasks being realized by a given device, such as communication task. That issue is analyzed in the following paper in case of Raspberry Pi and Beaglebone Black embedded devices and Ethercat and Modbus communication protocols.

Keywords: Embedded device · CODESYS · Raspberry Pi ·
BeagleBone Black · Real-time system · Modbus · TCP · EtherCAT

1 Introduction

There are many single-board microcontroller, so called System on a Chip SoC, available on the market nowadays and or great variety, too. One may choose a device suitable from the point of view of the size of build-in memory, processor power, number of ports or ready-to-use extension boards. There are for example Raspberry Pi, Blackberry, Banana Pi, Android, Odroid or Sparky, to name only a few. The most popular hardware platform seems to be the first mentioned above – Raspberry Pi, which is used in many applications described not only in the Internet, but also in scientific papers. The applications applies to many fields like computer networks [1], micro-machining [2], agriculture [3], smart homes systems [4] and many others. Another thing that tells about the wide use of Raspberry devices is the number and diversity of extension boards available, which for Raspberry Pi are numerous.

Wide use of Raspberry devices may be even greater considering that from two years it may be programmed with CODESYS development platform. CODESYS

© Springer Nature Switzerland AG 2019
P. Gaj et al. (Eds.): CN 2019, CCIS 1039, pp. 156–167, 2019.
https://doi.org/10.1007/978-3-030-21952-9_12

is dedicated to Programmable Logic Controllers PLCs used for industrial purposes. Programs in that environment may be implemented in typical PLC languages, compliant with international standard IEC 61131-3, which are: LD, FBC, ST, IL, SFC [5]. Now it is possible to implement applications for embedded devices using these specific languages. Moreover, the user application that is executed on an embedded device programmed with CODESYS is being executed in the same way as on industrial PLC – it runs in a cycle, preceded by inputs acquisition and followed by outputs update.

This kind of applications based on embedded devices are more and more common. There are some examples of Raspberry Pi device implemented for realization of electrical power system [6] or power management systems [7]. At this point the temporal characteristics of Raspberry Pi devices may become crucial. Finding no publication considering that case, in previous work authors measured the cycle time of Raspberry Pi [8]. It was not as a feasible task as one could expect. There are many different system configuration that should be taken into account. First of all, the user program may be executed in two basic modes: *cyclic* and *freewheeling*. Besides, other important tasks influence the cycle time of the device like data exchange through communication interfaces [9]. Apart from that, user may run on the embedded device another task such as visualization tasks, which influences the user program execution. The Raspberry Pi however, lacks in fact the possibility of creating a real-time applications, because the Raspbian operating system, which is dedicated for Raspberry, is not a system of the real-time type. The real-time operation may be obtained on another popular single-board microcontroller platform – Beaglebone Black. That issue is described in the following sections.

The fact that Raspberry Pi and BeagleBone are not conventional RealTime devices is not a reason not to use that devices in project where hard real-time is not the requirement, eg. solutions like Home Automation or traffic lights do not have strict time constraints. In the market there are solutions where RealTime clock is an option and where RealTime clock is not present at all. Using embedded device with CodeSys user may implement communication utilizing different protocols to exchange data with parts of industrial instalations, may create visualization both mobile and computer PC based applications. What is also important, remplacement costs or implementation of new devices is relatively low. In the figure below we presents schema of building automation network.

The system example shown in Fig. 1. consists of three embedded device which integrated site devices – light controller, thermostat, door controller. The first of them get information about light from user control panel and transmits it to the EtherCAT I/O module. The second device controls temperature in a building according to parameters sent from visualization to temperature controller via Modbus TCP. Visualization runs on every embedded device and is available on a PC computer. Additionally, the embedded devices prepare data to smartphone applications (not shown in the figure) which get data through MQTT protocol. Embedded devices can help to integrate several systems into one and prepare our data from older device to connect to the Industrial Internet of Things.

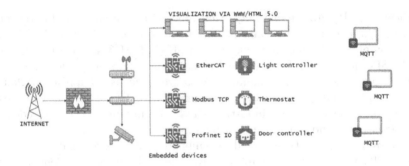

Fig. 1. Automation network with embedded devices example

The next section presents the testbed with details about the experimental research concerning temporal characteristics of Raspberry Pi and Beaglebone Black embedded devices. It is followed by analysis of results obtained during the experimental research and some conclusions.

2 Testbed

The research work has been performed on two single-board microcontrollers: Raspberry Pi[1] RPi and Beaglebone Black[2] BBB, latter called Embedded Devices EDs. Both devices were programmed in CODESYS development platform. Therefore, their operation becomes similar to Programmable Logic Controllers PLC operation, i.e. the user program was executed in a cyclic manner. Each cycle consisted of input acquisition, user program execution and outputs update.

The goal of the research work was to check how changes made in EDs configuration influence their temporal characteristics. In other words, how different configurations changes the EDs operation cycle. In both cases, it was done by observing the state of one of the outputs of both devices, which was being constantly toggled. A given state of the output corresponded to current cycle execution time.

The toggled output observation was performed by another ATmega based microcontroller and HP Logic Analyser[3] as presented in Fig. 1. The toggled output was connected to the ATmega input and the Logic Analyzer. ATmega measured and recorded the duration of every logical state of the observed output. For every ED configuration minimum 50,000 measurements were taken. Logic Analyzer was used for basic validation of correctness of the measurements.

---

[1] Raspberry Pi: 1,2 GHz quad-core ARM-8 Cortex-A53 CPU, 1 GB RAM, Raspbian Operating System, version 3.

[2] Beablebone Black: TI Sitara AM335x 1 GHz ARM®Cortex-A8, 512 MB RAM, Debian Operating System.

[3] Logic Analyser: HP/Agilent 54620C Logic Analyzer, maximum sample rate: 500 MSa/s.

The output state duration time was checked on the Logic Analyzer and compared to the most common value registered by ATmega (Fig. 2).

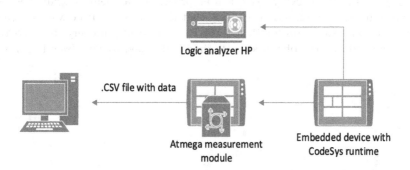

Fig. 2. The testbed for obtaining temporal characteristics of Embedded devices

In the experimental research two user program execution modes were considered: *cyclic* and *freewheeling*. There are also other execution possibilities (*event triggered, status triggered*), but they are considered as less common and are out of the scope of the paper.

In the *cyclic* execution mode the time between every execution of the user program is constant and may be defined by the user. In *freewheeling* mode, the program execution starts one after another, i.e every execution starts after the previous execution is finished.

Modbus TCP and EtherCAT communication protocols were used for test of the influence of the communication tasks on the operation cycle of the EDs. Both protocols are well known in industrial applications. There is also possibility to communicate using Profinet IO protocol [10], however, it will be taken into consideration during future research.

3 Results

The test were executed for several configurations:

- user program execution set as *cyclic* without communication tasks,
- user program execution set as *freewheeling* without communication tasks,
- *cyclic* program execution with Modbus TCP communication,
- *cyclic* program execution with EtherCAT communication.

Following subsections includes descriptions of experimental results. Firstly, operation of a standalone Raspberry Pi (RPi) and Beaglebone Black (BBB) devices were considered. i.e. without communication tasks running. For each device 50,000 program execution cycles were measured in two operating modes: *cyclic* with *freewheeling*. Results are presented in the next 2 subsections. The latter two sections includes description of EDs operation while realization of ModbusTCP and EtherCAT communication tasks.

3.1 Cyclic Program Execution Without Communication Tasks

Results of EDs operation in the *cyclic* mode with cycle set to 2 ms interval are presented in Fig. 3. In every cycle a user program was being executed with a constant number or instructions causing that the 2 ms cycle time wasn't enough for the program execution for Raspberry Pi. The result on every user program call was missed and the observed cycle time is doubled to 4 ms (see: Fig. 3a).

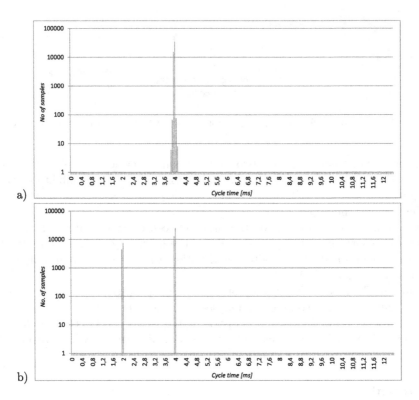

a)

b)

Fig. 3. *Cyclic* program execution without communication tasks: (a) Raspberry Pi, (b) Beaglebone Black Raspberry

In the same configuration BBB appears to be more capable of realization of the user program in one cycle. That is the reason why on the Fig. 3(b) significant part of the measurements of the cycle time are about 2 ms.

3.2 Freewheeling Program Execution Without Communication Tasks

In the *freewheeling* operation mode, the user program is being realized one time after another, i.e. next program execution is realized just after the previous realization is finished. It should be expected that as soon as realization of one program call is finished, the next program call will be executed without additional delay. In other words, this kind of program execution should effect in more frequent program calls than the *cyclic* operation mode for the same user program. As shown in Fig. 3 it is not so. For a very short user program, i.e. program which execution should last about 80–100 μs, the program is in fact being executed every 3.1 or every 6.2 ms. It is common for both RPi and BBB. It is much more than in *cyclic* operation mode for the same user program.

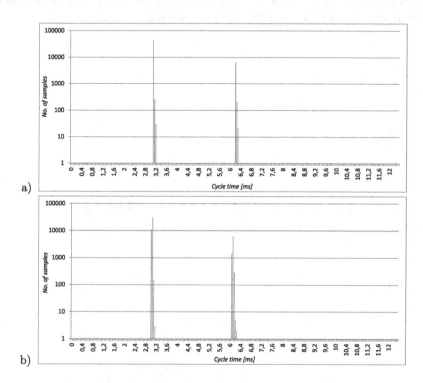

a)

b)

Fig. 4. *Freewheeling* program execution without communication tasks for k = 10: (a) Raspberry Pi, (b) Beaglebone Black Raspberry

The lack of cycle time measurements between the values of about 3 ms and 6 ms suggests, that consecutive user program calls are not performed immediately one after another, but with some delay. It seems that CODESYS calls the user program about every 3 ms in the presented example, i.e. first call after 3 ms, next after 6 ms etc. It means that when one program execution is not finished after 3 ms, then next call is planned to be in next 3 ms. It is important to notice again, that the user program realization for the above example was much less than 3 ms. It is not known to authors, why the call cycle was every 3 ms.

During our research we introduced parameter k, which determines how many basic arithmetic operations are realized in every cycle. In Figs. 4, 5 and 6 we present results to different values of the k parameter, while the user program consists of $k*(16$ sine or cosine$)$ calculation. After setting the k parameter to the value of 81 only one single group of measured cycle time are visible. That means that the user program execution was done about every 4.8 ms for RPi device, and about every 5.5 ms for BBB device. The value 81 was specifically found in order to show that this kind of user program execution also occurs, i.e. it is not always as shown in Fig. 3 when jitter of program calls is so great.

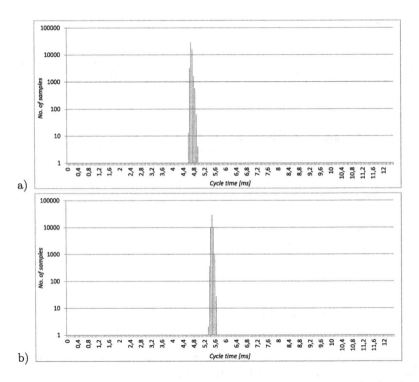

a)

b)

Fig. 5. *Freewheeling* program execution without communication tasks for k = 81: (a) Raspberry Pi, (b) Beaglebone Black Raspberry

After increasing the k parameter to the value of 230 (Fig. 5), in case of RPi device some program cycles were about 5.4 ms long, some about 6.7 ms long and some 8.0 ms long. It means that the program was tried to be called about every 1.3 ms. When at that time the previous user program call hadn't been finished, then the new program call wasn't performed and there was another 1.3 ms delay.

Concerning the BBB device with the k parameter equal 230 (Fig. 6b) the program was called and executed about every 8.7 ms, so with much less jitter.

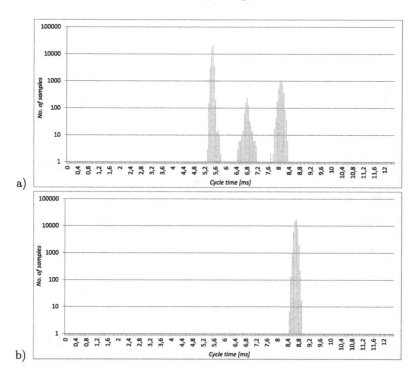

Fig. 6. *Freewheeling* program execution without communication tasks for k = 230:
(a) Raspberry Pi, (b) Beaglebone Black Raspberry

3.3 Cyclic Program Execution with Modbus TCP Communication

The first case of networked system considered communication with Modbus
TCP protocol. In the testbed we had four ATmega microcontrollers with Ethernet interfaces. In our test we set embedded device cycle time to 2 ms and the
embedded device exchange data with ATmegas every: 1 ms, 2 ms, 3 ms and 5 ms.
The results are shown in Fig. 7a and b for Raspberry Pi and BeagleBone Black
accordingly.

Realization of the user program itself was reported by CODESYS to be
done in 100 us. Even though, communication tasks caused the embedded device
to be unable to finish the user program execution in 2 ms, which is the time
given for one cycle. What is interesting, jitter in case of BBB device (Fig. 7b) is
significantly smaller than in case of RPi device.

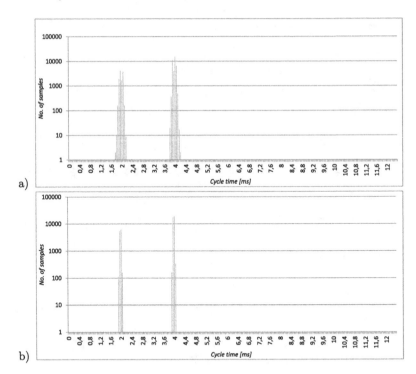

Fig. 7. *Freewheeling* program execution with Modbus TCP communication tasks: (a) Raspberry Pi, (b) Beaglebone Black Raspberry

3.4   Cyclic Program Execution with EtherCAT Communication

Second network configuration applies to EtherCAT protocol. The EDs exchange data with remote I/O stations based on Beckhoff EK1100 modules. The Ether-CAT communication was in two configurations with two different user program cycle set to 2 ms and 4 ms. In both cases the EtherCAT communication task was set to 4 ms (Figs. 8 and 9 accordingly). In our research EEDs exchanged 4 digital I/O states and 2 analogue values from and to our module.

The figures show that there is significant influence of the EtherCAT communication task to the embedded device user program cycle. The 2 ms cycle wasn't enough to perform both user program and data exchange in a timely manner and about half of the program cycles were exceeded. Here again it has to be noticed that the user program execution itself (without communication) took about 100 μs.

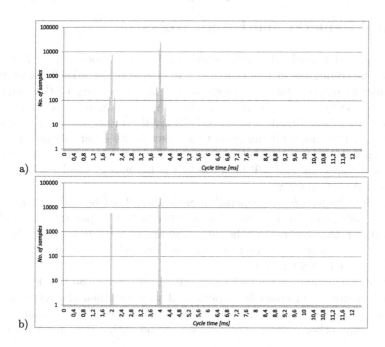

Fig. 8. *Cyclic* (interval 2 ms) program execution with EtherCAT communication task: (a) Raspberry Pi, (b) Beaglebone Black Raspberry

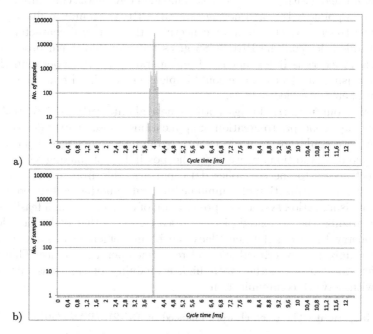

Fig. 9. *Cyclic* (interval 4 ms) program execution with EtherCAT communication task: (a) Raspberry Pi, (b) Beaglebone Black Raspberry

4   Final Remarks and Future Works

As we expected embedded devices which we use work correctly as PLC controllers. The results could be summarized in couple of remarks. First of all, in *cyclic* operation mode the user program is called according to the frequency defined by the cycle time. Therefore, when the cycle time is set to 2 ms, the user program will be called every 2 ms, but only if previous program realization is finished. For example, if the cycle time is set to 2 ms but the user program takes 2.5 ms, the user program won't be called every 2.5 ms but every 4 ms, so every second program call will be missed. It is likewise in regular PLC devices programmed by CODESYS.

Secondly, the *freewheeling* operating mode should not be understood as calling of the user program one time after another as fast as possible and without breaks. It is rather like calling the user program in a *cyclic* manner. However, the cycle time is hard to be described and authors hasn't found any information on that in CODESYS manuals.

Moreover, communication tasks may have great impact on the device cycle, and even greater than could be expected. It may lengthen the response time significantly. Timely operation of the devices may be even more volatile when there are some other processes running on the operating system on the device itself.

The above results are common for Raspberry Pi and Beaglebone Black device, no matter if the operating system of the embedded device was configured as real-time. At this point, it is worth noting that the PRU-ICSS unit available on the Beaglebone Black may change a lot concerning the temporal characteristics of that embedded device. The PRU-ICSS gives the user additional processors in the device which runs independently from operating system operation [11]. It seems that using this feature, one could implement even hard real-time systems on off-the-shelf embedded devices.

Although our research brings nothing new about work PLC controller, it gives some new concepts to creation of applications for embedded device. Using CODESYS we could create projects in programming languages dedicated for control systems. Additionally we have the possibility of making visualization with direct access to the controller data and available from internet browser level. In all makes realization of communication with another embedded devices, creation of visualization systems or preparation of control system a feasible task.

Summarizing, the presented piece of research shows that embedded devices like Raspberry Pi or BeagleBone Black can be implemented in systems where time determinizm is not critical and soft real-time operation suffice. That hardware offers low costs together with high availability of solutions, also those related with network communication.

Acknowledgment. The research work financed by BK-213/RAU2/2018.

References

1. Caldas-Calle, L., Jara, J., Huerta, M., Gallegos, P.: QoS evaluation of VPN in a Raspberry Pi devices over wireless network. In: 2017 International Caribbean Conference on Devices, Circuits and Systems (ICCDCS), pp. 125–128, June 2017
2. Akash, K., Kumar, M.P., Venkatesan, N., Venkatesan, M.: A single acting syringe pump based on Raspberry Pi - SOC. In: 2015 IEEE International Conference on Computational Intelligence and Computing Research (ICCIC), pp. 1–3, December 2015
3. Harish Kumar, B.: WSN based automatic irrigation and security system using Raspberry Pi board, pp. 1097–1103, September 2017
4. Alhasnawi, B.N., Jasim, B.H.: Scada controlled smart home using Raspberry Pi3. In: 2018 International Conference on Advance of Sustainable Engineering and its Application (ICASEA), pp. 1–6, March 2018
5. IEC61131-3: Programmable controllers-part 3: Programming languages (2014)
6. John, A., Varghese, R., Krishnan, S.S., Thomas, S., Swayambu, T.A., Thasneem, P.: Automation of 11 kv substation using Raspberry Pi. In: 2017 International Conference on Circuit, Power and Computing Technologies (ICCPCT), pp. 1–5, April 2017
7. Jaleel, A.H.A., Devassy, G., Vincent, M.C., Rose, N., Raphel, R.: Power management system in vessel. In: 2016 Online International Conference on Green Engineering and Technologies (IC-GET), pp. 1–4, November 2016
8. Stój, J., Smołka, I., Maćkowski, M.: Determining the Usability of Embedded Devices Based on Raspberry Pi and Programmed with CODESYS as Nodes in Networked Control Systems, pp. 193–205, January 2018
9. Jamro, M., Rzońca, D.: Impact of communication timeouts on meeting functional requirements for IEC 61131-3 distributed control systems. Automatika **56**, 499–507 (2015)
10. Gaj, P., Kwiecień, B.: Useful efficiency in cyclic transactions of profinet IO. Studia Informatica **31**(1), 29–41 (2010)
11. Molloy, D.: Exploring BeagleBone: Tools and Techniques for Building with Embedded Linux, pp. 23–53. Wiley, Hoboken (2015)

Ateb-Modeling Method Application for High Quality Video Traffic Modeling

Olga Fedevych[1(✉)], Ivanna Droniuk[1], Danylo Lizanets[2], Yurii Klishch[1], and Volodymyr Polziukov[1]

[1] ACS Department, Lviv Polytechnic National University,
12 Bandera Street, Lviv, Ukraine
olha.fedevych@gmail.com, ivanna.droniuk@gmail.com,
chlichch@gmail.com, pozyakrusher@gmail.com
[2] Division of Microengineering and Photovoltaics,
Wroclaw University of Science and Technology,
27 Wybrzeze Wyspianskiego, Wroclaw, Poland
danylo.lizanets@pwr.edu.pl
http://www.lp.edu.ua/asu
http://www-old.wemif.pwr.wroc.pl/zmf/

Abstract. This paper presents the procedure of implementing Ateb-modeling method, used for high definition video traffic samples. The subject of the study were high definition video traffic samples with different characteristics. It was shown, that those traffic samples have a self-similar nature, so the developed method could be used for modeling of future traffic values. Simulation tests were prepared to show the adequacy of used Ateb-modeling method for chosen high definition video samples. The obtained research results indicate the need to adapt the proposed method of modeling traffic values for 8K high definition video samples.

Keywords: Computer network · Ateb-modeling ·
High definition video · 4K · 8K

1 Introduction

In the conditions of the development of modern society, the amount of information exchanged by the world community is constantly increasing. According to recent forecasts, globally, clear video traffic will account for 55% of traffic by 2022 [1].

If take a look at other statistics, the growth of streaming video views on the most popular custom services for viewing such a video can be seen. According to these data, already today 58% of downstream traffic of Internet is video traffic from YouTube, Netflix, HTTP Media Stream, Amazon Prime, Facebook, Raw MPEG-TS and other services [2].

If prepare a select of statistics country by country and consider one of the largest consumers of world-wide Internet traffic - the United States, it can be

© Springer Nature Switzerland AG 2019
P. Gaj et al. (Eds.): CN 2019, CCIS 1039, pp. 168–177, 2019.
https://doi.org/10.1007/978-3-030-21952-9_13

seen exactly how the number of unique streaming video viewers grows on the most popular network services and platforms (statistical data is provided per month), among which Facebook, Vevo, Comcast and others can be highlighted [3].

Given the analyzed statistics, a steady increase in the load on computer networks and their components can be seen. The self-similar nature of network traffic is one of the most important reasons for the risk of the need for additional network equipment supply of computer networks, which in turn can lead to errors in its operation and loss of throughput and useful information [4].

Therefore, the analysis and forecasting of traffic varioforms in computer networks provide an opportunity: to provide adaptive control of the intensity of downloading buffers of network equipment of the computer network, to reduce the pulsations of delays in the transmission of traffic in the computer network and to reduce the transmission delay of data packets in the computer network due to forecasting these pulsations. If the process of data transfer in a computer network from node to node can be considered, then that the best and most successful stage for implementing optimal management decisions is part of the process of data transmission at intermediate nodes located between dispatch and delivery points. That is why it would be advisable to predict the intensity of traffic at the level of network equipment. To reduce the likelihood of overload and collapse of this equipment, it is necessary to investigate and evaluate the state of its load at a particular time point. Given the situation that occurs in the network equipment at a specific time point, there is a high probability of congestions occurring during further data transmission. It is this situation that generates narrowed capabilities in used network equipment. Therefore, it would be advisable to implement and develop new solutions to improve load balancing in such computer networks.

In particular, taking into account the diversity of modern architecture and the structure and configuration of network equipment, it would be advisable to develop effective methods for predicting traffic values, and taking into account its structure and nature, and taking into account its own architectural diversity of equipment, the best solution would be to develop a software product to achieve these goals.

2   Proposed Solution

Taking into account the aforementioned approaches, described in [5], the purpose of this article was to investigate the adequacy of the developed Ateb-modeling methodology for video traffic samples in order to further identify and assess the reduction of time delays in the transmission of video data in computer networks. The objectives of the study in this article were formulated as follows: it was necessary to analyze the statistics of traffic on various Internet services in the global computer network, as well as assess the magnitude and dynamics of changes taking place with it at this point in time; to explore the method of Ateb-modeling on various samples of video trafficking; provide a comparative description of various applied modeling parameters; to improve the software according to the obtained results; submit conclusions and recommendations for further research. That is why for the theoretical modeling of video traffic a nonlinear oscillating system with a single degree of freedom with a small perturbation was chosen, the motion of which is described by the ordinary differential equation of the second order with a small parameter ε. This is the first equation

$$\ddot{x} = \alpha^2 x^n = \varepsilon \bar{f}(x, \dot{x}, t) \tag{1}$$

where $x(t)$ – number of packets in computer network at time moment t; α – constant to determine the value of the period of fluctuation of traffic values, $f(x, \dot{x}, t)$ – any analytic function used to describe and simulate minor variations in traffic values from the main component of oscillations n – a number that determines the degree of nonlinearity of the equation and affects the period of the main component of the oscillations.

An analytical solution of this equation is the system of relations (2), which is written using the Ateb-functions

$$\begin{cases} \xi = aCa(n, 1, t) + \varepsilon f(\xi, \zeta, t) \\ \zeta = a\frac{1+n}{2}hSa(1, n, t) + \varepsilon g(\xi, \zeta, t) \end{cases} \tag{2}$$

where a – oscillation amplitude, $Ca(n, 1, t), Sa(1, n, t)$ – Ateb-cosine and Ateb-sine accordingly, ε – small parameter.

For modeling, a simplified solution of Eq. (2) in the form (3) was constructed:

$$\begin{aligned} f(\xi, \zeta, t) &= \sum_{i=1}^{N_1} a_i \delta(t_i), \\ g(\xi, \zeta, t) &= \sum_{i=1}^{N_2} b_i \delta(t_i), \end{aligned} \tag{3}$$

where N_1, N_2 – range of oscillations of functions f and g accordingly, a_i - amplitude of -th oscillation for function f at time moment $t + i$, bn_i – amplitude of -th oscillation for function g at time moment t_i, δ – Dirac function.

The mathematical apparatus of the Ateb-functions allowed to solve the differential Eq. (1) which describes essentially nonlinear processes in systems with one degree of freedom.

Ateb functions are a reversal to the Beta-functions. An incomplete Beta function is determined by equation

$$B_x(p, q) = \int_0^x t^{p-1}(1 - t)^{q-1} dt \tag{4}$$

where p and q – real numbers.

In the partial case, if $x = 1$, Eq. (4) takes the form of the Euler integral of the first kind

$$B_1(p, q) = \int_0^1 t^{p-1}(1 - t)^{q-1} dt \tag{5}$$

or complete Beta function.

For all x from interval $[0,1]$ functions $B_x(p,q)$ and $B_1(p,q)$, given by formulas (4), (5), are positive and satisfying conditions

$$0 \le B_x(p,q) \le B_1(p,q)$$
$$B_x(p,q) = B_1(p,q) - B_{1-x}(p,q)$$

It is accepted to consider two options, namely

$$p = \frac{1}{n+1}, \; q = \frac{1}{m+1} \tag{6}$$

$$p = \frac{1}{n+1}, \; q = \frac{m}{m+1} - \frac{1}{n+1} \tag{7}$$

where m and n are determined as $n = \frac{2\theta_1'+1}{2\theta_1''+1}, m = \frac{2\theta_2'+1}{2\theta_2''+1}, (\theta_1', \theta_1'', \theta_2', \theta_2'' = 0,1,2,\dots)$.

The work uses periodic Ateb functions to ensure that oscillations and traffic fluctuations are taken into account.

On the basis of Eq. (3), a short-term forecast of traffic intensity was developed.

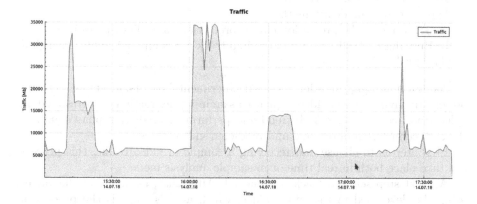

Fig. 1. The sample of traffic is reflected in the developed software

On the basis of the described mathematical apparatus, the Ateb-modeling method was developed, which made it possible to evaluate and model the value of the traffic on the background of a certain number of pre-obtained values and to verify the fact that the obtained values are adequate. Developed software solution interface with test traffic sample is presented in Fig. 1.

In this article, a number of model experiments were conducted for various video traffic samples involving Ateb-functions with different parameters, for example $Sa(1/7,3), Sa(0.01,0.1), Sa(1,1/3)$. Ateb-functions with different parameters were considered, but this article shows the three best options for the Ateb-function parameters (based on the criterion of the correlation value between simulated and real traffic values).

3   Video Traffic Analysis

Considering different types of videos viewed on PC users of the World Wide Web from around the world, it can be seen that each year there is a tendency to high definition videos views increasing. It is clear that first of all this is due to increasing number of users with access to high-speed connection that can rise certain problems during load balancing in any subnets backbone nodes [6–8].

That is why four formats were chosen as study samples of traffic from those that become widely distributed among users and gaining popularity.

4   Simulation Procedure

The process of constructing and testing received forecast of video traffic values was developed in order to evaluate results of applied method of Ateb-forecasting. Simulation was carried out by developed software.

This process consists of several phases, including:

– Selecting and receiving a video file
– Selection (playback) of the track;
– Applying Ateb-modeling method;
– Construction of the prediction curve (prediction (modeling) phase);
– Testing phase (Traffic values in original track are compared to obtained simulated values).

As shown in Fig. 2, in order to obtain experimental results, five main steps are presented: having chosen video sequence as an input, extraction of corresponding video trace, constructing Ateb-trend and performing prediction phases and, at the end, comparison of original and predicted video traces.

First step is to choose and receiving a sample of video traffic, this process can take place both in real time or in sample analysis mode.

As next step selected track is launched. After that, modeling of the values using the developed method take place, which at this stage is the process of selecting and adapting the Ateb-function with the best parameters. During forecasting phase, predicted values are calculated based on the selected forecasting parameters. At last phase, original video track is tested in comparison with predicted track.

5   Description of Conducted Experiments

Sample of traffic in this article is defined as short part (with duration less than one hour in real time) of computer network video traffic with defined characteristics. Samples characteristics is provided in Table 1.

Parameters used to select and construct predicted values for developed software are shown in Table 2. This table shows video traffic samples types that were chosen for experiments, three Ateb-function types that have shown themselves

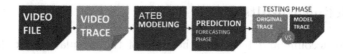

Fig. 2. Stages of video traffic forecast (modeled) values construction and testing processes.

Table 1. Studied video traffic samples characteristics.

Video type	Sample length	Sample receive time	Sample source	Quality, pixels	Bitrate, Mbps	Sample time duration, min	Browsing speed, Mbps
1080p60	30 min	18:00 CET 22/12/2018	YouTube	1920 × 1080	12	10	100
1080	30 min	18:30 CET 22/12/2018	Twitch.tv	1920 × 1080	3.5	10	100
4K	30 min	19:00 CET 22/12/2018	YouTube	3480 × 2160	35	10	100
8K	30 min	19:30 CET 22/12/2018	YouTube	7680 × 4320	75	10	100

Table 2. Video traffic trends simulation parameters using for developed method of Ateb-modeling

Video type	Ateb-function types	Interval Step, s	Offset Step, s	Delta
1080p60	Sa (1/7,3), Sa (0.01, 0.1), Sa (1, 1/3)	10	5	0.1
1080	Sa (1/7,3), Sa (0.01, 0.1), Sa (1, 1/3)	10	5	0.1
4K	Sa (1/7,3), Sa (0.01, 0.1), Sa (1, 1/3)	10	5	0.1
8K	Sa (1/7,3), Sa (0.01, 0.1), Sa (1, 1/3)	10	5	0.1

as the best ones among other, interval step as a step of period change between maximum and minimum, offset step as a period shift of the modeling function.

To ensure the purity and reliability of the study results, sampling of video traffic was done out using Wireshark software. Formulas of calculations are shown in the articles [9,10].

6   Modeling Results

Taking into account and based on the data of the calculations of the Hurst parameter, it is possible to apply the Ateb-modeling method to the selected video traffic samples.

Table 3. Calculated results of performed experiments

Data	Function	Correlation	Hurst approx	Hurst traffic
1080p60	Sa(1/7,3)	0.753815	0.777182	0.738174
	Sa(0.01,0.1)	0.702475	0.760986	0.738174
	Sa(1,1/3)	0.72702	0.77381	0.738174
4k	Sa(1/7,3)	0.94084	0.770557	0.776768
	Sa(0.01,0.1)	0.882845	0.756193	0.776768
	Sa(1,1/3)	0.924781	0.767806	0.776768
Twitch	Sa(1/7,3)	0.876081	0.763977	0.758839
	Sa(0.01,0.1)	0.834676	0.742042	0.758839
	Sa(1,1/3)	0.85499	0.76025	0.758839
8k	Sa(1/7,3)	0.344267	0.890392	0.773683
	Sa(0.01,0.1)	0.337655	0.884663	0.773683
	Sa(1,1/3)	0.34424	0.889599	0.773683

Experiments results are shown in Table 3. Results comparison of performed calculations, in particular correlation value, self-similarity and Hurst approximation for 4 different types of video traffic samples, are shown in Figs. 3, 4,

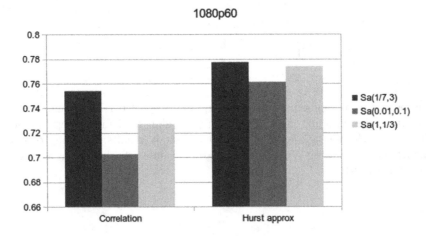

Fig. 3. Model calculations results for a sample of 1080p60 traffic.

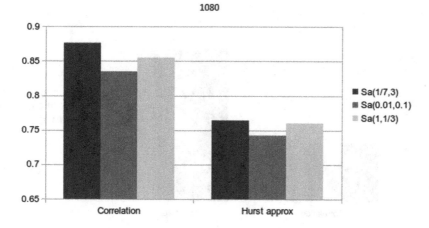

Fig. 4. Model calculations results for a sample of 1080p traffic.

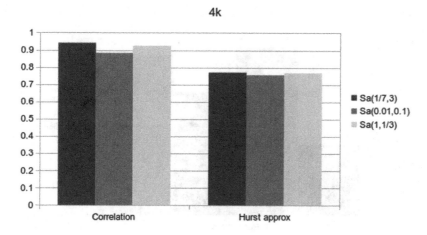

Fig. 5. Model calculations results for a sample of 4k traffic

5, 6 and 7. Correlation coefficient was calculated between real and predicted traffic in Table 3. In Hurst traffic column, Hurst parameter is calculated for the sample of real video traffic via R/S method. Hurst approx column contains calculated Hurst parameter using R/S method for predicted traffic with corresponding parameters [11, 12].

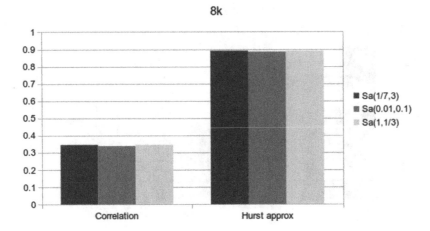

Fig. 6. Model calculations results for a sample of 8k traffic

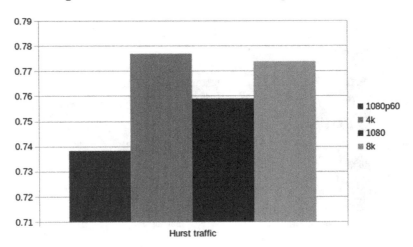

Fig. 7. Hurst parameter calculations results for all 4 samples of traffic.

7   Conclusions

Traffic forecasting trend simulation results have shown that analyzed traffic is self-similar and therefore developed method can be successfully used for trends prediction for 1080p60, 4k, 1080p traffic types but for 8k traffic type models have to be improved. Total traffic on a telecommunications network is divided into different components. One of the most important parts is video traffic. Conducted studies show that the behavior of this component determines the behavior of traffic in general, regardless of the type of video traffic. Our research shows that video traffic is self-similar, which allows us to apply certain known methods for optimizing the operation of network equipment. In the next research, it is planned to further explore the 8k video traffic to find out its features.

References

1. Cisco website: Index, Cisco Visual Networking. Forecast and methodology, 2015–2020 white paper. Retrieved 1st June (2017). Accessed 20 Dec 2018
2. Sandvine website: Downstream traffic statistic. https://www.sandvine.com/. Accessed 18 Oct 2018
3. Statista website: Most popular online video properties in the United States. https://www.statista.com/. Accessed 15 Sep 2018
4. Domański, A., Domańska, J., Czachórski, T., Klamka, J., Marek, D., Szyguła, J.: The influence of the traffic self-similarity on the choice of the non-integer order PI^{α} controller parameters. Commun. Comput. Inf. Sci. **935**, 76–83 (2018). https://doi.org/10.1007/978-3-030-00840-6_9
5. Bagchi, S.: Computational analysis of network ODE systems in metric spaces: an approach. J. Comput. Sci. Sci. Publ. **13**(1), 1–10 (2017). https://doi.org/10.3844/jcssp.2017.1.10
6. Tanwir, S., Perros, H.G.: VBR Video Traffic Models. Wiley, Hoboken (2014). 148 P
7. Kastrinakis, M., Badawy, G., Smadi, M.N., Koutsakis, P.: Video frame size modeling for user-generated traffic in an enterprise-like environment. Comput. Commun. **109**, 24–37 (2017). https://doi.org/10.1016/j.comcom.2017.05.008
8. Liew, C.H., Kodikara, C., Kondoz, A.M.: Video traffic model for MPEG4 encoded video. In: 62nd IEEE VTS Vehicle Technology Conference, vol. 3. pp. 1854–1858 (2005). https://doi.org/10.1109/VETECF.2005.1558427
9. Fedevych, O., Dronyuk, I., Lizanets, D.: Researching measured and modeled traffic with self-similar properties for ateb-modeling method improvement. In: Gaj, P., Sawicki, M., Suchacka, G., Kwiecień, A. (eds.) CN 2018. CCIS, vol. 860, pp. 13–25. Springer, Cham (2018). https://doi.org/10.1007/978-3-319-92459-5_2
10. Demydov, I., Dronyuk, I., Fedevych, O., Romanchuk, V.: Traffic fluctuations optimization for telecommunication SDP segment based on forecasting using ateb-functions. In: Kryvinska, N., Greguš, M. (eds.) Data-Centric Business and Applications. LNDECT, vol. 20, pp. 71–88. Springer, Cham (2019). https://doi.org/10.1007/978-3-319-94117-2_4
11. Pathan, A.-S.K., Monowar, M.M., Khan, S.: Simulation Technologies in Networking and Communications. CRC Press, Boca Raton (2014)
12. Cajueiro, D., Tabak, B.: The rescaled variance statistic and the determination of the Hurst exponent. Math. Comput. Simul. **70**, 172–179 (2005). https://doi.org/10.1016/j.matcom.2005.06.005

A Comparison of Request Distribution Strategies Used in One and Two Layer Architectures of Web Cloud Systems

Krzysztof Zatwarnicki$^{(\boxtimes)}$ and Anna Zatwarnicka

Institute of Computer Science, Opole University of Technology,
76 Prószkowska Street, 45-758 Opole, Poland
k.zatwarnicki@gmail.com, anna.zatwarnicka@gmail.com

Abstract. Web Cloud systems are becoming more and more popular. In the article we want to examine HTTP request distribution strategies that can be used in one and two layer architectures of Web cloud systems. In particular, we want to compare our intelligent solutions with each other, and with popular and most commonly used in Web cloud systems. We describe modern solutions, present the test-bed and results of conducted experiments. At the end, we discuss results and present final conclusions.

Keywords: Web cloud · HTTP request distribution ·
Cloud computing · Fuzzy-Neural modeling · Web systems simulation

1 Introduction

For the last several years we have come across the statements that we live in the time of digital revolution and that, the revolution is as important to humanity as the industrial one. After expansion of the Internet infrastructure and the popularization of mobile devices, information and entertainment services moved to web systems allowing people to use services where they are and also, at any time.

In recent years, many companies have moved away from the classic model of keeping IT resources at their headquarters to the using of the specialized services provided by professional businesses. The main advantage of this approach is that those companies no longer have to be involved in the installation, maintenance of infrastructure and devices, supplying power and Internet connections.

The three main types of cloud computing solutions can be distinguished:

- Software as a Service (SaaS) – where the consumer is able to use and manage the provider's applications that run on a cloud infrastructure but not cloud infrastructure and software;
- Platform as a Service (PaaS) – in this approach the consumer does not control the structure of the lower cloud infrastructure level but has control over the deployed applications;

© Springer Nature Switzerland AG 2019
P. Gaj et al. (Eds.): CN 2019, CCIS 1039, pp. 178–190, 2019.
https://doi.org/10.1007/978-3-030-21952-9_14

- Infrastructure as a Service (IaaS) – in case of using IaaS cloud, the consumer has control over deployed applications as well as operating systems and storage.

Gartner forecasts that in 2019 the use of the public cloud will grow by 17.3% – from 175.8 to 206.2 billion of dollars [3]. Until now, real values are compatible with forecasts. Additionally, there are predictions that by 2022, 90% of organizations will use the integrated platform hosted in cloud [4]. The forecasted rapid development of cloud systems will require the development of new solutions, algorithms and architectures.

In the article, we concentrate on request distribution strategies in a Web cloud environment. This approach is consistent with the need to provide Service Level Agreement (SLA) and adequately end-user satisfaction.

The aim of our work is to examine HTTP request distribution strategies that can be used in one and two layer architectures of Web cloud systems. Especially we want to focus on comparing our intelligent solutions with each other and with popular and most commonly used in Web cloud systems.

We examine our Two-Layer Cloud-base Web System (TLCWS) that works in a two layer architecture, with one layer Fuzzy-Neural Request Distribution (FNRD) strategy. Additionally, both systems are compared with the most popular request distribution strategies: Round-Robin, Last Loaded and Path-based routing, that also work in one and two layer architectures.

TLCWS and FNRD systems are intelligent, and distributes HTTP requests in SaaS based cloud architecture. Both systems use neuro-fuzzy approach to minimize request response times and were presented in our previous papers [17,19].

The rest of the paper is structured as follows: Sect. 2 contains related work, Sect. 3 describes request distribution strategies to be examined. In Sect. 4 the test-bed, results of experiments and discussion is presented. Lastly, Sect. 5 concludes the paper.

2   Related Work

The proper and efficient load balancing is one of the main challenges in cloud environments [5]. A lot of research has been done in recent years to improve performance of web systems – from web clusters to web systems based on cloud environment. Various approaches were examined, from simple and fast load balancing algorithms, through more complicated, to algorithms using Artificial Intelligence mechanisms.

The simple and classical Round-Robin algorithm, that is often used in Web clusters as well as in Web clouds, was modified by Xu Zongyu and Wang Xingxuan [15]. Modified Round Robin (MRR) algorithm is more intelligent than the classic one and takes into account the load of servers.

An interesting and relatively easy to implement solution was proposed by Ramana and Ponnavaikko. Their architecture named Global Dispatcher-Based Load Balancing (GDLB) uses Domain Name System Mechanism and dispatcher to distribute requests over nodes [11]. A similar approach was presented in [9].

Authors use two layer decision making algorithm in specialized, high-scalable system based on cloud architecture. The algorithm uses synergy of DNS and internal dispatcher to maintain a high level of performance.

Many articles address intelligent load balancing strategies including Genetic Algorithms, Ant Colony Optimization, Artificial Bee Colony and Neuro-Fuzzy Inference System (ANFIS) [6]. K.R. Remesh Babu and P. Samuel have noticed that dynamic nature of cloud computing environment needs a dynamic algorithms for efficient scheduling and load balancing among cloud elements. The static load balancing will only work when there is a small variation of the workload. Their throttle algorithm used the behavior of bee colonies [12]. In [14] authors present Particle Swarm Optimization (PSO) based on heuristic method to schedule applications to particular cloud resources. This method takes into account the computational cost as well as the cost of data transmission.

To group of intelligent Web cloud systems belongs also systems which in decision algorithms takes into account HTTP request response time. Two of such systems have already been mentioned (FNRD and TLCWS), but there are also systems that distribute HTTP request globally, like GARDiB [16] and guarantees the web page response time (MLF) [18].

Regardless of the research, in practice, simple and non-intelligent request distribution algorithms are the most commonly used. The most popular and frequently chosen by customers cloud system is AWS (Amazon Web Services) [1]. Over 1 billion of end users from 190 countries use services hosted on AWS. Amazon Web brokers use following strategies: Round-Robin, Last Loaded and Path-based routing.

3   Two and One Layer Architectures of Cloud Web Systems

Most of the cloud systems consist in elements called regions, placed in different locations and continents around the world [1]. A region is placed in one particular location and houses many cooperating server rooms called availability zones (in short zones). The regions and zones consist of brokers that distribute requests/tasks. Brokers in cloud Web systems are distributing HTTP requests. Broker can be located in region and distribute requests among zones. Zones can also contain brokers that distributes request among servers not only in zone where they are placed (cross-zone distribution). When the region contains broker and, additionally zones that contain brokers distributing requests, we call it two layer architecture (Fig. 1a). If there is only one broker than we call it one layer architecture (Fig. 1b).

3.1   Request Distribution Strategies

Brokers can have implemented many distribution strategies letting to achieve different goals. In this article we deal with problem of load balancing and load

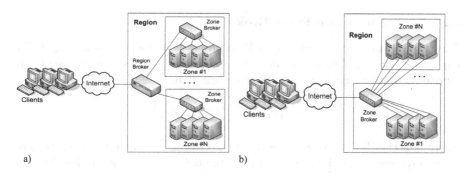

Fig. 1. Web cloud systems: (a) two layer architecture, (b) one layer architecture.

sharing, and consequently, from now on, we will assume that all of Web servers in the cloud can provide the same content.

The most popular strategies that are implemented in Amazon Web Services are [1]:

- Round Robin (RR) – forward HTTP request to subsequent Web server,
- Last Loaded (LL) – forward request to server with the least number of currently serviced requests,
- Path-based routing (P) – forward requests based on the URL in the request. In our experiments we are using a modification of this algorithm, making it to behave similarly like LARD algorithm [10]. During the first service, a given type of request is forwarded to the server with the last load. During the next service, the request is redirected to the same server as long as the server is not overloaded, and the number of requests serviced concurrently by the server is not much higher than on other servers. If the load of the server is too high the request is redirected to server with least load.

In the experiments we compared presented algorithms with the two strategies/systems proposed by us:

- Fuzzy-Neural Request Distribution (FNRD) [17] – system using one layer architecture and distributing requests taking into account HTTP requests response time,
- Two-Layer Cloud-based Web System (TLCWS) [19] – system similar to the FNRD that uses a two layer architecture.

Both of the systems presented, work in this way to minimize the response time for each individual HTTP request. The response time is estimated with the use of fuzzy-neural mechanism placed in an intelligent Web broker. In TLCWS system, there is a broker working in the region layer and it cooperates with the zone brokers. Region and zone brokers are constructed, and work in the exact same way as broker in FNRD system. FNRD system consists of only one broker that distributes load among all the servers working in the zones.

3.2   Intelligent Neuro-Fuzzy Broker

The main purpose of the intelligent Web broker is to forward incoming HTTP requests r_i, $i = 1, ...I$ to servers or zones (in short executors), which are able to service requests in shortest time. We can formulate it as follows:

$$z_i = \min_w \left\{ \hat{s}_i^w : w \in \{1, 2, ..., W\} \right\} \quad , \tag{1}$$

where z_i is the chosen executor out of W executors, $\hat{s}_i^w$ is an estimated response time to request r_i for the wth server, i is the index of the request (Fig. 2).

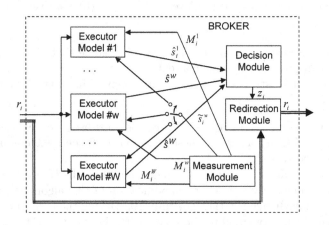

Fig. 2. Broker design.

The broker is composed of the following modules: executor models, a decision module, a redirection module and a measurement module. The executor model estimates the response time $\hat{s}_i^w$ taking into account load $M_i^w = [e_i^w, f_i^w]$ of the executor, where e_i^w is the number of all requests being concurrently serviced by the executor and f_i^w is the number of dynamic request. Executor model can adapt to the environment using actual response time $\tilde{s}_i^w$. This module contains neuro-fuzzy model of executor. There are W executor modules in the broker. Decision module makes decision according to formula 1. The redirection module forwards incoming request r_i to the chosen executor (Fig. 3).

The executor module is complex and consists in two modules: classification and neuro-fuzzy. Classification module classify incoming requests to class k_i, where $k_i \in \{1, ..., K\}$. Objects belonging to the same class have similar response times. For example static objects (files) are classified by their sizes, while dynamic objects (created while the service) get a separate class for each of them. For each of class k_i neuro-fuzzy model have different parameters $Z_i = [Z_{1i}, Z_{ki}, ..., Z_{Ki}]$, where $Z_{ki} = [C_{ki}, D_{ki}, S_{ki}]$ are parameters of input $C_{ki} = [c_{1ki}, ..., c_{lki}, ..., c_{(L-1)ki}]$, $D_{ki} = [d_{1ki}, ..., d_{mki}, ..., d_{(M-1)ki}]$ and output $S_{ki} = [s_{1ki}, ..., s_{jki}, ..., s_{Jki}]$ fuzzy set functions (Fig. 4a). We can say that there

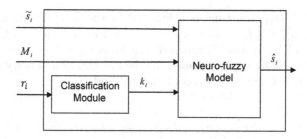

Fig. 3. Executor model.

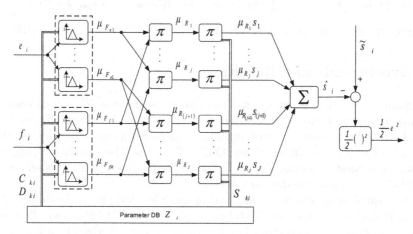

a)

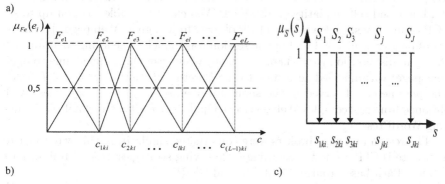

b)   c)

Fig. 4. Server model: (a) neuro-fuzzy model, (b) input fuzzy sets functions, (c) output fuzzy sets functions.

are K neuro-fuzzy networks in each executor module. Fuzzy set functions for inputs $\mu_{F_{el}}(e_i)$, $\mu_{F_{fm}}(f_i)$, $l = 1, ..., L$, are triangular (Fig. 4b), while fuzzy sets functions for outputs $\mu_{Sj}(s)$ are singletons (Fig. 4c). We have conducted preliminary experiments during which we determined the optimal number of input fuzzy sets $L = M = 10$, while the number of output fuzzy sets is equal to

$J = M \cdot L$. The response time is calculated in the defuzzification phase and $\hat{s}_i = \sum_{j=1}^{J} s_{jki}\mu_{R_j}(e_i, f_i)$, where $\mu_{R_j}(e_i, f_i) = \mu_{F_{el}}(e_i) \cdot \mu_{Ffm}(f_i)$ After the service of the HTTP request is finished, the broker gets the response time $\tilde{s}_i^w$ and can adapt to changing environment by the modification of fuzzy sets parameters with the use of the Back Propagation Method:

$$s_{jk(i+1)} = s_{jki} + \eta_s \cdot (\tilde{s}_i - \hat{s}_i) \cdot \mu_{R_j}(e_i, f_i),$$

$$c_{\phi k(i+1)} = c_{\phi ki} + \eta_c (\tilde{s}_i - \hat{s}_i) \sum_{m=1}^{M} \left(\mu_{Ffm}(f_i) \sum_{l=1}^{L} \left(s_{((m-1) \cdot L + l)ki} \ \partial\mu_{F_{el}}(e_i) / \partial c_{\phi ki} \right) \right),$$

$$d_{\gamma k(i+1)} = d_{\gamma ki} + \eta_d (\tilde{s}_i - \hat{s}_i) \sum_{l=1}^{L} \left(\mu_{F_{el}}(e_i) \sum_{m=1}^{M} \left(s_{((l-1) \cdot M + m)ki} \ \partial\mu_{Ffm}(f_i) / \partial d_{\gamma ki} \right) \right),$$

where η_s, η_c, η_d are adaptation ratios, $\phi = 1, ..., L - 1$, $\gamma = 1, ..., M - 1$ [19].

4   Experiments and Results

The main goal of our experiments was to compare efficiency of Web cloud systems working in one and two layer architectures under control of an intelligent Web broker. In order to conduct experiments we prepared a simulation application with use of discrete event simulator OMNET++ [8]. The simulation program contained: request generator, region and zone brokers, as well as Web servers in availability zones and the database servers (Fig. 5).

The request generator was composed of modules that model the behavior of typical Web clients (which are Web browsers), and the HTTP request flow was compatible with the flow observed on the Internet, characterized by a high variety, burst and self-similarity [2,13]. Each Web client was able to open up to six TCP connections, to download HTTP objects from a given Web page. The number of pages downloaded within one session was modeled according to an Inverse Gaussian distribution ($\mu = 3.86$, $\lambda = 9.46$) while user think time (time to open next page) was modeled according to the Pareto distribution ($\alpha = 1.4, k = 1$). The parameters of served in the simulator Web service (Web page sizes) were obtained from real complex Web service of Opole University of Technology [7] running WordPress.

The region and zone brokers were constructed similarly, in this way to let to distribute HTTP requests according to following strategies: Round Robin, Last Loaded, Path-based routing, TLCWS and FNRD.

The zones in the simulator contained modules acting in similar way to both Web and database servers. In Web server CPU, RAM and SSD drive were modeled as queues, while database server was modeled as single queue. The service times we acquired in preliminary experiments with use of a computer running a Wordpress service. Both of the servers contained Intel Core i7 7800X CPU, a Samsung SSD 850 EVO driver and 32 GB of RAM. In the Web server module, the operating memory was also modeled, acting as the cache memory for the file system. The size of the memory for each of the servers was 1 GB.

We prepared two versions of simulator. In the first version, the simulator contained region and zone broker in a two layer architecture. Experiments had

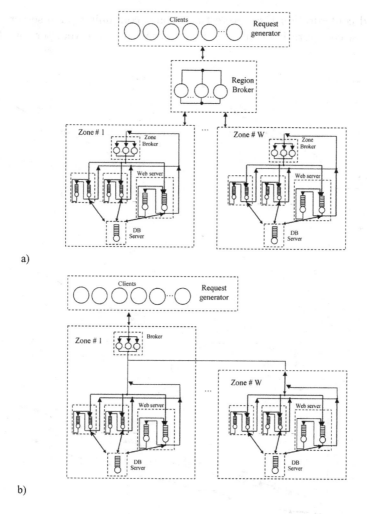

a)

b)

Fig. 5. Simulation model: (a) two layer architecture, (b) one layer architecture.

been conducted for both brokers running Last Load strategy (marked as LLLL), Round Robin strategy (RRRR), Partitioning (marked as PP) and TLCWS. The second version of the simulator consisted in the system with a one layer architecture containing only one broker placed in the zone, and distributing HTTP requests among servers in all of the zones. The broker ran the following strategies: LL, RR, P, FNRD.

In the experiments, we measured the mean HTTP requests response time. Experiments were conducted for different loads of the system measured as the number of clients sending requests. We used three configurations of Web cloud system for each version of simulator. In the first configuration the system contained two zones with five Web servers and one DB server in each zone (2 x 5).

In second configuration there were three zones containing three servers (3×3) and the last configuration contained five zones with two servers per zone (5×2).

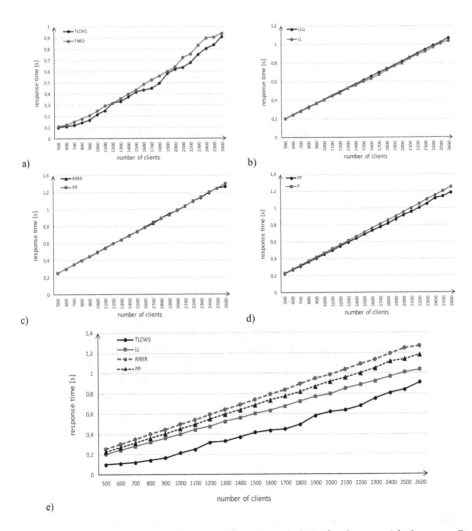

a)

b)

c)

d)

e)

Fig. 6. Response time in load function (number of clients), cluster with 2 zones, 5 servers in each zone, (a) strategy TLCWS vs. FNRD, (b) LLLL vs. LL, (c) RRRR vs. RR, (d) PP vs. P, (e) comparison of the best results.

Figures 6, 7 and 8 present the results of experiments conducted for three different cloud systems working in one and two layer architectures. Pictures from (a) to (d) presents the results for the same distribution strategy for one and two layer architectures. Picture (e) shows the best results (lower response times) taken from charts (a), (b), (c) and (d).

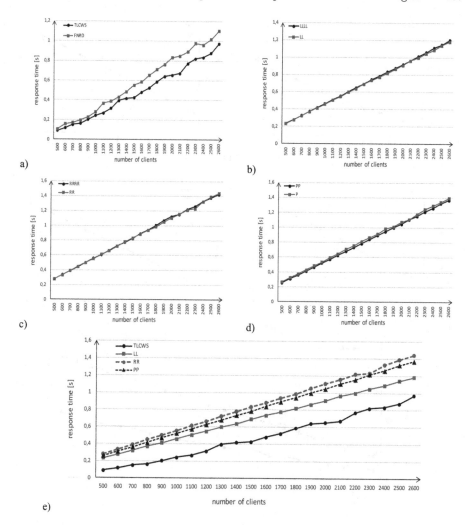

Fig. 7. Response time in load function (number of clients), cluster with 3 zones, 3 servers in each zone, (a) strategy TLCWS vs. FNRD, (b) LLLL vs. LL, (c) RRRR vs. RR, (d) PP vs. P, (e) comparison of the best results.

As it can be observed for all of the cluster systems response times for LL and LLLL strategies are very similar. The same conclusion can be drawn for RR and RRR strategies. The results obtained for P and PP strategies show that for two layer architecture response times are shorter, the difference, however, is not significant.

We obtained very interesting results for the TLCWS and FNRD strategies. Both of the strategies are significantly better than other non-intelligent strategies. The response times in some cases are even 50% shorter than for other strategies. In all of the experiments two layer architecture is better for the neuro-fuzzy

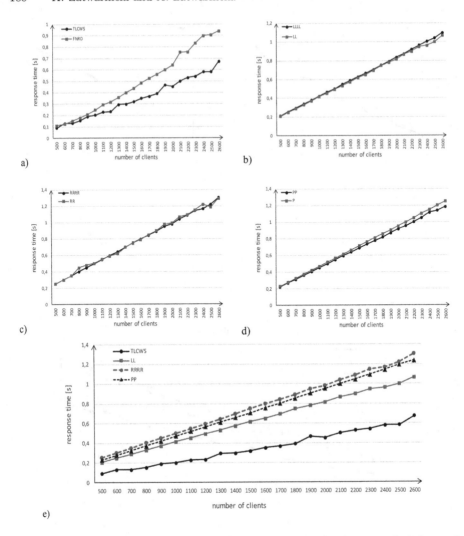

Fig. 8. Response time in load function (number of clients), cluster with 5 zones, 2 servers in each zone, (a) strategy TLCWS vs. FNRD, (b) LLLL vs. LL, (c) RRRR vs. RR, (d) PP vs. P, (e) comparison of the best results.

strategy than one layer architecture. We have observed that when the number of zones increases (Figs. 6a, 7a and 8a) the difference between TLCWS and FNRD is becoming greater in favor of the two layer strategy (especially observed on Fig. 8a). It may be due to the fact that when the number of zones is low, then the algorithms behave similarly, like in a one layer architecture. When the number of zones is greater, neuro-fuzzy algorithms in region and zone brokers become to cooperate and adapt better to behaving mutually.

Based on the results of the experiments it can be concluded that cloud Web systems working in one and two layer architectures with use non-intelligent strategies get similar results. When using intelligent neuro-fuzzy strategies it is worth using two layer architectures taking advantage of the cooperation of Web brokers.

5 Summary

The aim of the article was to compare HTTP request distribution strategies working in one and two layer architectures of Web cloud systems. In particular, we wanted to compare our intelligent solutions with each other, and with popular and most commonly used in Web cloud systems strategies.

The results of the experiments revealed that Web cloud systems using non-intelligent strategies get similar results for one and two layer architectures. Additionally, one and two layer intelligent strategies get significantly shorter response times than not intelligent ones. We also observed that intelligent, two layer TLCWS strategy works better than the one layer FNRD strategy, especially when the number of zones is greater.

Based on the results of the experiments, it can be concluded that TLCWS strategy is a good solution for modern Web cloud systems. However we should take into account that making distribution decisions that use intelligent algorithms are time consuming. The region Web broker, that takes on the whole load, can become a bottleneck in the system. For this reason, in our further studies, we will examine the possibility of a cooperation of non-intelligent brokers with intelligent ones.

References

1. AWS documentation, How Elastic Load Balancing Works. https://docs.aws.amazon.com/elasticloadbalancing/latest/userguide/how-elastic-load-balancing-works.html. Accessed 21 Dec 2018
2. Cao, J., Cleveland, W.S., Yuan, G., Jeffay, K., Smith, F., Weigle, M.: Stochastic models for generating synthetic HTTP source traffic. In: Proceedings of Twenty-Third Annual Joint Conference of the IEEE Computer and Communications Societies, INFOCOM 2004, Hong-Kong, pp. 1547–1558 (2004)
3. Estimation of cloud computing market. https://www.forbes.com/sites/louiscolumbus/2018/09/23/roundup-of-cloud-computing-forecasts-and-market-estimates-2018/. Accessed 02 Jan 2019
4. Gartner forecasts on cloud computing development. https://www.gartner.com/en/newsroom/press-releases/2018-09-12-gartner-forecasts-worldwide-public-cloud-revenue-to-grow-17-percent-in-2019. Accessed 02 Jan 2019
5. Jadeya, Y., Modi, K.: Cloud computing - concepts, architecture and challenges. In: Proceedings of International Conference on Computing, Electronics and Electrical Technologies (ICCEET) 2012, IEEE, Kumaracoil, India (2012)
6. Lee, S.-P., Nahm, E.-S.: Development of an optimal load balancing algo-rithm based on ANFIS modeling for the clustering web-server. In: Communications in Computer and Information Science, CCIS, South Korea, vol. 310, pp. 783–790 (2012)

7. Main page of Opole University of Technology. https://www.po.opole.pl/. Accessed 02 June 2018
8. OMNeT++ Discrete Event Simulator. https://www.omnetpp.org/. Accessed 01 Jan 2019
9. Opiola, L., Dutka, L., Wrzeszcz, M., Slota, R., Kitowski, J.: Two-layer load balancing for Onedata system. In: Proceeedings of Computing and Informatics, Slovak Academy of Sciences, vol. 37, no. 1, pp. 1–22, Slovakia (2018)
10. Pai, V.S., et al.: Locality-aware request distribution in cluster-based network servers. ACM SIGPLAN Not. **33**(11), 205–215 (1998)
11. Ramana, K., Ponnavaikko, M., Subramanyam, A.: A global dispatcher load balancing (GLDB) approach for a web server cluster. In: Kumar, A., Mozar, S. (eds.) ICCCE 2018. LNEE, vol. 500, pp. 341–357. Springer, Singapore (2019). https://doi.org/10.1007/978-981-13-0212-1_36
12. Remesh Babu, K.R., Samuel, P.: Enhanced bee colony algorithm for efficient load balancing and scheduling in cloud. In: Snášel, V., Abraham, A., Krömer, P., Pant, M., Muda, A.K. (eds.) Innovations in Bio-Inspired Computing and Applications. AISC, vol. 424, pp. 67–78. Springer, Cham (2016). https://doi.org/10.1007/978-3-319-28031-8_6
13. Suchacka, G., Wotzka, D.: Modeling a session-based bots' arrival process at a web server. In: ECMS (2017)
14. Suraj, P., Linlin, W., Siddeswara, M., Rajkumar B.: A particle swarm optimization-based heuristic for scheduling workflow applications in cloud computing environments. In: 2010 24th IEEE International Conference on Advanced Information Networking and Applications, Perth, WA, Australia, pp. 20–23 (2010)
15. Xu, Z., Wang, X.: A predictive modified round robin scheduling algorithm for web server clusters. In: Proceedings of 34th Chinese Control Conference (CCC) 2015, IEEE, Hangzhou, China (2015)
16. Zatwarnicka, A., Zatwarnicki, K.: Adaptive HTTP request distribution in time-varying environment of globally distributed cluster-based web system. In: König, A., Dengel, A., Hinkelmann, K., Kise, K., Howlett, R.J., Jain, L.C. (eds.) KES 2011. LNCS (LNAI), vol. 6881, pp. 141–150. Springer, Heidelberg (2011). https://doi.org/10.1007/978-3-642-23851-2_15
17. Zatwarnicki, K.: Adaptive control of cluster-based web systems using neuro-fuzzy models. Int. J. Appl. Math. Comput. Sci. (AMCS) **22**(2), 365–377 (2012). Zielona Gora
18. Zatwarnicki, K., Borzemski, L.: Guaranteeing the quality of service in cluster-based Web systems. In: Grana, M., et al. (eds.) Advances in Knowledge-Based and Intelligent Information and Engineering Systems, pp. 1141–1150. IOS Press, Amsterdam (2012)
19. Zatwarnicki, K., Zatwarnicka, A.: Two-layer cloud-based web system. In: Borzemski, L., Świątek, J., Wilimowska, Z. (eds.) ISAT 2018. AISC, vol. 852, pp. 125–134. Springer, Cham (2019). https://doi.org/10.1007/978-3-319-99981-4_12

Communications

Mobile Voice Traffic Load Characteristics

Zagroz Aziz[✉] and Robert Bestak

Czech Technical Universiy in Prague, Technicka 2, Prague 6, Czech Republic
{azizzagr,robert.bestak}@fel.cvut.cz

Abstract. Knowledge of traffic load evolution in time is essential to properly configure and dimension a mobile network. Moreover, it is a key parameter to indicate the network performance and quality of service. In this paper, we use interarrival time (time between arrivals), waiting time, and service time parameters to investigate outgoing and incoming traffic of an international voice traffic carrier. Both types of traffic are analyzed for the previously mentioned parameters by taking into account short and long-distance international call scenarios. The obtained results follow the expected Poisson and Exponential distributions for these parameters. In addition, the traffic load of neighboring countries shows a long-term stability and consistency.

Keywords: Mobile network · Voice traffic · Queueing theory ·
Interarrival time · Service time · Waiting time

1 Introduction

Mobile networks generate enormous amount of data every moment, such as signaling and user data (paging, location update, Call Detail Records (CDR), Short Message Service (SMS), etc.). By analyzing such data, the network performance can be improved, and a better Quality of Services (QoS) to users can be ensured. Additionally, the data can be also utilized for other purposes, such as public transportation planning, public health analysis, social behavior studies, crime investigations, etc. [1].

One of data usage in mobile networks is teletraffic. The teletraffic in telecommunications represents an application of probability theory that is used to support network planning, dimensioning, performance evaluation, operation and maintenance [2]. The objective is to make the traffic measurable, and to quantify the relation between grade-of-service and system capacity [2].

In the teletraffic theory, the word "traffic" is usually denoted for the traffic intensity, i.e. the number of calls carried by network per a time unit [3]. International Telecommunication Union (ITU-T) defines the traffic intensity as "the instantaneous traffic intensity in a pool of resources, which is the number of busy resources at a given instant of time" (ITUT B.18).

This study is based on data that are collected from an international voice traffic carrier. The topology of studied scenario is illustrated in Fig. 1. The topology consists of home network (hereafter referred as the reference network) that

© Springer Nature Switzerland AG 2019
P. Gaj et al. (Eds.): CN 2019, CCIS 1039, pp. 193–207, 2019.
https://doi.org/10.1007/978-3-030-21952-9_15

is located in one country while a destination network can be located in a country wherever around the world. Both, reference and destination networks, are mobile operators. A third network can be involved in a communication, and it is called as a carrier network. The carrier network interconnects two networks and can be part of a mobile network operator.

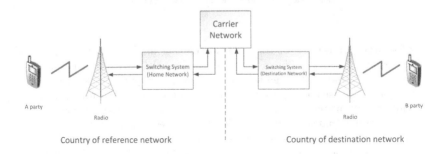

Fig. 1. International voice call flow through a carrier network

Our analyses are based on CDRs, which contain detailed information about a phone call connection (e.g., connection time, release time, call duration, called and calling number, etc.). A CDR is created with a unique ID number for each call session, including both successful or unsuccessful call session.

A CDR file (which is composed of several CDRs) is created in the system every 15 min, and it consists of several thousands of lines of recorded CDRs. Each line begins with an element that indicates the current call status.

Having a 3-month CDR file dataset of outgoing and incoming international voice traffic, we investigate characteristics of interarrival time, waiting time, and service time. The analyses are provided as for the total global traffic, as for selected destinations (countries), covering short and long-distance international call scenarios.

The rest of paper is organized as follows. Section 2 summarizes the related works. In Sect. 3, we detail the analyzed data set and we discuss the data processing procedure. The call flow process is explained in Sect. 4 and the theory background is summarized in Sect. 5. Section 6 demonstrates and discusses the obtained results. Finally, Sect. 7 concludes our work.

2   Related Work

Traffic characteristics and their modeling are still very hot topics in the mobile network community. Especially as newer technologies being deployed and the amount of carried out data in networks is exponentially growing.

In [4], authors focus on the lognormal distribution of interarrival time and service time with exponential distribution in the queueing system. The paper shows the influence of keep-alive mechanism on the queue process. Authors declare that the waiting time and the queue length are shorter than what is proposed by 3rd Generation Partnership Project organization (3GPP).

In [5], Carrier Sensing Multiple Access (CSMA) mechanism of MAC layer of IEEE 802.15.4 is analyzed. The paper investigates the interarrival time distribution in order to analyze a network queue model and to detect anomalies in the network. The model is based on the discrete time Markovian chain and the proposed analytic model closely predicts the behavior of simulated network.

A traffic congestion model for computer network environment is presented in [6]. The studied model is based on the continuous-time finite-state homogenous bivariate Markov chain. The model is used to evaluate the performance of protocols and applications in a network. Possible evaluated network features include random path delays and packet losses, which can occur due to a traffic congestion.

In [7], the performance of mobile cellular networks is mathematically analyzed and numerically evaluated by using Markov Modulated Poisson Process call arrivals (MMPP). There are three different analytical approaches that are assessed based on birth and death processes with different dimensions. The accuracy of numerical results based on the analytical approach is compared with the results that are based on handoff call requests. The presented results follow the Poisson process.

In [8], an exponential model is assumed for the traffic variance. Using this assumption, authors propose an analytical expression to estimate the byte loss probability for a single server queue system alongside with multifractal traffic arrivals. Based on the results, authors propose a novel admission control strategy.

The problem of modelling voice traffic in mobile networks, considering interarrival time and holding time characteristics, is addressed in [9]. Authors define a mathematical model that complied with experimental data provided by a mobile operator.

Authors in [10] use the queueing theory to design a model simulating the IP Multimedia Subsystem (IMS) service, including the Voice over IP (VoIP). The developed model includes a single server following the Poisson distribution of arrivals and the exponential distribution of service time. The designed algorithm is declared to improve the IMS performance.

In [11], the authors present a scheme to transmit a message by means of a queue timing channel. The message is encoded in a sequence of additional delays in service time. The paper concludes that the proposal scheme gives the same results to the Bits through Queues (BTQ).

To study the QoS of base stations in cellular networks, Poisson distribution is used in [12]. Firstly, authors evaluate a natural class of typical-user service characteristics (including path-loss, interference, signal-to-interference ratio, spectral efficiency). Outcome of the first part is then used in the next phase to optimize the mean value of energy efficiency in cellular networks.

In [13], traffic characteristics of cellular data network based on a 2.2 billion traffic record set is presented. The paper proposes a Zipf-like model to characterize the distributions of traffic, subscribers, and requests among service providers.

In comparison to the previously mentioned works, we focus on and analyze the outgoing and incoming international voice traffic distribution and characteristics based on a CDR file dataset.

3   Data Acquisition and Processing

Discovering useful information in large datasets is always a big challenge for data analysts and statisticians.

Data in a CDR are typically represented through attributes and terminologies (i.e. Session Initiation Protocol (SIP), Time Division Multiplexing (TDM), etc.) that detail call scenario steps and their status. A CDR provides one of the most important information about how a call session is accomplished, which is denoted as a cause value (i.e. successfully or unsuccessfully). The cause value is an indicator of successful call session establishment and release between calling and called parties. We process the attributes and terminologies by transforming them into demanded knowledge.

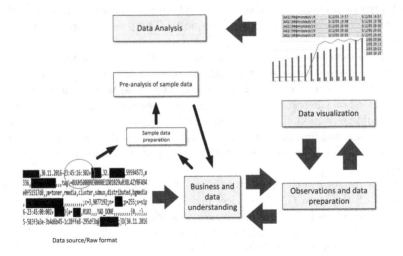

Fig. 2. Data analysis process

In our work, the data processing (see Fig. 2) is initially carried out on a dataset sample, instead of on the whole studied dataset. The data sample comprises of one-week traffic for four selected countries all over the world to represent short and long-distance international call scenarios. The chosen countries cover Europe, Asia and Middle East regions.

The reason of working firstly with the dataset sample is to get a preliminary understanding and view on outcome. It helps us to get an insight how the traffic evolves over one week for different regions. Additionally, having preliminary outputs, we can relatively easily detect and identify errors due to the data processing or/and due to inconsistency in the dataset itself.

Once satisfied with the results based on the data sample, we process the whole dataset (3-month data set). This provides us with a detailed view on the network traffic evolution in time and worldwide. On the other hand, to detect errors and inconsistency in the data set, it becomes a much more challenging task as the amount of processed data importantly grows.

Data selection and preparation steps are typically the most time-consuming phases in the data processing. We have a universe pool of attributes in the dataset, and only attributes leading to required goals and knowledge have to be adequately selected from the vast pool. In addition, there are many attributes and terminologies per every line that have to be jointly interconnected with attributes and terminologies in other lines. Furthermore, missing, inconsistent or erroneous records in an analyzed dataset are another typically issues that we have to face when processing data, and which take a lot of time to satisfyingly solve them. In real-world, there are many reasons why the stored raw data in a network system can be incorrect, invalid, or certain attributes even missing (e.g. incorrect data recording, fraud calls, network misconfiguration, device malfunctioning, etc.). Finally, the data have to be transformed into a structure which is suitable for the final presentation and visualization.

4   Call Flow Process

Figure 3 illustrates a SIP phone call scenario. The scenario is generally similar as for an international as for a local phone call. A completed process of call, starting from the *Connecting* status to the *End of Call* status, is declared as a successful call. A call is established through a SoftSwitch (SSW). The SSW is basically a computer specialized SW, representing the IMS part in the core network, i.e. the switching system element in the network. The SSW typically handles the call processing control and call routing [14]. Among others, it also converts a voice bit stream into packets and recto-verso.

At the initial stage, a *Calling party* (denoted as 'A' party) sends an INVITE message to a *Called party* ('B' party) through the network. The INVITE message initializes the *Connecting* status of the call, which means that the A party is trying to setup a connection and reaching the B party.

In the next phase, the SSW creates an incoming-leg (in other words an incoming channel) for the A party and tries to find out a route that would interconnect the A and B parties. Afterwards, Media Gateway (MGW) inside the SSW is activated and it begins to handle the SIP call and to create an outgoing-leg for the B party.

A channel for both the A and B parties is assigned, and both incoming-leg and outgoing-leg of the call are specified and identified in the system via a unique ID [15].

As soon as the INVITE message reaches the B party, 18x ringing response message (*Alerting*) is sent back to the SSW and to the A party. The SIP 18x message is usually sent to indicate the status of B party. If the B party is ready to accept the call, it sends back to the A party the 200 OK message to establish the

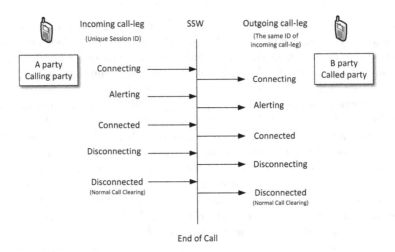

Fig. 3. Phone call scenario

call [15]. Once the call connection is established, a *Connected* status is assigned to both sides of call. The created CDR file details the whole call scenario (e.g., the time and date of call, the connection type: SIP or TDM, the channel ID, etc.).

The call is terminated once the SSW receives another 200 OK message. As next, the BYE message and *Disconnecting* request status are issued to the A and B parties by network. The call session is ended, and the *Disconnected* status is regularly sent by network till the MGW is deactivated [16]. The call session profile is successfully released if all aspects of the call are correctly recorded in the CDR.

5   Theoretical Approach

In this section, we discuss the theoretical background. Figure 4 illustrates different time phases of a call. Once a call request arrives, denoted as the arrival time t_A the waiting time period, Δt_w is initialized. The Δt_w represents a time interval between the *Connecting* and *Connected* status of a call as described in Fig. 3. At this phase, the A party, denoted as x in Fig. 4 is in the idle status while waiting to setup the call. Once the party A is connected to the B party, at a moment denoted as connection time t_C, the connection enters to the active state. The active call continues till one of the parties clears the call at the moment t_D (disconnect time). The interval between t_C and t_D is referred as a serving time, Δt_s. Thus, the whole call time period, Δt_x, consists of two periods Δt_w and Δt_s [17]. The time t_A and period Δt_s are ones of key parameters when analyzing the performance of a queueing system.

Having two consecutive call arrivals, x_1 and x_2, the call arrival x_2 at t_2 does not depend on the call arrival x_1 at t_1 and vice versa (Fig. 5). The time

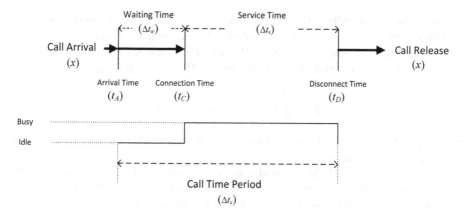

Fig. 4. Waiting and service time phases

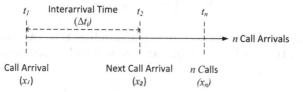

Fig. 5. Interarrival time process

period between two consecutive calls is denoted as the interarrival time Δt_i, i.e.
$\Delta t_i = t_2 - t_1$.

The interarrival time basically captures the importance of destination; the lower the average time between two consecutive calls, the more important the destination is from the point of traffic amount. In our study, the system is considered to be ideal, i.e. a call arrival, x, starts to be served once arriving to the system. In other words, the call does not have to wait in a queue to be served by system. A delay is only due to the ringing period and the network call processing.

The ringing period in networks varies from a few seconds and can go as high as tens of seconds and has a strong influence on the waiting time. Typically, a maximum ringing time period threshold is configured by network operator; for example, in our reference network, the ringing threshold is set to 30 s. The ringing time period depends on several aspects such as the calling party patience, the called party awareness of incoming call, the importance of call, the daytime of call, the culture of given country, etc.

The network call processing time consists of all partial periods before the ringing period itself starts; e.g., paging, processing, transmission and propagation delay, etc. The network may also perform other tasks such as the security check, the user credential check, the B party availability checking, finding the route, etc., which increase the total network call processing time.

The number of calls within a time unit follows the Poisson distribution, i.e. the time between 2 consecutive calls follows the exponential distribution and call arrivals are independent to each other [18][19]. The Poisson distribution indicates the probability that a call arrival will occur in the next given time unit:

$$P_j = \frac{\lambda^j}{j!}\, e^{-\lambda} \qquad (j = 0, 1, ...). \qquad (1)$$

where j is the number of call occurrences in a time unit, and λ represents the mean value of call arrivals.

The service time Δt_s is modeled using the exponential distribution with the service time mean value represented as β. The probability density function of an exponential distribution is given by following equation [20]:

$$f(t) = \begin{cases} \frac{1}{\beta} e^{-\frac{1}{\beta}n} & \text{if } n \geq 0 \\ 0 & \text{if } n < 0 \end{cases} \qquad (2)$$

where n is the number of call occurrences in a unit of time. As shown in Sect. 6, the interarrival time and service time are exponentially distributed.

6   Results

As mentioned above, we analyze 3-month CDR dataset files, which cover the period 10/2016–12/2016. The dataset was collected via a CDR mediation server in the switching center and consists of about 9 million records.

For the analyses, we consider incoming and outgoing traffic of one short and three long-distance international call scenarios: (i) one neighboring country to the reference network (country C in the figures below), (ii) one European country (country B in the figures), and finally (iii) two Asian countries (countries A and D in the figures). A detailed study of traffic distribution in time and among different countries are presented in [21]. Additionally, results for the whole world are provided as well, including the measured and theoretically calculated values. Notice that the incoming traffic consists of network traffic originating from the whole world while the outgoing traffic only includes traffic originating from one country, i.e. the reference network. In other word, the incoming user pool is much larger than the outgoing one.

Table 1 indicates mean values theoretically calculated for the outgoing and incoming traffic for the world case scenario. The outgoing traffic shows a higher mean value of waiting time with a lower mean value of service time than the incoming traffic. One of the main reasons is the cost of an outgoing international call, which is generally higher than a local call. Moreover, internet services are nowadays globally available, which can be used as an alternative to traditional voice call networks. As to the interarrival time, the mean value of outgoing traffic is higher as more traffic is generated in the world comparing to the amount of traffic originating in the reference network.

Table 1. Mean values of the analyzed parameters fpr the world scenario

Traffic direction	Waiting time(λ)[s]	Service time(β)[s]	Interarrival time(β)[s]
Outgoing	16.26	132.29	0.62
Incoming	13.46	154.06	0.52

Figure 6 shows the probability of occurrences for the service time. On the right (resp. left) y-axis, the probability of occurrences for the world scenario (resp. selected countries) is presented. The graphs illustrate the frequency of occurrences for the given service time, i.e. the sum of all probabilities when the service time goes to infinity equals to 1 for each country, resp. the world.

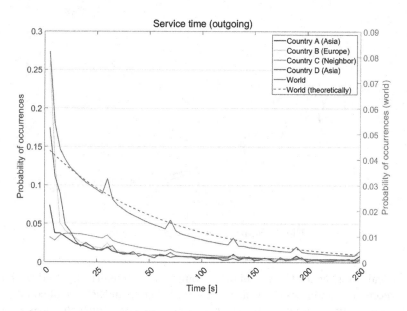

Fig. 6. Service time, outgoing traffic

As can be observed from Fig. 6, the service time is exponentially distributed for all considered scenarios; the figure also indicates the theoretical result for the world scenario. Majority of considered scenarios show the maximum for the value around 5 s and then the probability rapidly decreases as the service time increases. In our case, the service time is affected by many factors, such as the type of calls (private, business calls), the network operator price policy (distant calls are typically cheaper), the distance between the reference and destination network, etc.

Comparing to the long-distance call scenarios, the probability of occurrences for the short-distant scenario (country C) is smaller, but the probability is relatively constant up to around 60 s. In other words, the probability of having a longer call duration is higher for the short-distance call scenarios than for the long-distance ones. This is typically due to the strong business and/or personal relationships among the neighboring countries.

In Fig. 7, the waiting time outcome is shown. Correspondingly to Fig. 6, the right y-axis illustrates the probability of occurrences for the world scenario case, whereas the left y-axis, for the 4 selected countries. Based on the scenario, the peaks occur between 6 to 21 s.

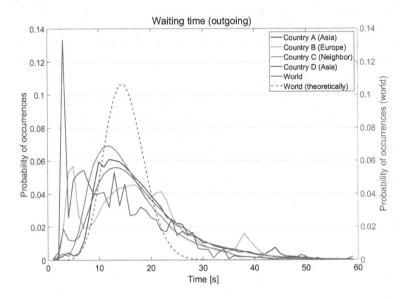

Fig. 7. Waiting time, outgoing traffic

The waiting time for the neighboring country case (country C), and the country A, more or less follows the world scenario. Whereas in the case of countries B and D, a slightly different curve progresses are observed, with the peak around 3 s. This is possibly due to:

- A user intentionally clears the call after the called party is being rung, as the ringing is only used as a sign between these two parties.
- The called party relatively quickly hangs up as the party knows in advance about the incoming call and expects the ringing.

The theoretical world scenario shows a higher waiting time in the interval 10 to 20 s and a lower one in the interval 30 to 60 s comparing to real values. Such variations between the real and theoretical values are due to the nature of people behavior around the world; for instance, lifestyle, network policies, time zone difference, etc.

Figure 8 shows the interarrival time, where the description of x-y axes is analogous to the previous figures. We can see in the graph that the peak of interarrival time for the world scenario is very small, close to 0s. According to Table 1, the probability of having 2 consecutive calls in the world scenario within 1 s is about 0.6.

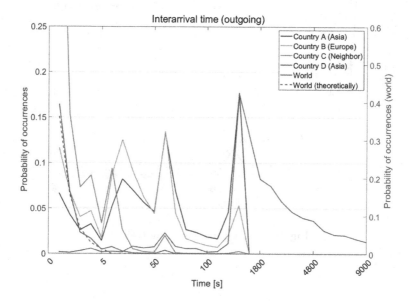

Fig. 8. Interarrival time, outgoing traffic

The lowest interarrival time manifests the neighboring country C, which is due to the strong relationships between the neighboring countries as explained previously. The interarrival time grows as destinations are farther and farther (see countries A, B and D). For example, the interarrival time between two calls in case of a faraway destination could reach even as high values as 10^6 s.

Additionally, based on the daily/weekly traffic profiles, we can deduce the country relationship nature, i.e. if the relationship is much more business or personal oriented.

Figure 9 illustrates the service time for the incoming traffic case. In general, the probability of occurrences is higher for the incoming traffic than for the outgoing one (shown in Fig. 6). As the incoming calls are from a much larger pool of users (the whole word, or all network operators in a given country), a call with a longer service time typically occurs more frequently.

The waiting time of incoming traffic is illustrated in Fig. 10. The waiting time of incoming traffic for countries A-D manifests higher peaks compared to the outgoing traffic case (Fig. 7). In case of world scenario, the theoretical values are, similarly to the outgoing scenario, higher than the real ones. The reason is the same as explained in Fig. 7.

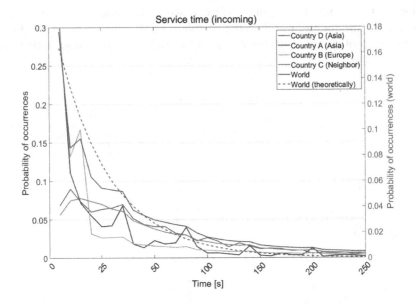

Fig. 9. Service time, incoming traffic

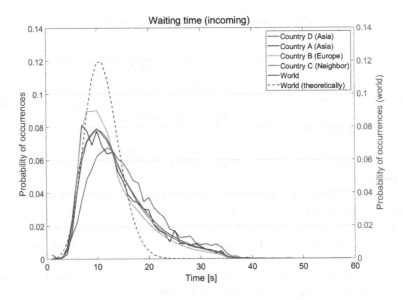

Fig. 10. Waiting time, incoming traffic

As to the interarrival time, the probability of occurrences shows similar behavior for the outgoing (Fig. 8) and incoming traffic (Fig. 11), where the probability is higher for the incoming interarrival time (due to the larger pool of users). However, comparing to Fig. 8, the incoming interarrival time peaks are shifted to right, resp. left (around 500 s, resp. 10 s) for the countries A and D, resp. B and C.

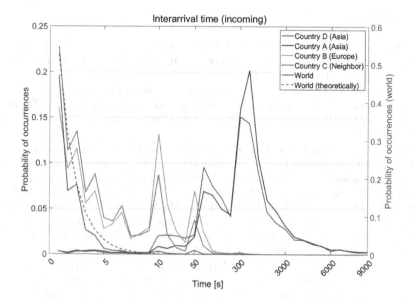

Fig. 11. Interarrival time, incoming traffic

7   Conclusion

In this paper, we study characteristics of the interarrival, waiting, and service time for outgoing and incoming traffic of an international voice traffic carrier. The analyzed CDR dataset covers 3 months (October-December 2016).

For both traffic, outgoing and incoming, the interarrival, and service time follow the exponential distribution. The interarrival time increases as the destination to the reference network becomes less important. A longer service time can be observed from the nearby countries to the reference country.

The waiting time follows the Poisson distribution, and it varies based on the network configuration, the user resilience, or the call importance.

In the world scenario case, the theoretical values show higher peaks than the real measurements. Such difference is due to the nature of people behavior around the world; for instance, lifestyle, network policies, time zone difference, etc.

In our future work, we plan to utilize mobile signaling data to investigate network and/or mobile terminals configuration inconsistency and the impact on the network performance.

Acknowledgments. This research work was supported by the Grant Agency of the Czech Technical University in Prague, grant no. SGS18/181/OHK3/3T/13.

References

1. Sultan, K., Ali, H., Zhang, Z.: Call detail records driven anomaly detection and traffic prediction in mobile cellular networks. IEEE Access **6**, 41728–41737 (2018). https://doi.org/10.1109/ACCESS.2018.2859756
2. Zonoozi, M.M., Dassanayake, P., Faulkner, M.: Teletraffic modelling of cellular mobile networks. In: Proceedings of Vehicular Technology Conference (VTC 1996), vol. 2, 1274–1277. IEEE Press, Atlanta (1996). https://doi.org/10.1109/VETEC. 1996.501517
3. Rico-Paez, A., Cruz-Perez, F.A., Hernandez-Valdez, G.: Teletraffic analysis formulation based on channel holding time statistics. In: IEEE International Conference on Wireless and Mobile Computing, Networking and Communications, pp. 326–330. IEEE Press, Marrakech (2009). https://doi.org/10.1109/WiMob.2009.62
4. Fan, L., Zhao, Z., Qi, Ch., Li, R., Zhang, H.: A revisiting to queueing theory for mobile instant messaging with keep-alive mechanism in cellular networks. In: IEEE International Conference on Communications (ICC), 1–6. IEEE Press, Paris (2017). https://doi.org/10.1109/ICC.2017.7996707
5. Mehr, K.A., Nobar, S.Kh., Niya, J.M.: Inter-arrival time distribution of IEEE 802.15.4 under saturated traffic condition. In: 23rd Iranian Conference on Electrical Engineering, pp. 2164–7054. IEEE Press, Tehran (2015). https://doi.org/10.1109/ IranianCEE.2015.7146211
6. Mark, B.L., Ephraim, Y.: On modeling network congestion using continuous-time bivariate Markov chains. In: 45th Annual Conference on Information Sciences and Systems, pp. 1–6. IEEE Press, Baltimore (2011). https://doi.org/10.1109/CISS. 2011.5766118
7. Castellanos-Lopez, S.L., Cruz-Perez, F.A., Hernandez-Valdez, G., Miranda-Tello, J.R.: Performance analysis of mobile cellular networks with MMPP call arrival patterns. In: 9th IFIP International Conference on New Technologies, Mobility and Security (NTMS), pp. 2157–4960. IEEE Press, Paris (2018). https://doi.org/ 10.1109/NTMS.2018.8328711
8. Stenico, J.W.G., Lee, L.L., Vieira, F.H.T.: Queuing modeling applied to admission control of network traffic flows considering multifractal characteristics. IEEE Latin Am. Trans. **11**, 749–758 (2013). https://doi.org/10.1109/TLA.2013.6533964
9. Pattavina, A., Parini, A.: Modeling voice call inter-arrival and holding time distributions in mobile networks. In: ITC19/Performance Challenges for Efficient Next Generation Networks, pp. 729–738 (2005)
10. Nagy, L., Tombal, J., Novotny, V.: Proposal of a queueing model for simulation of advanced telecommunication services over IMS architecture. In: 36th International Conference on Telecommunications and Signal Processing (TSP), pp. 326–330. IEEE Press, Rome (2013). https://doi.org/10.1109/TSP.2013.6613945

11. Ferrante, G.C., Quek, T.Q.S., Win, M.Z.: Timing capacity of queues with random arrival and modified service times. In: IEEE International Symposium on Information Theory (ISIT), pp. 370–374. IEEE Press, Barcelona (2016). https://doi.org/10.1109/ISIT.2016.7541323

12. Błaszczyszyn, B., Karray, M.K., Keeler, H.P.: Using Poisson processes to model lattice cellular networks. In: Proceedings IEEE INFOCOM, pp. 773–781. IEEE Press, Turin (2013). https://doi.org/10.1109/INFCOM.2013.6566864

13. Jun, L., Tingting, L., Gang, C., Hua, Y., Zhenming, L.: Mining and modelling the dynamic patterns of service providers in cellular data network based on big data analysis. China Commun. 10, 25–36 (2013). https://doi.org/10.1109/CC.2013.6723876

14. Stallings, W.: Data and Computer Communications, 8th edn., Chap. 10, pp. 307–308 (2007)

15. Forouzan, B.A.: TCP/IP Protocol Suite, 4th edn., Chap. 25, pp. 748–751 (2010)

16. Hartpence, B.: Packet Guide to Voice over IP, Chap. 3, pp. 70–75 (2013)

17. Tunnicliffe, G.W., Murch, A.R., Sathyendran, A., Smith, P.J.: Analysis of traffic distribution in cellular networks. In: VTC 1998. 48th IEEE Vehicular Technology Conference. Pathway to Global Wireless Revolution (Cat. No. 98CH36151), vol. 3, pp. 1984–1988. IEEE Press, Ottawa (1998). https://doi.org/10.1109/VETEC.1998.686103

18. Schay, G.: Introduction to Probability with Statistical Applications, Chap. 6, pp. 183–184 (2007)

19. Cooper, R.B.: Introduction to Queueing Theory, 2nd edn., Chap. 2, pp. 50–56 (1981)

20. Ross, Sh.M.: Introduction to Probability and Statistics for Engineers and Scientists, 4th edn., Chap. 5, pp. 176–182 (2009)

21. Aziz, Z., Bestak, R.: Analysis of call detail records of international voice traffic in mobile networks. In: Tenth International Conference on Ubiquitous and Future Networks (ICUFN), pp. 475–480. IEEE Press, Prague (2017). https://doi.org/10.1109/ICUFN.2018.8436669

Vehicle in Motion Weighing
Based on Vibration Data Collected
from Sensor Network

Marcin Bernas[1]($\boxtimes$), Wojciech Korski[2], Bartłomiej Płaczek[3],
and Jarosław Smyła[2]

[1] Faculty of Mechanical Engineering and Computer Science,
University of Bielsko-Biala, Bielsko-Biala, Poland
`marcin.bernas@gmail.com`
[2] Institute of Innovative Technologies EMAG, Katowice, Poland
`{wojciech.korski,jaroslaw.smyla}@ibemag.pl`
[3] Institute of Computer Science, University of Silesia, Katowice, Poland
`placzek.bartlomiej@gmail.com`

Abstract. This paper introduces a method for estimating weight of moving vehicle based on vibration measurements registered by sensor network. The sensor network consists of sensor nodes installed on roadside and curbs. The sensor nodes calculate statistical features of vibration measurements and verify them using a proposed quality function. Then statistical features are used to create a vibration model, which allows the sensor nodes to estimate vehicle weight. Vibration models were created in this study using linear regression and fully connected neural network. A final assessment of vehicle weight is performed at the network sink, by taking into consideration the estimations and data quality indicators provided by particular sensor nodes. Several machine learning methods were compared with the proposed approach. The proposed method has low computational requirements, thus it can be adapted to sensor networks with computational and memory constrains. Advantages of the introduced method were demonstrated in real-world experiments, using various modelling approaches and installation types of sensor nodes. The experimental results confirm that the introduced method can be adapted for the node, which can weight vehicles independently or in cooperation with other nodes (as an ensemble).

Keywords: Sensor network · Accelerometers · Neural networks · Road traffic · Vehicle weight estimation

1 Introduction

The overloaded vehicles are a common issue, which causes deterioration of roads and possibility of accidents [1]. The most tangible example of this problem is a rapid degradation of road surface by overloaded and technically inefficient

© Springer Nature Switzerland AG 2019
P. Gaj et al. (Eds.): CN 2019, CCIS 1039, pp. 208–219, 2019.
https://doi.org/10.1007/978-3-030-21952-9_16

trucks. In some extreme situations, the overloaded truck can be a direct cause of bridge collapse. Moreover, the overloaded trucks have much longer braking distance, which directly impacts on traffic safety and can lead to serious accidents [2]. Therefore, vehicle weighing systems are necessary to support traffic safety, protect road infrastructure and maintain transportation systems in a good condition.

Several vehicle weighing systems have been proposed over past decades. The existing vehicle weighing techniques fall into two main categories: static weighing techniques and weigh-in-motion (WIM) techniques. On the one hand, the static weighing techniques are very accurate, however the weighing procedure is costly and time-consuming. On the other hand, the WIM technologies, which have been developed since the 1960s, allow determining weight of vehicles without stopping them at the cost of lower accuracy [3]. The up to date WIM solutions use sensors installed in pavement, such as: bending plates, load cells, capacitance mats, and strip sensors. To reduce the cost of installation in pavement, several WIM systems were proposed in the related literature, including a wireless sensor network and an array of sensors with strain gauges and optic fibers [4].

The above mentioned pavement-based solutions cannot be easily moved and can be avoided by drivers that know the locations where vehicles are weighed. Therefore, in this paper a portable sensor network was proposed, where nodes are installed on curbs or on roadside. The novelty of proposed solution lays in the network portability and in the dedicated weight estimation algorithm with data quality measure, which was tested in real world conditions.

The paper is organized as follows. Section 2 includes a survey of related literature and describes contribution of this work. The proposed two weighing methods are discussed in Sect. 3. Results of experiments are described in Sect. 4. Finally, conclusions and further work directions are given in Sect. 5.

2   Related Works

The vehicle weighing is an active research topic. Methods of rough vehicle weight estimation via WIM technique for bridges were discussed in [5]. This survey shows that it is possible to distinguish the axes of moving vehicle and estimate its weight by using sensors placed on the bottom of bridge elements. Further research [6] was conducted to determine the levels of gross weight accuracy for bridge weighting systems. One of the considered systems was ranked in C class and the others in D class and bellow, according to COST323 specification.

There is a considerable amount of research related to vehicle weighing on the road. In [7] a traffic monitoring system was developed to monitor road damage status, traffic volume, vehicle speed and weight. The system utilises embedded sensors: asphalt strain gauges, vertical strain gauges, load cells and a moisture sensor. This solution requires the sensors to be placed symmetrically along the centre of the road. The sensors measure both horizontal (7 sensors) and vertical (2 sensors) strain. The sensors are placed in such a way that vehicle wheels are moving as close to sensor as possible.

A sensor network (over 100 sensors) measuring dynamic response of concrete pavement under the travelling vehicles was proposed in [8]. The network includes strain gages measuring linear displacement, vertical accelerometers and thermocouples. To identify vehicle weight the data from sensor network were correlated with a finite element model [9]. The proposed solution was able to detect dynamic changes with high accuracy.

Similar method using embedded strain sensors was proposed in [7]. The authors have used three nodes with five strain sensors installed in pavement at centre of a road. The experimental results proved that it is possible to measure the axle weight with embedded strain sensors. The promising results were obtained using mean value of multiple sensors samples. Further research was conducted to establish a relationship between the signal value and the weight measurement [10].

Specific tire configuration and distress development was investigated in [11], however major impact was put on strain gauges sensors and road model. In another work along these lines [4] a precise numerical simulation method was presented to analyze the relationship between sensor location and the accuracy of static weight estimation. The considered model has shown that vibrations propagate not only into road structure but also on its surface.

All the above mentioned solutions were installed in pavement to reduce the influence of external factors and enable precise weighing. According to the aforementioned model, the weighting precision drops with distance.

It should be also noted that several attempts were made to detect overloaded vehicles based on vibrations and currently available accelerometers [12]. The accelerometers were used in [13] to measure vibrations caused by heavy vehicles. The 3185D S/N 2723 accelerometer (Dytran Instruments) was mounted on a 1.2 m long steel bar with diameter of 20 mm and put in the ground to absorb the energy of vibrations. This precise sensors has enabled distinguishing personal cars and trucks via frequency analysis.

In [14] the accelerometers were placed on the road surface and an extended Kalman filter was used for estimating parameters of passing vehicles. The experimental results have confirmed the possibility of vehicle detection and velocity estimation. A disadvantage of that method is a dependence of the results on unknown scaling factors.

The utilisation of accelerometers and magnetometers for vehicle detection was discussed in [15]. The accelerometers were used to detect axle locations, and the magnetometers were applied to estimate vehicle speed. The authors have obtained detection accuracy close to 99% for sensors placed on the road.

The magnitude of pavement vibrations caused by vehicles was analysed experimentaly and theoreticaly in [16]. The authors have proposed a model of tuned wave propagation on pavement using a system identification approach. The introduced model confirms that vibrations can be detected both in the pavement as well as on its surface.

A multi-sensor network with piezoelectric (MEMS) accelerometers was proposed in [17]. This type of sensors allows the energy of vibrations to be harvested and utilized to power the sensor nodes. Dedicated algorithms have been proposed to enable vehicle detection, recognition of driving direction, and speed estimation based on amplitude and frequency analysis of the registered vibrations.

Finally, a high quality seismic sensors (three-axis geophone and single-channel seismometer) were used for vehicle detection in [18]. The authors have managed to recognize vibrations produced by passing vehicles from distance, based on time-frequency analysis approach. The method was accurate on distance up to 3 m on both paved and asphalt roads.

The up-to-date solutions allow precise weighting of vehicles with use of build-in-pavement sensors and detect vehicles from distance. This paper introduces a method to estimate the weight of moving vehicle based on accelerometer readings. The novelty of the proposed solution lies in measuring weight of vehicle with use of sensors that are placed on road side or on curb. Details of the proposed system are described in next section.

3   Proposed Method

The proposed method is based on observations made during initial research [19], where usefulness of data provided from various accelerometers was evaluated. Movement of vehicle, which was passing the sensor, did not influence the sensor readings for X an Y axis, thus only Z axis (the vertical one) was considered in this research.

The data for this study were collected from a sensor network, which was placed on the road side and on the curbs. Two types of sensors were used. Both sensors provide $+/-2G$ sensitivity range, however they have different resolutions (14-bit and 16-bit). Figure 1a, b, c presents the data from sensor node placed on the roadside, while Fig. 1d, e, f shows data registered on the curb.

Vertical red lines in Fig. 1 indicate the time period when a vehicle (above 3.5 ton) was passing. The accelerometers were installed in sensor node (b and e) or outside it and directly glued to the road surface (a, c, d and f). As it was mentioned earlier in this section, the vibrations are registered for Z axis. The measurements confirmed that the vibration from vehicles are relatively small, thus more precise accelerometer (16-bit) is necessary. The best results were obtained for sensors connected directly to the road surface. The results for different types of connection are presented in Fig. 2.

The observation of sensors readings allows us to identify this sensor, which is not installed (glued) correctly. The incorrectly installed sensor collects vibration values with higher average standard deviation and minimal/maximal value within measurement period. E. g., the range of values for a period, when vehicles are not present on the road, does not exceed 50 units in case of correctly installed sensor and is above 100 otherwise. Additionally, the single high values that exceeds 300 units are registered by the incorrectly installed sensor at random. Finally, the moving vehicles do not change the amplitude of readings significantly. Based on the above observations and previous research [1,19] the novel rough weighting method using sensor network was introduced (Algorithm 1). The presented algorithm is executed for each sensor separately.

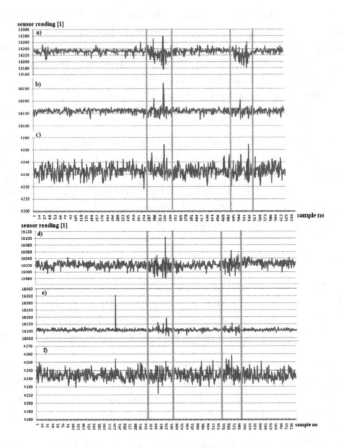

Fig. 1. Data from sensor node placed on the roadside (a, b, c) and on the curb (d, e, f)

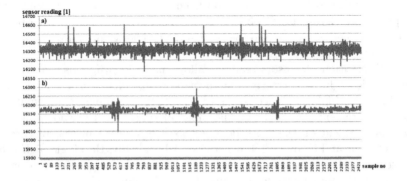

Fig. 2. Accelerometer readings for two types of connection: (a) sensor put on road surface without glue, (b) sensor glued to the road surface

1. collect sensor readings $X(t)$,
2. build feature representation for $X(t)$ and the analysis window dt:
   (a) $F_{std}(X, t, dt) = standard\_deviation(X(t), t \in (t - dt, t))$,
   (b) $F_{mm}(X, t, dt) = max(X(t), t \in (t - dt, t)) - min(X(t), t \in (t - dt, t))$,
   (c) $F_{pp}(X, t, dt) = 90'perc.(X(t), t \in (t - dt, t)) - 10'perc.(X(t), t \in (t - dt, t))$,
3. evaluate data quality (q) using Algorithm 2 and parameter t_h,
4. if ($q > p_{thres}$) then
   (a) calculate weight $W(X, t, dt)$ using Eq. 2.
   (b) send values W and q to the sink.

Input data of the proposed algorithm include accelerometer readings for the Z axis. Three features are calculated based on the input data: standard deviation, difference between maximum and minimum, and difference between 90'th and 10'th percentile. Percentile measure was used to reduce the influence of outliers. The feature values are used at the first step to estimate the quality of data registered by the sensor (Algorithm 2). If the calculated quality exceeds a predefined threshold (p_{thres}) then the weighting procedure is performed (Eq. 1). Two parameters have to be adjusted for the proposed method: size of the aggregation window (dt) and width of the history window (t_h).

The weight value (W_{fin}) is calculated at sink of the sensor network in accordance with Eq. 1, where values (W_i) are obtained form i-th sensor node with corresponding quality measure (q_i). In order to suppress unnecessary data transmissions, only the weight values with quality above p_{thers} are reported by the sensor nodes to the sink.

$$W_{fin}(t, dt) = \frac{\sum_i W_i(X_i, t, dt) * q_i(X_i, t, dt, t_h)}{\sum_i q_i(X_i, t, dt, t_h)} \tag{1}$$

The quality evaluation algorithm (Algorithm 2) was proposed to minimize the effect of incorrectly installed sensors on vehicle weighing result (noise presented in Fig. 2). Algorithm 2 is defined as follows:

1. if ($t == dt$) then // algorithm initialisation
   (a) $prot1_{std} = prot2_{std} = F_{std}(X, t, dt)$, $prot1_{cnt} = 0$,
   (b) $prot1_{mm} = prot2_{mm} = F_{mm}(X, t, dt)$, $prot2_{cnt} = 0$, step $= 0$;.
2. if ($t >= dt$) then
   (a) $prot1_{val} = prot1_{std}^2 + prot1_{mm}$; // measure of prot1
   (b) $prot2_{val} = prot2_{std}^2 + prot2_{mm}$; //measure of prot2
   (c) $curr_{val} = F_{std}(X, t, dt)^2 + F_{mm}(X, t, dt)$; //measure of current reading
   (d) if ($t < t_h$) then //initialization phase
       i. if $|curr_{val} - prot1_{val}| > |(prot1_{val} - prot2_{val}|$ then
          A. $prot2_{std} = F_{std}(X, t, dt)$,
          B. $prot2_{mm} = F_{mm}(X, t, dt)$.
       ii. if $|curr_{val} - prot2_{val}| > |(prot1_{val} - prot2_{val}|$ then
          A. $prot1_{std} = F_{std}(X, t, dt)$,
          B. $prot1_{mm} = F_{mm}(X, t, dt)$.
   (e) if ($t >= t_h$) then //normal mode

    i. if $|curr_{val} - prot1_{val}| > |(curr_{val} - prot2_{val}|$ then
      A. $prot1_{std} = (1\text{-}\alpha)*prot1_{std} + \alpha*F_{std}(X, t, dt),$
      B. $prot1_{mm} = (1\text{-}\alpha)*prot1_{mm} + \alpha*F_{mm}(X, t, dt),$
      C. $prot1_{cnt} + +.$
    ii. else
      A. $prot2_{std} = (1\text{-}\alpha)*prot2_{std} + \alpha*F_{std}(X, t, dt),$
      B. $prot2_{mm} = (1\text{-}\alpha)*prot2_{mm} + \alpha*F_{mm}(X, t, dt),$
      C. $prot2_{cnt} + +.$
(f) $step + +;$
(g) $q = \min(1.0 - \frac{min(prot1_{val}, prot2_{val})}{max(prot1_{val}, prot2_{val})} - |\beta - \frac{min(prot1_{cnt}, prot2_{cnt})}{step}|, 0)$

Algorithm 2 is divided into initialisation stage ($t < t_h$) and normal stage ($t > t_h$). During first stage two most distinctive profiles are found: $prot1$ and $prot2$. Then, in normal stage each new value is assigned to one of two profiles. The selected profile is modified. Magnitude of the profile modification depends on learning parameter (α). The data quality measure q is calculated by taking into consideration the distinctiveness of the two profiles and number of elements within profiles. The value $\beta = 1.5\%$ parameter is the percentage share of truck traffic in overall traffic. The calculated profiles $prot1$ and $prot2$ are also used to adjust the weighting model for separate sensors.

The weight in Algorithm 1 is calculated using Eq. 2. During experiments it was noticed that characteristic of accelerometer readings are changing with time (due to weather or temperature changes), therefore the weighting model M with profiles ($prot1$, $prot2$) are saved during training process and then it is scaled to current accelerometer reading using Eq. 3 (a_{thers}).

$$W(X_t, t, dt) = a_{thres} * M(X, t, dt) \tag{2}$$

$$a_{thres} = \frac{max(prot2_{std}^M, prot1_{std}^M)}{2 * max(prot2_{std}, prot1_{std})} + \frac{max(prot2_{mm}^M, prot1_{mm}^M)}{2 * max(prot2_{mm}, prot1_{mm})}, \tag{3}$$

where: $prot1^M$, $prot2^M$ are profiles build based on historical data and used to tune the model (M).

The method can be simplified for short time readings (fixed - $a_{thres} = 1$), or it can adapt to changing weather conditions (dynamic - a_{thres} calculated for each weighting procedure). The weighting model M used for weight calculation by the sensor node should be computationally simple and not memory consuming. Therefore, several machine learning methods were considered [21]. Based on previous studies, two models were selected. The first model is based on fully connected neural networks with linear activation function [20]. Second model uses linear regression, which represents a baseline simple approach for comparison purposes. The regression model is described by Eq. 4.

$$M(X, t, dt) = \frac{\sum_{i=1}^{3} a_i * F_i(X, t, dt) + b_i}{3}, \tag{4}$$

where a_i and b_i are parameters of the lineal relation, F_i are feature functions : ($F_1 \equiv F_{std}$, $F_2 \equiv F_{mm}$ and $F_3 \equiv F_{pp}$).

4 Experiments

Experiments were conducted on a two lane road. The sensor network was deployed close to existing WIM system, which provided a reference data for testing and calibration. Six sensor nodes were installed on roadside and on the curb.

Parameters of the proposed method were set based on initial experiments. The time window dt was set to 1 second, which is equal to average vehicle passing time. The initialisation time (t_h) was set to 600 s to ensure passing of at least one vehicle. The learning parameter α was set to 0.05 to neglect fast changing of profile. The initial research shows that $\alpha > 0.1$ provides unstable results. Finally, the data quality q was evaluated for various data sets to estimate p_{thres} parameter. The values of q calculated for real sensor readings as well as synthetic data (random with Gaussian and uniform distribution) were presented in Table 1.

Table 1. Data quality analysis

Data set	Sensor low noise	Sensor moderate noise	Sensor high noise	Random Gaussian	Random uniform
Data quality (q)	0.84	0.65	0.56	0.505	0

Based on the calculated values of q the p_{thres} was set to 0.6 to take under consideration only those sensors, for which the registered data are reliable.

The next task was to train the model M for weight calculation. The model was build using 80 randomly selected vehicles. The least square method was used to evaluate the parameters of regression model. The approximation is presented in Fig. 3 both for total weight of vehicle and for the heaviest axis. According to model defined in Eq. 4, the regression parameters were defined as $a_i = (900, 555, 225)$ and $b_i = (-18, -42, -120)$. To verify the model a single sensor node was used and the results of weighing are presented in Fig. 4.

The regression model provides a rough weight estimation with mean average weighting error (MAE) equal 3029 kg.

In a similar manner the FCNN model was trained, however in this case the PSO method was used as discussed in [20]. The obtained MAE error was comparable and equal to 2983 kg. The proposed model was compared with state-of-the-art approaches [21] denoted as M_{MLP}, M_{PNN}, and M_{fuzzy}. The models are tuned version of WEKA models. Each sensor was tuned separately (fixed configuration - a_{thres}) using stratified sampling. The results of weighting for one sensor were presented in Table 2 for measurements performed on the roadside and on the curb for 100 vehicles with various weights.

The proposed models proved to be superior in comparison with the other classical models. The FCNN model was generating the lowest error ratio in case

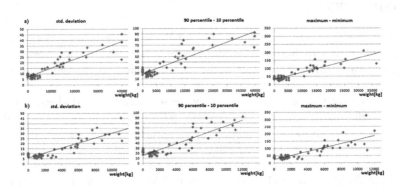

Fig. 3. Model M approximation for (a)weighting vehicles and (b) heaviest vehicle axis

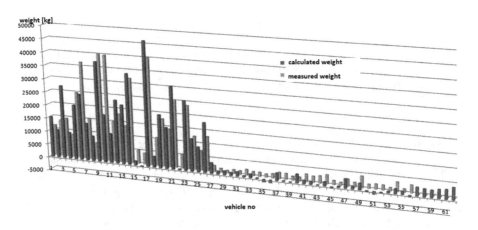

Fig. 4. The comparison of vehicle weight using linear model M

Table 2. Weighing model comparison [MAE]

Model type	M	M_{FCNN}	M_{MLP}	M_{PNN}	M_{fuzzy}
Sensor1 on roadside [1000 kg]	3.5	3.3	3.8	4.1	3.87
Sensor1 on curb [1000 kg]	4.01	3.96	4.1	5.2	4.7
Sensor2 on roadside [1000 kg]	3.73	3.7	4.2	5.0	4.3
Sensor2 on curb [1000 kg]	4.3	4.2	4.4	4.5	4.7
Sensor3 on roadside [1000 kg]	4.8	5.0	5.1	6.3	5.9
Sensor3 on curb [1000 kg]	5.2	5.6	5.2	5.8	6.6

of low noise (sensor 1 and 2), while the regression model was more adequate for noised data (sensor 3).

The values presented in Table 2 were obtained after tuning the model for each sensor node separately. Further analysis was conducted for the scenario, where the model was tuned for one sensor node and the other sensor nodes have used the sane model with dynamic a_{thres} parameter. The results are presented in Table 3. In this case the regression model was used, because it is unaffected by random factors.

Table 3. Fixed and dynamic training comparison [MAE]

Model used	Sen.1 (fixed)	Sen.1 (dynamic)	Sen.2 (fixed)	Sen.2 (dynamic)	Sen.3 (fixed)	Sen.3 (dynamic)
Sensor 1 [1000 kg]	3.6	3.5	4.1	3.8	8.3	5.8
Sensor 2 [1000 kg]	3.8	3.7	3.9	3.7	7.7	6.4
Sensor 3 [1000 kg]	5.8	4.8	6.8	5.1	5.8	5.6

The results in Table 3 show that when using a dynamic a_{thres} value it is possible to maintain error ratio and even decrease it in some cases. The result is visible for sensors with comparable level of data quality (sensor 1 and sensor 2). In case of sensor 3 (most noised) the correct model can improve the result, however model based of noised data increase the error even further.

Finally, the results obtained from three sensor nodes were used to estimate vehicle weight with and without quality function. The results are presented in Table 4. As borderline value the simple averaging was considered.

Table 4. Results obtained using 3 sensors data [MAE]

Model type	M	M_{FCNN}	M_{MLP}	M_{PNN}	M_{fuzzy}
Roadside (averaging) [1000 kg]	5.2	4.7	4.8	4.1	6.1
Curb (averaging) [1000 kg]	5.36	4.9	5.0	3.9	6.17
Roadside (with Q function) [1000 kg]	3.55	3.5	3.7	4.1	4.3
Curb (with Q function) [1000 kg]	4.3	4.2	4.3	4.3	4.4

The results in Table 4 show that for sensor network with nodes installed in various locations, the weighing error can increase if all sensors readings are equally taken under consideration. Using the proposed quality function, the most noised sensor readings are ignored. The experimental results proved that by using the quality function it is possible to obtain MAE comparable to that of the best sensor and at the same time discard the faulty sensors. Moreover, using data from multiple sensors provides more stable results. All considered weighing models provided better and more stable results than in case of one sensor node.

5   Conclusion

The proposed vehicle weighing system uses the data collected by sensor network to roughly estimate weight of passing vehicles. The sensor network is composed of sensor nodes that are installed on road surface or on curbs to register vibrations caused by the vehicles. Results of the experiments show that major role in weighing takes the position of a sensor and type of connection to the surface. The comparison of two sensor types showed that 14-bit accelerometer provided insufficient precision to correctly estimate the vehicle weight. Additionally, type of installation also influences the estimation error (increasing error by 17%).

The proposed weighing system allows the estimation model to be trained for particular sensor nodes independently or trained for one sensor node and shared with the other nodes. The proposed weighting method based on FCNN proved to be superior in comparison with state-of-the-art methods. The considered machine learning methods provided comparable results with those of the linear regression model, which has lower computational complexity. The proposed models, after training phase, has low computational requirements and can be implemented into sensor nodes with limited CPU and memory resources.

It is worth to note that proposed quality function allows us to reduce the impact of incorrectly installed nodes and make the results more stable and reliable. Moreover, it allows the system to reuse a trained weighting model for multiple nodes, which significantly simplifies a training process.

As for further work it would be useful to install sensor nodes on a longer road section for reducing the influence of vehicle speed on weighing results. Another interesting possibility is to use more features to describe sensor data and obtain more precise weighing results.

Acknowledgements. The research was supported by The National Centre for Research and Development (NCBR) grant number LIDER/18/0064/L-7/15/NCBR/2016. The authors would like to thank the GDDKiA in Poland and APM Pro for providing the reference data sets from existing WIM system.

References

1. Fu, G., Hag-Elsafi, O.: Vehicular overloads: load model, bridge safety, and permit checking. J. Bridge Eng. **5**(1), 49–57 (2000)
2. Jacob, B., Beaumelle, V.F.L.: Improving truck safety: potential of weigh-in-motion technology. IATSS Res. **34**(1), 9–15 (2010)
3. Richardson, J., Jones, S., Brown, A., et al.: On the use of bridge weigh-in-motion for overweight truck enforcement. Int. J. Heavy Veh. Syst. **21**(2), 83–104 (2014)
4. Qin, T., Lin, M., Cao, M., Fu, K., Ding, R.: Effects of sensor location on dynamic load estimation in weigh-in-motion system. Sensors (Basel) **18**(9), 3044 (2018)
5. Yu, Y., Cai, C.S., Deng, L.: State-of-the-art review on bridge weigh-in-motion technology. Adv. Struct. Eng. 1–17. https://doi.org/10.1177/1369433216655922.
6. Lydon, M., Robinson, D., Taylor, S.E., et al.: J. Civil Struct. Health Monit. **7**, 325 (2017). https://doi.org/10.1007/s13349-017-0229-4

7. Zhang, W., Suo, C., Wang, Q.: A novel sensor system for measuring wheel loads of vehicles on highways. Sensors **8**, 7671–7689 (2008). https://doi.org/10.3390/s8127671

8. Darestani, M.Y., Thambiratnam, D.P., Nataatmadja, A.: Experimental study on structural response of rigid pavements under moving truck load. In: Proceedings of the 22th ARRB Conference-Research into Practice, Canberra, Australia. 29 October-2 November 2006

9. Darestani, M.Y., Thambiratnam, D.P., Nataatmadja, A., Baweja, D.: Structural response of concrete pavements under moving truck loads. J. Trans. Eng. **133**, 670–676 (2007). https://doi.org/10.1061/(ASCE)0733-947X(2007)133:12(670)

10. Xue, W., Weaver, E., Wang, L., Wang, Y.: Influence of tyre inflation pressure on measured pavement strain responses and predicted distresses. Road Mater. Pavement Des. **17**, 328–344 (2016). https://doi.org/10.1080/14680629.2015.1080180

11. Zhang, Z., Huang, Y., Bridgelall, R., Lu, P.: Optimal system design for weigh-in-motion measurements using in-pavement strain sensors. IEEE Sens. J. **17**, 7677–7684 (2017). https://doi.org/10.1109/JSEN.2017.2702597

12. Lombaert, G., Degrande, G.: The experimental validation of a numerical model for the prediction of the vibrations in the free field produced by road traffic. J. Sound Vibr. **262**, 309–331 (2003)

13. Profaska, M., Gora, M.: Badania emisji drgan z ciagu komunikacyjnego-studium przypadku. Systemy Wspomagania w Inzynierii Produkcji (2013, polish)

14. Hostettler, R., Birk, W.: In analysis of the adaptive threshold vehicle detection algorithm applied to traffic vibrations. In: Proceedings of the 18th IFAC World Congress, Italy, Milano, pp. 100–105 (2011)

15. Ma, W., et al.: A wireless accelerometer-based automatic vehicle classification prototype system. IEEE Trans. Intell. Transp. Syst. **15**, 104–111 (2014)

16. Hostettler, R., Birk, W., Nordenvaad, M.L.: Feasibility of road vibrations-based vehicle property sensing. IET Intell. Transp. Syst. **4**, 356–364 (2010)

17. Rivas, J., Wunderlich, R., Heinen, S.J.: Road vibrations as a source to detect the presence and speed of vehicles. IEEE Sens. J. **17**, 377–385 (2017)

18. Ghosh, R., Akula, A., Kumar, S., Sardana, H.: Time-frequency analysis based robust vehicle detection using seismic sensor. J. Sound Vibr. **346**, 424–434 (2015)

19. Bernas, M., Płaczek, B., Korski, W., Loska, P., Smyła, J., Szymała, P.: A survey and comparison of low-cost sensing technologies for road traffic monitoring. Sensors **18**(10), 3243 (2018)

20. Bernas, M., Płaczek, B.: Fully connected neural networks ensemble with signal strength clustering for indoor localization in wireless sensor networks. Int. J. Distrib. Sens. Netw. **11**(12), 403242 (2015)

21. Berthold, M.R., et al.: KNIME-the Konstanz information miner: version 2.0 and beyond. AcM SIGKDD Explor. Newsl. **11**, 26–31 (2009)

Optimal Spectrum Sharing in RF Energy Harvesting Cognitive Femtocell Networks

Jerzy Martyna[✉]

Institute of Computer Science, Faculty of Mathematics and Computer Science,
Jagiellonian University, ul. Prof. S. Lojasiewicza 6, 30-348 Cracow, Poland
jerzy.martyna@uj.edu.pl

Abstract. A cognitive femtocell network is considered as a highly promising solution to tackle communications in indoor environments. Currently, the spectrum allocated in the femtocells is a licensed macrocell spectrum and is additionally provided by the mobile operator. The allocated spectrum is subject to interference, as well as limited by the possibilities of femtocells. In this paper, the model of dynamic spectrum sharing in cognitive femtocell network with RF energy harvesting is proposed. Thus, the cognitive radios are able to access licensed bands not only from macrocells but also from other licensed systems. Additionally, they gain RF energy from the licensed transmitters, which makes them extremely important for minimizing energy consumption. The obtained simulation results confirm that the presented model and its solution are able to achieve greater capacity and energy savings than the normally employed femtocell networks.

Keywords: Cognitive radio · Femtocell networks ·
Radio frequency energy harvesting · Spectrum sharing · Power control

1 Introduction

Cognitive radios (CRs) [1] are considered as a key technology for future wireless communications and mobile computing. A computer network built with their use, called cognitive radio networks (CRNs), is intelligent and can automatically sense the environment. It allows this network to adapt the communication parameters to the environment. The CRN may have the ability to optimize a waveform to one or many criteria. For example, the radio may be able to optimize for data rate, for packet success rate, for service cost, for battery power minimization, etc. The main purposes for the use of a CRN is usage of frequency bands that are owned by their licensed users. Therefore, one of the most significant requirements of CR radio is that the interference caused by cognitive devices to licensed users remains at a negligible level.

The CR system enabled femtocells are able to access spectrum bands not only from macrocells but also from other licensed systems (e.g. TV systems) provided

P. Gaj et al. (Eds.): CN 2019, CCIS 1039, pp. 220–232, 2019.
https://doi.org/10.1007/978-3-030-21952-9_17

the interference from femtocells to the existing systems. They are called cognitive femtocell radio networks [2,3]. These networks could achieve much higher capacity than the femtocell networks which do not employ agile spectrum access. Moreover, these networks can avoid interference to macrocell and the other femtocell user equipment. The secondary users (SUs) are allowed to transmit their data concurrently with primary users (PUs), which is possible thanks to the cooperation of SUs by deferring their transmissions when PUs are busy. It allows users to operate in licensed bands without license.

Radio frequency (RF) energy harvesting has emerged as a promising technique to supply energy to wireless networks [4]. The RF energy harvesting does not depend on the sun, wind or sea wave motion, etc., which are renewable energy sources. This is particularly important for cognitive networks which, if their nodes are equipped with RF energy harvesting capability, can receive RF energy from licensed radio transmitters. This can significantly extend the lifetime of such a network. The technology is expected to emerge in the near future, which will increase the use of RF energy.

Most of the devices used for RF energy harvesting in CRNs take energy from a single band. Recently, devices have also been designed for those networks that use either single antenna using a dual-band circuit using single antenna [5] or multiple antennas [6] for energy consumption. Moreover, a multiband RF energy harvesting system with triple antennas [7] has been developed. This means that such devices can simultaneously receive energy from radio waves and work in the CRN.

There is some research studying different issues in RF energy harvesting in CRNs. A Markov decision process to obtain an optimal channel selection to maximise the long-term average throughput of the SU for a multiple-channel in the CRNs has been presented by X. Lu et al. [8]. In the paper presented by D. T. Hoang et al. [9] an online method that allows an optimal solution to be found to maximise average throughput of the SU when the SU has no information about the channels. In the paper by W. Wang et al. [10] an optimal power allocation for underlay-based CRNs with PU's statistical delay QoS provisioning is presented. A Stackelberg game for distributed time scheduling in RF-powered backscatter CRNs was introduced by W. Wang et al. [11]. Despite this, the issue of RF harvesting in cognitive femtocell networks has not been studied. In the near future, such a technology will be widely developed that will increase the use of RF energy.

The main purpose of this paper is to develop a model of spectrum sharing in. The second goal is to maximize the capacity of the cognitive femtocell network with RF energy harvesting, and furthermore, to maximize the harvested RF energy for each RF band. In addition, the article presents the sharing spectrum algorithm for acquiring RF energy in cognitive femtocell networks.

The rest of the paper is as follows. In the Sect. 2 the model of the cognitive femtocell system with RF energy harvesting is presented. Section 3 provides a model of spectrum sharing in the cognitive femtocell network with RF energy harvesting. Section 4 presents spectrum sharing problem. In Sect. 5 an algorithm

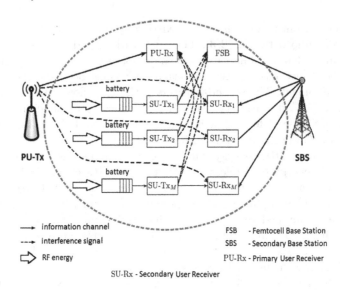

Fig. 1. System model of cognitive femtocell network with RF energy harvesting.

for the dynamic spectrum sharing in femtocell network with RF energy harvesting is given. The performance of the proposed scheme for spectrum sharing in femtocell network with RF energy harvesting is presented. The conclusion ends this paper.

2   System Model

In this section, the system model is presented in detail. For simplicity, it concerns a single cognitive femtocell system.

Let the system be created by a single PU receiver (PU-Rx) and M SUs (see Fig. 1). Each SU has an RF energy harvesting (EH) device. It is assumed that both primary and secondary networks use the OFDMA scheme. Additionally, it is assumed that one subcarrier can only be used by one SU at each time slot, but each SU can use multiple subcarriers at each time slot. For simplicity, the interference between individual SUs is omitted.

It is assumed that the Rayleigh fading channel can be modeled by a two-state Markov chain (see Fig. 2) [12]. It can be seen that the state B states active PU, while the state F denotes that the PU is inactive. For a time slot $k, k \in \{1, \ldots, K\}$ it is possible to define the state of the n-th channel, namely

$$x_k^n = \begin{cases} 1, & \text{if the channel is in the state } B \\ 0, & \text{if the channel is in the state } F \end{cases} \tag{1}$$

Let N be the number of subcarriers in the femtocell and L be the number of subcarriers occupied by PU receiver. Thus, the number of random subcarrier state can be expressed by

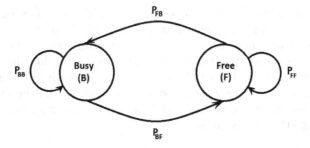

Fig. 2. Markov channel model [12].

$$I = \binom{N-M}{L} \tag{2}$$

Let Q_k be the set of states of all channels at the k-th time slot. Then states of all channels available in cognitive femtocell at the time slot k can be given by

$$y_k = \{x_k^1, x_k^2, \ldots, x_k^N\}, \quad i \in \{1, 2, \ldots, K\} \tag{3}$$

For the transition matrix of PU receiver is defined the occupation state as $\mathbf{P}^o$. The state transition probability of the n-th channel can be given by

$$P_{BF}^n = \Pr\{x_{k+1}^n = 0 \mid x_k^n = 1\} \tag{4}$$

$$P_{FB}^n = \Pr\{x_{k+1}^n = 1 \mid x_k^n = 0\} \tag{5}$$

The transition probability of $\mathbf{P}^o$ can be described as follows

$$p_{ij}^o = \Pr\{O_{k+1} = y_j \mid O_k = y_i\} \tag{6}$$

After transformation

$$p_{ij}^o = \prod_{n=1}^{N} \Pr\{x_{k+1}^n \mid x_k^n\} \tag{7}$$

3 Model of Cognitive Femtocell with RF Energy Harvesting

RF harvesting in wireless networks is a technology making these networks self-sufficient. But cognitive radio additionally allows for additional efficiency increase in energy harvesting, which results from the use of active licensed devices being sources of energy in these networks.

Each SU has a built-in power conversion circuit that can extract DC power from the received electromagnetic waves [13]. For each SU the harvesting zone, which is a disk with radius r_h centered at each PU-Tx with the radius r_P, is determined. The radius r_h is determined by the sensitivity of circuit of RF energy harvesting for a given transmission power level of PU-Tx as P_P.

It is assumed here that each SU within a harvesting zone inside the femtocell can obtain energy larger than the energy harvesting threshold, which is denoted by $P_P r_h$, where $\alpha > 2$ is the path-loss exponent. The received energy power by a SU outside any harvesting zone is too small to activate the energy harvesting circuit which means it can be omitted. The probability mass function (PMF) inside the disk SU(Y, r_h) is given by

$$\Pr\{W = w\} = e^{-\pi r_h^2 \lambda_p} \frac{(\pi r_h^2 \lambda_p)^w}{w!}, \quad w = 0, 1, 2, \ldots \tag{8}$$

The probability that the SU lies in femtocell within radio range of PUs-Tx, p_h, is given by

$$p_h = \Pr\{SU \in \mathcal{F}\} \tag{9}$$
$$= \Pr\{W \geq 1\} \tag{10}$$
$$= \sum_{w=1}^{\infty} e^{-\pi r_h^2 \lambda_p} \frac{(\pi r_h^2 \lambda_p)^w}{w!} \tag{11}$$
$$= 1 - e^{-\pi r_h^2 \lambda_p} \tag{12}$$

In practice, values λ_p and r_h are both small. Thus, it is assumed that $\pi r_h^2 \lambda_p \ll 1$. This allows approximation of Eq. (12) by ignoring the higher-order terms with $w > 1$. It indicates that if SU is inside the harvesting zone of one single PU-Tx most probably most probably, which equivalently means that the harvesting zones of different PUs-Tx do not overlap at most time. Thus, the amount of average power harvested by SU in femtocell in a time slot can be lower-bounded by $\eta p_p R^{-\alpha}$, where $R \leq r_h$ indicates the distance between SU and and its nearest PU-Tx, η is the harvesting efficiency.

It is assumed that each SU transmitter has energy queue. This means that the harvested battery is of finite capacity. Let packets composed of harvested energy be saved as E_k^m, $m \in \{1, \ldots, M\}$. It is assumed that the process of energy-harvesting is denoted as the Poisson process with mean e_λ [14]. Thus, the probability density function can be given by

$$Pr(E_k^m = e_i) = e^{e_\lambda} \frac{(e_\lambda)^i}{i!}, \quad i = 1, \ldots, J \tag{13}$$

It is necessary to define the remaining battery energy for each m-th SU in the k-th time slot. It can be assumed that the battery energy is available at the next slot $k + 1$ in the n-th channel can be defined as:

$$B_{k+1}^m = \min\{B_k^m - p_k^{m,n} T + E_k^m, B_{max}\}, \quad k \in \{1, \ldots, K\}, \; m \in \{1, \ldots, M\} \tag{14}$$

where $p_k^{m,n}$ is the transmission power allocated in the m-th SU in the n-th channel at time slot k, T denotes the duration of one time slot and B_{max} denotes the maximum energy battery capacity.

For a cognitive femtocell signal-to-interference-plus-noise ratio (SINR) $SINR_k^{m,n}$ of the m-th SU in the n-th channel at the time slot k can be defined as

$$SINR_k^{m,n} = \frac{h_k^{m,n} p_k^{m,n} \cdot x(k,m,n,\omega)}{\sum_{\omega=1}^{\Omega} h_k^{\omega,n} p_k^{\omega,n} + \sum_{j=1,j\neq m}^{M} h_k^{j,n} p_k^{j,n} + h_k^{F,n} p_k^{F,n} + \sigma^2},$$
$$k \in \{1,\dots,K\}, \quad j,m \in \{1,\dots,M\}, \quad n \in \{1,\dots,N\} \quad (15)$$

where $p_k^{m,n}$ and $p_k^{F,n}$ denote the transmission power allocated in the m-th SU in the n-th channel at time slot k of the ω PU-Tx and femtocell BS (FBS), respectively, at the time slot k; $h_k^{m,n}$ is the channel coefficient at the m-th SU in the n-th channel, $x(k,m,n)$ is a binary indicator. If $x(k,m,n,\omega) = 1$, the m-th SU in femtocell ω works on channel n in the k-th slot; zero otherwise. $h_k^{\omega,n}$, $h_k^{F,n}$ are the channel coefficients at the ω-th PU-Tr and FSB, σ^2 denotes the background noise power.

The downlink capacity of the m-th SU in cognitive femtocell ω can be presented as follows:

$$C^{down}(\omega,m,n) = x(k,m,n,\omega)B \cdot log_2(1 + SINR_k^{(m,n)}),$$
$$k \in \{1,\dots,K\}, m \in \{1,\dots,M\}, \omega \in \{1,\dots,\Omega\} \quad (16)$$

where B denotes the channel bandwidth. Thus, the downlink capacity of cognitive femtocell ω is given by:

$$C(\omega,n) = \sum_{m=1}^{M} C^{down}(\omega,m,n), \quad m \in \{1,\dots,M\} \quad (17)$$

The spectrum sharing problem in cognitive femtocell network is to maximise the downlink capacity of all femtocell BSs while guarantee the channel and power constraints, namely

$$\max \sum_{n=1}^{N} \sum_{\omega=1}^{\Omega} \left(C(\omega,n) + E^H(\omega,n)\right) \quad (18)$$

subject to

$$x(k,m,n,\omega) \in \{0,1\}, \quad \forall k \in \{0,\dots,K-1\}, \quad \forall m \in \{1,\dots,M\},$$
$$\forall n \in \{1,\dots,N\}, \quad \forall \omega \in \{1,\dots,\Omega\} \quad (19)$$

$$\sum_{k=0}^{K-1} \left(\sum_{m=1}^{M} \sum_{n=1}^{N} \left(\sum_{\omega=1}^{\Omega} h_k^{\omega,n} p_k^{\omega,n} + \sum_{j=1}^{M} h_k^{j,n} p_k^{j,n} + h_k^{F,n} p_k^{F,n} \right) \right) + \sigma^2 \leq I^{TH}$$
$$\forall k \in \{0,1,\dots,K-1\} \quad (20)$$

$$\sum_{n=1}^{N} p_k^{m,n} \leq \frac{B_k^m}{T}, \quad k \in \{0, 1, \ldots, K-1\} \tag{21}$$

$$p_k^{m,n} \geq 0, \quad \forall m \in \{1, \ldots, M\}, \quad \forall k \in \{0, 1, \ldots, K-1\}, \quad \forall n \in \{1, \ldots, N\} \tag{22}$$

$$R_k^m \geq R_{min}^m, \quad k \in \{0, 1, \ldots, K-1\}, \quad \forall m \in \{1, \ldots, M\} \tag{23}$$

where E_ω^H is the RF energy harvesting rate (in watts) by the ω SU from the RF transmitter in a fading channel and is given by [12]:

$$E^H(\omega, n) = \frac{\tau \beta P_S g_m}{d_m^\alpha} \tag{24}$$

where β is the RF-to-DC power conversion efficiency of the device, P_S is the transmit power of the PU transmitter, α is the path-loss exponent.

4   Spectrum Sharing Problem

Notice that the formulated optimisation problem in Eqs. (18–23) is a mixed-integer nonlinear problem (MINLP). For simplifying the problem is proposed here a scenario concerning the worst case when all SUs are at the cell boundary and have a similar chain gain. Thus, the FBBS does not distinguish among its SUs users in its distance range. Then, the downlink SINR of channel is given by

$$SINR^{down}(\omega, n) = \frac{H(d_\omega, n)P(\omega, n)}{N_o + I_s(\omega, n)} \tag{25}$$

where $H(d_\omega, n)$ defines the gain on channel n in the femtocell ω through the distance d_ω. $P(\omega, n)$ denotes the transmission power for the ω-th FBS on channel n, $I_s(\omega, n)$ denotes the interference on channel n to all SUs within femtocell ω. Thus, the downlink capacity of femtocell ω can be obtained as

$$C_\omega = n_\omega B \sum_{n=1}^{N} x_B(\omega, c) \log_2 \left(1 + SINR^{down}(\omega, n)\right) \tag{26}$$

where $x_B(\omega, n)$ is a binary variable, whereby $x_B(\omega, n) = 1$ represents that channel n is allocated to femtocell ω, zero otherwise. The inter-cell spectrum sharing problem can be formulated as follows:

$$\max_{x_B(\omega, n) \in \{0, 1\}, P(\omega, n) \geq 0} \sum_{n=1}^{N} \left(C(\omega, n) + E^H(\omega, n)\right) \tag{27}$$

subject to:

$$\sum_{n=1}^{N} x_B(\omega, n) = n_\omega, \quad \omega \in \{1, \ldots, \Omega\} \tag{28}$$

$$SINR^{down}(\omega, n) \geq x_B(\omega, n) \cdot SINR_{min}^{down} \tag{29}$$

$$\sum_{n=1}^{N} P(\omega, n) \le P_{max} \tag{30}$$

$$\sum_{n=1}^{N} E^{H}(\omega, n) \le E^{H}_{max} \tag{31}$$

Constraint given by Eq. (28) represents the total number of channels can be used in one femtocell in one femtocell and is equal to the number of users in that femtocell. Constraint given by Eq. (29) means that the channel n is allocated to SU m in femtocell ω for downlink transmission, the SINR received on the m-th SU should be higher than a defined threshold. allocated to femtocell ω. Constraint given by Eq. (30) means the total transmission power of all transmission channels can not exceed the maximum power P_{max}. Constraint given by Eq. (31) indicates that the total harvested energy on channels can not exceed the maximum energy battery capacity.

4.1 Optimal Spectrum Sharing in Cognitive Femtocell Network with RF Energy Harvesting

Optimal spectrum sharing in cognitive femtocell networks with RF energy harvesting belongs to multi-objective optimisation and has an infinite number of solutions. The maximisation of both objective functions, i.e. capacity and energy harvesting, requires their definition of thr relative weight of the objectives. If both have the same importance, then the following function can be minimised

$$F(\omega, n) = (C(\omega, n) - K)^2 + (E^{H}(\omega, n) - T)^2 \tag{32}$$

where K and T are the known maximal values of downlink capacity and energy harvesting, respectively.

To find the maximum value of $F(\omega, n)$ can be used function values of F at a given point (ω, n). This is know as derivative-free optimisation [15]. The most of these use approximate derivatives based on finite differences or derivatives of interpolating functions. Then, it can be constructed for a point $p^k = (\omega^k, n^k)$, $k = 0, 1, 2, \ldots$, a local quadratic model, namely

$$m_k(p^k + d) = F(p^k) + (y^k)^T d + \frac{1}{2} d^T \mathbf{H}^k d \tag{33}$$

In the above equation, it is assumed that $p^{k+1} = p^k + d^k$ an minimised d^k. Then, for a small (but not too small) $\epsilon > 0$ it is obtained:

$$g^k = (g_1, g_2)^T, \quad g_i = \frac{F(p^k + \epsilon e_i) - F(p^k - \epsilon e_i)}{2\epsilon^2} \tag{34}$$

with $e_1 = (1, 0)^T$ and $e_2 = (0, 1)^T$, is the approximate gradient and the Hessian matrix, $\mathbf{H}^k$, is given as follows:

$$\mathbf{H}^k = \begin{pmatrix} h_{11} & h_{12} \\ h_{21} & h_{22} \end{pmatrix}, \quad h_{ij} = \frac{F(p^k + \epsilon e_i + \epsilon e_j) - F(p^k) - F(p^k + \epsilon e_j) + F(p^k)}{\epsilon^2}$$

$$\tag{35}$$

is a Taylor approximation of the Hessian.

5 Heuristic Algorithm for Power Allocation Ib Cognitive Femtocell Network with RF Energy Harvesting

The algorithm for power allocation presented here maximizes the downlink capacity of all SUs, while maintaining guaranteed interference from the PU-Tx below a threshold value. Its affect is limited to a finite time interval. It was assumed that the algorithm uses a reward function. It has been assumed here that the reward function is defined as maximum of the sum downlink capacity at the future time slot from the current system state. The capacity of the m SU at the time slot k can be presented as follows:

$$C_k^m = \sum_{n=1}^{N} \log_2(1 + g_k^{m,n} \cdot p_k^{m,n}), \quad m \in \{1, \ldots, M\} \tag{36}$$

where $g_k^{m,n}$ means the channel gain distribution of the m-th SU in the subcarrier-occupied state of time slot k in the n-th channel.

It is assumed that the selected n channels of BS of femtocell will be allocated to its SU after maximising the energy value. This is due to the reception that the current reward at time slot k is a function of the current energy budget B_k^m and the current system state S_k of each SU, namely

$$V_k(B_k^1, B_k^2, \ldots, B_k^M; S_k) = \max_{p_k^{m,n}} E\{\sum_{v=k}^{K-1} \sum_{m=1}^{M} \sum_{n=1}^{N} log_2(1 + g_k^{m,n} p_k^{m,n})\}$$

$$k \in \{1, \ldots, K\} \tag{37}$$

Details of the algorithm are provided in Fig. 3.

6 Simulation Results

This part will present the results of cognitive femtocell network simulation research.

It was assumed that the simulation model consists of a single femtocell with BS and a single PU-Rx inside the femtocell. The maximum transmission power for femtocell BS is $10\,$dBm [16]. It was assumed that the number of available channels is 20. The path loss is equal to d^{-2}, where d is the distance between a transmitter and its receiver. Derivative-free optimization was carried out using the NEWUOA software [17].

Figure 4 shows the dependence of average capacity per femtocell versus the energy budget. For comparison, the optimal allocation algorithm in the cognitive network with energy harvesting [18]. It can be seen from the figure that the total energy budget is much higher for the optimal algorithm than the presented algorithm.

Figure 5 shows the total average of harvested energy versus the number of channels of primary channel. As illustrated by increasing the number of channels

---

Algorithm 1 Power allocation in CRFN system for energy harvesting

---

1: **procedure** PA IN CFN FOR ENERGY HARVESTING
2:     **for** $m \leftarrow 1, M$ **do**
3:         **if** $\sum_{n=1}^{N} \sum_{m=1}^{M} (a_k^{m,n} p_k^{P,n} + b_k^{m,n} p_k^{F,n}) \leq I^{TH}$ **then**
4:             **while** $\sum_{n=1}^{N} p_k^{m,n} \leq \frac{B_k^m}{T}$ **do**
5:                 **for** $k \leftarrow 0, K-1$ **do**
6:                     **for** $k \leftarrow K-1, 0$ **do**
7:                         $V_k(B_k^1, B_k^2, \ldots, B_k^M; S_k)$
8:                         $= \max_{p_k^{m,n}} E\{\sum_{v=k}^{K-1} \sum_{n=1}^{N} \sum_{m=1}^{M} \log_2(1 + g_k^{m,n} p_k^{m,n})\}$
9:                     **end for**
10:                 **end for**
11:                 *Energy harvesting for the $m-th$ SU*
12:             **end while**
13:         **end if**
14:         **if** $\sum_{n=1}^{N} p_k^{m,n} > \frac{B_k^m}{T}$ **then**
15:             *Power allocation for the $m-th$ SU*
16:         **end if**
17:     **end for**
18: **end procedure**

Fig. 3. Algorithm for energy harvesting in cognitive femtocell network.

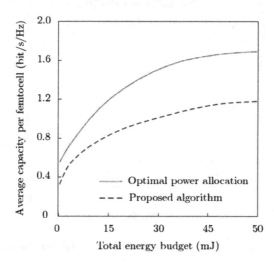

Fig. 4. Average capacity per femtocell versus total energy budget.

that the SUs are capable to energy harvest, the total average of harvested energy of the SUs increase.

Figure 6 shows the average capacity per femtocell versus the number of slots. In this case the proposed algorithm gives a slightly lower values of the average capacity per femtocell than the optimization method used here.

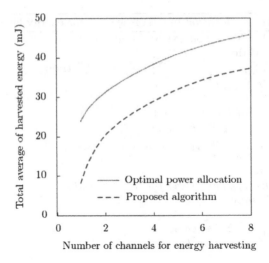

Fig. 5. Total average harvested energy versus the number of channels of primary channels.

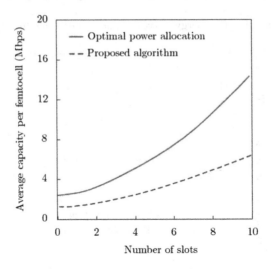

Fig. 6. Average capacity per femtocell versus the number of slots.

7 Conclusion

This article investigated the spectrum sharing problem in downlink in CRFN networks with RF energy harvesting. This problem was formulated as a mixed-integer problem. A multi-objective optimization with simultaneous maximisation of two functions without available derivatives. A free-derivative optimisation has been proposed. Alternatively, a distributed scheme was proposed with low complexity to jointly select downlink channels and allocate the transmission power for each channel with simultaneous RF energy harvesting by batteries of SUs in each femtocell. The numerical results obtained using simulation and the NEWUOA software show that the proposed method of simultaneous channel allocation method and RF energy harvesting can be successfully used in the CRFN networks.

References

1. Mitola III, J., Maguire Jr., G.: Cognitive radio: making software radios more personal. IEEE Pers. Commun. **6**(4), 13–18 (1999)
2. Xiang, J., Zhang, Y., Skeie, T., Xie, L.: Downlink spectrum sharing for cognitive radio femtocell networks. EEE Syst. J. **4**(4), 524–534 (2010)
3. Huang, L., Zhu, G., Zhong, H., Du, X.: Cognitive femtocell networks: an opportunistic spectrum access for future indoor wireless coverage. IEEE Wirel. Commun. **20**(2), 44–51 (2013)
4. Hasan, Z., Boostanimehr, H., Bhargava, V.R.: Green cellular networks: a survey some research issues and challenges. IEEE Commun. Surv. Tutorials **13**(4), 524–540 (2011)
5. Sun, H., Guo, Y.X., He, M., Zhong, Z.: A dual-band rectenna using broadband yagi antenna array for ambient RF power harvesting. IEEE Antenna Wirel. Propag. Lett. **12**, 918–921 (2013)
6. Kuhn, V., Lahuec, C., Seguin, F., Person, C.: A multi-band stacked RF energy harvester with RF-to-DC efficiency up to 84%. IEEE Trans. Microw. Theory Tech. **63**(5), 1768–1778 (2015)
7. Keyrouz, S., Visser, H.J., Tijhuis, A.G.: Multi-band simultaneous radio frequency energy harvesting. In: Proceedings of the 7th European Conference on Antennas and Propagation (EuCAP), Gothenburg, Sweden, pp. 3058–3061 (2013)
8. Lu, X., Wang, P., Niyato, D., Hossain, E.: Dynamic spectrum access in cognitive radio networks with RF energy harvesting. IEEE Wirel. Commun. **21**(3), 102–110 (2014)
9. Hoang, D.T., Niyato, D., Wang, P., Kim, D.I.: Opportunistic channel access and RF energy harvesting in cognitive radio networks. IEEE J. Sel. Areas Commun. **32**(11), 2039–2052 (2014)
10. Wang, Y., Ren, P., Du, Q., Sun, L.: Optimal power allocation for underlay-based cognitive radio networks with primary user's statistical delay QoS provisioning. IEEE Trans. Wirel. Commun. **14**(12), 6896–6910 (2015)
11. Wang, W., Hoang, D.T., Niyato, D., Wang, P., Kim, D.I.: Stackelberg game for distributed time scheduling in RF-powered backscatter cognitive radio networks. IEEE Trans. Wirel. Commun. **17**(8), 5606–5622 (2018)
12. Zhang, Q., Kassam, S.A.: Finite-state markov model for rayleigh fading channels. IEEE Trans. Commun. **47**(11), 1688–1692 (1999)
13. Ostaffe, H.: Power Out of Thin Air: Ambient RF Energy Harvesting for Wireless Sensors (2010). http://powercastco.com/PDF/Power-Out-of-Thin-Air.pdf
14. Usman, M., Koo, I.: Access strategy for hybrid underlay-overlay cognitive radios with energy harvesting. IEEE Sens. J. **14**(9), 3164–3173 (2014)
15. Conn, A.R., Scheinberg, K., Vicente, L.N.: Introduction to Derivative-free Optimization. SIAM-MPS, Philadelphia (2009)

16. Lopez-Perez, D., Valcarce, A., De La Roche, G., Liu, E., Zhang, J.: Access methods to wimax femtocells: a downlink system-level case study. In: 11th IEEE Singapore International Conference on Communication Systems (ICSS 2008), November 2008, pp. 1657–1662 (2008)
17. NEWUOA. http://mat.uc.pt/zhang/software.html#newuoa
18. Ho, C.K., Zhang, R.: Optimal energy allocation for wireless communications with energy harvesting constraints. IEEE Trans. Sign. Proc. **60**(9), 4808–4818 (2012)

Accuracy of UWB Location Tracking Devices When on the Move

Konrad Połys[1] , Krzysztof Grochla[1] , Artur Frankiewicz[2],
and Michał Gorawski[1](✉)

[1] Institute of Theoretical and Applied Informatics, Polish Academy of Science,
Bałtycka 5, 44-100 Gliwice, Poland
{kpolys,kgrochla,mgorawski}@iitis.pl
[2] AIUT sp z o.o., Wyczółkowskiego 113, 44-109 Gliwice, Poland
afrankiewicz@aiut.com
http://www.iitis.pl

Abstract. The Ultra-wideband location technology provides high accuracy of indoor positioning. However there is still little work on the accuracy of this solution, especially when monitoring moving objects. The paper shows the results of the measurement of the UWB indoor positioning system accuracy in stationary conditions and shows how the accuracy changes when the monitored object is on the move. It is also measured how the orientation of the antenna influences the accuracy. We show, that while the UWB provides very small error in the stationary scenario (within few centimeters), the accuracy significantly decreases when the device is moving and the average error grows to up to 20 cm. The experimental probability distribution function for different measurement scenarios is presented, showing how the location estimation variates in time and how this is correlated with the actual position of the device.

Keywords: UWB · Indoor positioning · IPS accuracy

1 Introduction and Motivation

The indoor positioning is a function needed in multiple applications, from indoor navigation for people and devices, to providing information services basing on the location of users. There are many implementations of this concept which had been developing in previous years in various fields of science, industry and critically important services e.g. military, fire department, police force, medical emergency services etc. The ability to pinpoint the position of soldiers, policemen, or fire-fighters inside the building they are operating gives invaluable data about their performance and also sends a constant feedback to external command center. Examples of civilian application like: the monitoring of people movement in museums, art galleries etc. to gather data about their movement

This research was partially funded by Polish National Center for Research and Development grant number POIR.04.01.04-00-0005/17.

P. Gaj et al. (Eds.): CN 2019, CCIS 1039, pp. 233–241, 2019.
https://doi.org/10.1007/978-3-030-21952-9_18

patterns and to monitor their position in the case of emergency or monitoring of vehicles and other mobile equipment in the industrial buildings. With the rise of new technologies and production of numerous IoT devices with sensing capabilities, the need for indoor positioning systems (IPS) grows.

There are multiple technologies used in IPS, such as infra-red, magnetism, radio waves, sound and other sensing data to pinpoint the object location and track its movement. The main problem is to choose the positioning technology that is adequate for a certain usage, taking into consideration the need for accuracy, availability, coverage area, scalability, cost, privacy and energy consumption [2]. The mostly used techniques are:

1. Zigbee
   Zigbee is a good choice for a short distance, low power and low throughput wireless communication, that can create stable, protected multi-hop mesh network to send short messages with low energy usage. It can be used for localization purposes by measuring signal strength and basing on this estimating the distance between its nodes [7].

2. Ultra Wideband (UWB)
   UWB is a technique with low-power, short range and high bandwidth. It bases on quick sending of short impulses lasting several picoseconds. It does not requite line of sight, and because of its low-power characteristics it does not interfere with other wireless stems. In theory it provides a high accuracy and is considered one of the most desirable and reliable means of indoor positioning nowadays [1, 11].

3. Radio Frequency Identification (RFID)
   RFID can be used to monitor almost any object with the RFID tag applied. The tags can be passive (without the power source, collects energy from a reader) or active (with power source, e.g. battery). Although the technology is quite universal and cheap (due to the use of passive tags), the position it gives is rather an information about proximity to the reader, not the adequate position in space

4. WiFi
   The WiFi was not designed as a positioning technology, however there are methods that can be used that gives object position basing on signal level [8], radio waves proximity detection or fingerprinting. The main pro of such approach is its low cost, as most indoor location already have WiFi infrastructure that can be use as a positioning system. The main drawback is low accuracy of a radio waves proximity detection positioning. Although the fingerprinting method can be more exact, it requires the precise reference measurements.

5. Bluetooth
   A wireless personal area network (WPAN). It can be used as a localization device by using the Bluetooth beacons and, triangulation of a mobile device (e.g. smartphone).

6. Ultrasound
   Ultrasound positioning systems use ultrasonic waves and the time of arrival as location means. The system uses smart devices, tags or badges to monitor

position. It is relatively low energy solution, however ultrasonic waves cannot penetrate walls, so highly obstructed environment will requite a lot of receivers to cover highly obstructed spaces, it can also be distracted by high frequency sounds, reflections and interference [15] and it is costly [6,12].

In comparison to the technologies such as WiFi, Bluetooth (BLE) the UWB provides higher accuracy. UWB is currently providing best balance between the low cost and high accuracy, especially indoors, where GPS may not be used. Although the UWB systems are less accurate than e.g ultrasound but it excels in range and coverage and can easily penetrate walls. Moreover it is not bound by line-of-sight, what is quite useful when dealing with large, cluttered industrial spaces like warehouses or production halls. Thus in this paper we concentrate on the analysis of UWB accuracy.

There are several commercial producers of the UWB systems, in [9,14] authors compare Ubisence, Bespoon and Decawave systems and pointing to Decawave as slightly better solution than others in terms of accuracy. We have used the Decawave equipment in our work.

There are multiple papers describing the location techniques and algorithms, such as e.g. [5,9,10], but very few evaluating the actual precision of the location measurements. This is the case especially for the location estimation for moving objects, however there are very few papers providing analysis of the IPS when the monitored object is on the move. In [13] authors analyze the accuracy of UWB in sports activities, showing estimation of the accuracy of the UWB IPS in the specific use case of monitoring the sportsman. Although the precision of UWB is evaluated, there are no full histograms showing the distribution of errors. In this paper we present the research on UWB accuracy and precision in locating a moving object, showing how the distribution of error changes when the device is on the move. We created a testbed environment in a closed space without external interference, we placed the UWB beacon on a moving device and preformed a series of measurements to research its accuracy and error.

The rest of paper is organized as follows: second chapter describes the testbed with used devices and location with four scenarios. Third and fourth chapter is about achieved results in stationary and mobile measurements. Paper is finished with short conclusion.

2   Testbed Description

We have used the DWM1001 Module produced by Decawave [4]. These modules are complete set of hardware which allow to test UWB in different scenarios. Our testbed consists of 5 devices in stationary measurement and 6 devices in moving scenario. Five devices is a minimum because there is a need of four anchors and one device which position is to be measured. Measuring device should be placed inside a region created by the anchors. Moving scenario had a one device more because it was hard to take a live data from device on the move because of wires. This additional device was connected to PC via USB and was sending live data from our measuring device. Testbed was placed on the table (Fig. 2) with dimensions 150 × 400 cm. The anchors was located on the corners of the table.

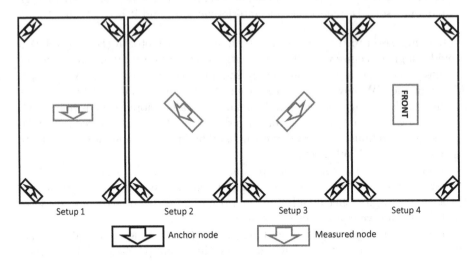

Fig. 1. Simple sketch showing orientation with respect to anchors

2.1 Stationary Measurements

During stationary measurements measuring node was in the middle of both axis. In this scenario four settings were considered. The first one setting was with device parallel to the anchors (Fig. 1, Setup 1). Second and third was with node turned 45° clockwise (Fig. 1, Setup 2) and anticlockwise (Fig. 1, Setup 3) from starting position (first scenario). Last setting is a measure with device laying on its back, with antenna pointing to the ceiling (Fig. 1, Setup 4).

2.2 Mobile Measurements

To conduct mobile measurements we have used a toy train. Dimensions of the train track are about 55 × 120 cm. We put measuring node on the train and record localization data during movement. To compute the accuracy of received position we took the following steps: we measured the exact dimensions of straight and curved pieces of track; we also conducted the series of measurement of speed ratio between straight and curved track; we designed and built an Arduino device, that recorded the moment when train passes defined point on the track (train passed through an end switch that marked full train route) and sent the data to a PC workstation. Having the proportions of moving speed depending on track shape and time periods between pushing end switch, we where able to calculate real position on the track. Lastly we approach the problem of synchronization between calculated and measured localization. We created a C#

Fig. 2. Testbed used for the evaluation of mobile device

application that simultaneously recorded readouts with timestamp from UWB node and the end switch. Mobile measurements were recorded in two setups. First one with track in the middle of both axis and second one with track moved by +162 mm on X axis.

3   Results of Stationary Measurements

We have started by analysis of the error for a stationary conditions, where the UWB device is placed in one location in the middle of 4 anchored modules. We have collected over 342 thousand measurements during approximately 10 h. The accuracy of the measurement was very high - the average error of calculated position was 11 mm and with the variance 42,67 mm^2. The Fig. 3 shows the experiment probability density function (EPDF) which presents on Y axis the probability that a variable will take a value within the interval shown on X axis. For the stationary measurement: over 50% of all measurements had error smaller than 1 cm, and there were no errors higher than 6 cm.

Next we have evaluated how the change of the orientation of antenna influences the accuracy of the measurement. We have rotated the UWB device by 45° clockwise and repeated the measurements. The accuracy was significantly worse - the average error was 89 mm and the variance 1447 mm^2. The Fig. 4 shows that the errors of up to 20 cm were observed.

We have observed that the direction of the rotation have significant influence on the accuracy. When the device was rotated counter-clockwise the average location error was just 25 mm with variance of 267 mm^2. The Fig. 5 shows that the error was not growing above 10 cm and more than 50% of measurement has an error of up to 3 cm.

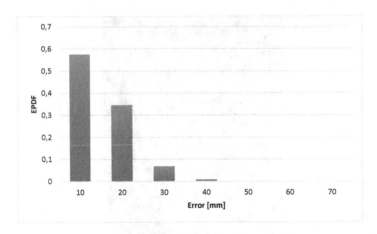

Fig. 3. EPDF of location error for stationary measurements

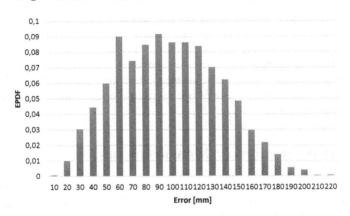

Fig. 4. EPDF of location error for stationary measurements with device shifted by 45°
clockwise.

We believe that the large influence of the device rotation on the errors is caused by the characteristic of the antennas. In theory, the UWB beacons use the omni-directional antennas printed on the circuit boards, but due to the limitations of the placement of the antenna element on one plane, the PCB antennas characteristic is not fully symmetric, as can be seen e.g. in [3]. The UWB device use the measurements of the distance to the 4 anchors, which has larger error when the antennas are placed in angle causing variance of the signal reception. Another source of errors are the deflection of radio signal from walls, which were received with higher level when the antenna was shifted by 45°, because the line of deflection was perpendicular to the PCB plane.

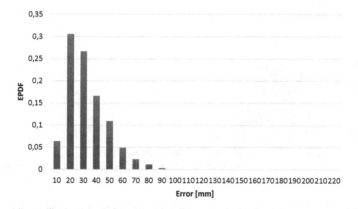

Fig. 5. EPDF of location error for stationary measurements with device shifted by 45° counter-clockwise.

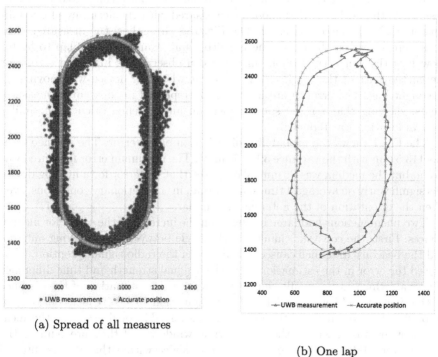

(a) Spread of all measures

(b) One lap

Fig. 6. Spread of measured locations for a moving UWB device.

4   Accuracy of UWB Location for Moving Objects

The next set of experiment considered the analysis of the accuracy of the UWB location measurements for a moving device. The UWB devices allowed to read

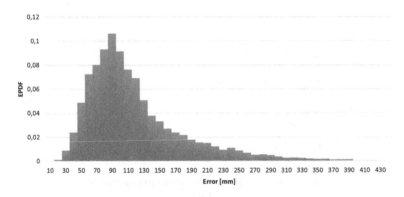

Fig. 7. EPDF of location error for measurements of moving device.

the location every 100 ms. For each measurement we have calculated the exact position of the device at that moment (measured with the accuracy of one millimeter) and calculated the error. The Fig. 6(a) shows how the measured locations were spread in the area. The Fig. 6(b) is an example of one lap to better show how the single track looks like. We can observe the increased errors for the curved part of the track and lower error when the device was moving on a straight line. The errors were in large part repeated in consecutive measurements, showing that for the specific location and antenna orientation similar error in location was reported.

The Fig. 7 shows the probability distribution of the error. The average error was 113,5 mm with the variance 3901,56 mm^2. The maximum error measured was 45 cm, but the median was 96 mm. We can see that the error for a moving device was significantly on average 4 times higher than in the stationary conditions, even when the orientation of the antenna was similar.

Two phenomenons had large influence on the increase of the errors for moving devices. First, the constant changes of the angle between the receiving antenna and the beacons' antennas caused variability in the radio signal reception, what caused the error in the estimation of received signal strength and time difference of arrival, which as the result cause the errors in the estimation of the distance to the anchors, which is used for the triangulation in UWB location system. Secondly, the movement of the device causes an additional error, as the measurement are taken within the 100 ms time window and they are send to the output every 100 ms. The movement of the device within this 100 ms interval adds to the error. But in our tests the speed of movement was relatively small: 0.2 m/s, what introduces an maximum change of position by 2 cm within one measurement window.

5   Conclusions

We have measured the accuracy of the UWB Indoor Positioning System. While the UWB is providing very high accuracy of the location measurements, we show that the accuracy is significantly degraded when the measurement are executed for a device which is on the move. The average error increases from approximately 1 cm for a device located in the optimal location (in the center between the anchored UWB beacons) to 11.4 cm when the device is moving. The orientation of the antenna also significantly influences the error: when the antenna was placed with an angle of 45 or 115° to the anchored devices the error up to 9 times higher comparing to the parallel placement of the antennas.

References

1. Alarifi, A., et al.: Ultra wideband indoor positioning technologies: analysis and recent advances. Sensors **16**(5), 707 (2016)
2. Brena, R.F., et al.: Evolution of indoor positioning technologies: a survey. J. Sensors **2017**, 21 pages (2017)
3. Chen, Z.N., See, T.S., Qing, X.: Small printed ultrawideband antenna with reduced ground plane effect. IEEE Trans. Antennas Propag. **55**(2), 383–388 (2007)
4. Decawave: Decawave MDEK1001 (2019). https://www.decawave.com/product/mdek1001-deployment-kit. Accessed 10 Jan 2019
5. He, S., Chan, S.H.G.: Wi-Fi fingerprint-based indoor positioning: recent advances and comparisons. IEEE Commun. Surv. Tutor. **18**(1), 466–490 (2016)
6. Ijaz, F., Yang, H.K., Ahmad, A.W., Lee, C.: Indoor positioning: a review of indoor ultrasonic positioning systems. In: 2013 15th International Conference on Advanced Communication Technology (ICACT), pp. 1146–1150. IEEE (2013)
7. Islam, T., Rahman, H.A., Syrus, M.A.: Fire detection system with indoor localization using ZigBee based wireless sensor network. In: 2015 International Conference on Informatics, Electronics & Vision (ICIEV), pp. 1–6. IEEE (2015)
8. Jekabsons, G., Kairish, V., Zuravlyov, V.: An analysis of Wi-Fi based indoor positioning accuracy. Sci. J. Riga Tech. Univ. Comput. Sci. **44**(1), 131–137 (2011)
9. Jimenez, A.R., Seco, F.: Comparing Decawave and Bespoon UWB location systems: indoor/outdoor performance analysis. In: IPIN, pp. 1–8 (2016)
10. Karbownik, P., Krukar, G., Shaporova, A., Franke, N., von der Grün, T.: Evaluation of indoor real time localization systems on the UWB based system case. In: 2015 International Conference on Indoor Positioning and Indoor Navigation (2015)
11. Marano, S., Gifford, W.M., Wymeersch, H., Win, M.Z.: NLOS identification and mitigation for localization based on UWB experimental data. IEEE J. Sel. Areas Commun. **28**(7), 1026–1035 (2010)
12. Qi, J., Liu, G.P.: A robust high-accuracy ultrasound indoor positioning system based on a wireless sensor network. Sensors **17**(11), 2554 (2017)
13. Ridolfi, M., et al.: Experimental evaluation of UWB indoor positioning for sport postures. Sensors **18**(1), 168 (2018)
14. Ruiz, A.R.J., Granja, F.S.: Comparing Ubisense, Bespoon, and Decawave UWB location systems: indoor performance analysis. IEEE Trans. Instrum. Meas. **66**(8), 2106–2117 (2017)
15. Sakpere, W., Adeyeye-Oshin, M., Mlitwa, N.B.: A state-of-the-art survey of indoor positioning and navigation systems and technologies. South Afr. Comput. J. **29**(3), 145–197 (2017)

Cells Interrelation in Mobile Networks

Iyad Khuder[✉] and Robert Bestak

Czech Technical University in Prague, Technicka 4, Prague, Czech Republic
{khudeiya,robert.bestak}@fel.cvut.cz

Abstract. To better optimize a mobile network, it's useful to have knowledge about the movement of users in the network. This can relatively easily be done via sending GPS coordinates calculated in a mobile terminal to network. However, this approach is, first of all, quite energy demanding, and secondly user dependent, as a user has to pose a mobile terminal supporting GPS and has to allow the usage of GPS. Another possibility is to make use of signaling data which is an essential and integral part of mobile network operations, plus it's more or less user independent. By combining the signaling data together with the network coverage map, we can estimate users' movements in the network. In this paper, we focus on cells interrelation in a network coverage map. We present a simplified cell graphical representation, using a so-called cell-vector, and we analyze the possible use of cell-vector position scenarios to predict whether a pair of cells are neighboring each other or not.

Keywords: Mobile network · Signaling · Coverage map · Cell · User's position

1 Introduction

For a decade, mobile technologies have been facing a rampant demand on both capacity and data rates [1]. The global mobile data traffic is expected to increase about eight times by the end of 2023 comparing to 2018 [2]. This enormous increase poses many challenges, including spectral and energy efficiency. This means, firstly to increase the information rates being transmitted over a given bandwidth, and secondly to optimize the operation of BSs base stations (i.e. the towers) according to a given request and traffic load at each BS. To support these aspects, users' movements in the network should be considered and investigated. The user mobility knowledge is not only useful to optimize the network performance, including both the radio and core network parts, but it can also be beneficial in other cases. For example, in urban planning, traffic control, transportation management, as well as in home and work place prediction [3–5], which in turn has many applications. Furthermore, the study of collective human mobility patterns can also be utilized in other areas such as epidemiology or sociology [6].

The detection/monitoring of users' motion is possible to be made without using the Global Positioning System (GPS) technology through the network signaling activity, which occurs as a part of the mobile network operation [7,8].

© Springer Nature Switzerland AG 2019
P. Gaj et al. (Eds.): CN 2019, CCIS 1039, pp. 242–255, 2019.
https://doi.org/10.1007/978-3-030-21952-9_19

Typically, a user terminal triggers about tens/hundreds of signaling messages per day (depending on supporting mobile network technologies and user activity). However, this network-based movement detection requires a relatively good know how of the mobile network principles, network topology and mainly the knowledge of the network coverage map; including relations among cells such as which cells are neighbors.

New mobile network technologies have been emerging and put in operation in parallel with their predecessors. Operators deploy those technologies together in the same areas for the purpose of achieving a good quality of service QoS and to meet the growing needs of capacity and data rates. Therefore, a nowadays coverage map typically includes several mobile network technologies (2G, 3G, 4G, etc.). Additionally, these technologies may use several frequency bands (e.g. LTE 800 MHz, LTE 1800 MHz, etc.) working together and overlapping each other in most areas. This results in a complex coverage map in many ways. First of all, the overlapping layers are usually not identical. Cells of different layers may be smaller or larger depending on the frequency bands assigned to these layers and covered territory. Furthermore, due to the capacity and traffic requirements, networks become densified, and the number of cells significantly rises. As a result, relations between cells become more complicated to be found out. Thus, the coverage map becomes a composite of overlapping layers which may only share in common the positions of BSs and may differ from each other in all other attributes (e.g., the number of sectors per BS, radii and azimuths of cells, etc.). Notice that in our paper we reciprocally use the terms "cell" and "sector" in the same meaning; the notation "sector" is mainly used when talking about the "number of sectors" per BS (tower).

An example of a network coverage map is shown in Fig. 1, where the crosses in the figure denote the BSs' positions. The illustrated map is representing only one network technology, i.e. one coverage layer. The complete coverage map consists of several layers overlapping each other.

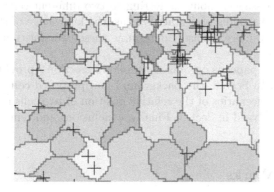

Fig. 1. A sample of a real coverage area.

Having a coverage map of high degree of complexity, it's impossible to detect and process the cells interrelations visually (i.e. by eyes). The approach has to be automated. However, the automation of this process implicates several challenges. Mainly, the coverage map typically consists of complicated visual data which are too large to be processed in a reasonable time interval for the practical use. For instance, the coverage map of one national operator may consist of thousands of cells, each of which is represented as a polygon of many vertices, which are given by their GPS coordinates. Therefore, it would be wise to reduce the amount of data as much as possible when investigating cell interrelations.

In this paper, we introduce a simplified cell graphical representation concept. As we are not interested in the real cell coverage (i.e. the cell polygon shape), but rather in cells interrelations, we represent the cell polygon as a simple vector, denoted as the cell-vector. The cell-vector of a certain cell is the vector whose starting point is the position of corresponding BS for the given cell, and the terminal point is the cell centroid. The latter is the "geometric center of gravity" of the cell polygon. The orientation of the cell-vector indicates the orientation of the signal beam emitted by the related BS antenna. Thus, the whole coverage map is simplified and only represented as a set of cell-vectors. We present and discuss possible scenarios of relative positions of two cell-vectors (cells) to predict whether the given cells are adjacent to each other or not. As a BS can serve a number of sectors, we introduce a representation of the cells related to one BS as a group of vectors, called a "vector cluster". The aim of this paper is to set the basis for a geometric concept which represents the coverage map in terms of cell-vectors and to analyze all potential scenarios of the relative position of two nearby cells.

The proposed concept could be useful in scenarios that require an analysis of the coverage, such as those related to the study of user mobility. An example on the applications is the study of user mobility using the Cell ID positioning technique (i.e. locating the user basing on the network signaling activity). In this study, there's a challenge which is to determine the user mobility status, as some signaling messages, though referring to two different cells, could indicate either a static or a moving user status. Thus, the knowledge of the cells interrelations allows us to more precisely determine these two status and the real user movement.

The rest of the paper is organized as follows. Section 2 provides an overview of related works. In Sect. 3, the background and proposed concept is discussed in detail. Possible scenarios of the relative position of two cells to each other are presented and analyzed in Sect. 4. Finally, conclusions and remarks are given in Sect. 5.

2   Related Works

There is a lot of work done in the field of mobile networks which uses some knowledge of the coverage map in many aspects of mobile networks, such as user mobility, network optimization, radio resource management, energy saving modes, spectrum efficiency, etc. However, to our best knowledge, there are no works on cells interrelations or the analysis of the coverage map of mobile network, due to confidentiality issues. Mobile network operators maintain the BSs' settings confidential and do not make their coverage maps publicly available.

There are several papers on the techniques which are used to estimate the coverage map by measuring the signal strength in many locations in the coverage area. Authors in [9] propose a crowdsourced measurement approach which can be employed by both network operators and service providers to predict the network coverage, whether before or after the deployment of network infrastructure. The proposed approach is based on spatial interpolation techniques using controlled measurements taken by a customized mobile application to obtain measurement information. This include the GPS/network location, general mobile network information, the location area code, the cell ID and the signal strength. In addition, the user manually inputs his/her exact location at each measurement point. Authors state that this approach is more cost-effective than the traditional drive testing technique.

In [10], authors demonstrate that the signal strength measurement doesn't give accurate results in estimating the coverage and could be misleading in many cases. For example, the same mobile phone may measure different signal strengths in different moments while being at the same position. As the signal strength at a certain point doesn't provide an enough indication of the network performance (QoS) at that point, the authors recommend estimating the coverage by measuring the TCP goodput, instead of the signal strength.

There are several papers on the representation of coverage maps using Voronoi diagrams. For example, authors in [11] present a customized version of Voronoi diagrams to better represent the coverage map. It shows that the basic definition of Voronoi diagrams doesn't really fit to the real cellular network coverage. Therefore, the authors propose a developed version of Voronoi diagrams' definition, called the "ordered order-k multiplicatively weighted Voronoi diagrams". The measured distance between a point and a BS is no longer the conventional Euclidean distance, but an adjusted version of the distance, which is the distance multiplied by a factor that is called the weight related to that BS. This model results in the edges of Voronoi cells being arcs, rather than straight segments.

A spatial model called "graph-cellular automata (graph-CA) of urban and regional change" to represent a network of cells is introduced in [12]. This model is mainly useful for dynamic cells, where each cell has more than one state that change over the time. However, it's still useful for static cells, as it proposes an extended definition of neighborhood relationship, which is not only based on adjacency, but also on a functional influence.

An example of an application which assumes knowledge of the coverage map is presented in [13]. The authors propose a model which predicts the user mobility in mobile networks, in order to proactively anticipate traffic hotspots.

The introduced system uses data (records) from OpenMobileNetwork[1], where a record consists of a pair of the Cell-ID and a timestamp (i.e. the time of record creation). However, a weakness point of the proposed approach is that it doesn't take the impact of the neighboring cells into account, as the authors consider this aspect to be unimportant. Nevertheless, the authors acknowledge the fact that the user's "oscillation phenomenon" has a negative impact on the prediction quality of user mobility.

In [14], authors propose a method to analyze users' mobility basing on wireless big data in a densely populated area. They take advantage of the wireless big data processing platform in order to construct user's trajectories from traffic data sources. They also introduce algorithms for identifying hotspots and discovering mobility patterns of users. These issues have theoretical and practical significance for urban planning, traffic control and mobile network resource optimization.

Different methods to estimate the mobile network coverage and their comparisons are presented in [15]. The methods are based on crowdsourcing approaches which aggregate measurements from disseminated smartphones. The authors introduce four approaches for estimating the BS position of a certain cell, graduating from the simplest to the most developed. The proposed approaches differ from each other by the algorithm utilized to deduce the BS position basing on the positions of measurement points and the values of these measurements.

When network data are available, they can be used as shown in [16], where the authors use these data to increase the efficiency of network operations in mobile operators. Authors use the "Base Station System Application Part" protocol to identify the location updates corresponding to handovers between two different technologies. Then, they use this information to detect discontinuities in a 3G network.

Notice that the coverage in practical applications are usually estimated and not very strictly defined, i.e. borders between cells are not absolute, but are based on the highest probability to connect through a certain BS. Hence, the main point of interest in our paper is cells interrelation, rather than the shape of each cell. Comparing to the above-mentioned works, we investigate a possible simplified coverage map representation, and possible cell interrelation scenarios that can occur when considering a pair of cells.

3   Cell Graphical Representation

In literature, the cell shape in a mobile network is typically simplified and represented as identical circles or hexagons [17,18]. Nevertheless, cells have in reality irregular shapes and different sizes (see Fig. 1). For this reason, other approaches have been proposed to represent the coverage of a mobile network such as representing the cells by means of polygons which are derived from the coverage map if available. Although the polygon representation provides a realistic illustration

---

[1] An open platform that provides approximated and semantically enriched mobile network and WiFi access point topology data. http://www.openmobilenetwork.org.

of the coverage, it's of high complexity as each cell is represented by a polygon of a high number of vertices and it requires storing the coordinates of all vertices of each cell. Another approach is to use Voronoi diagrams as they are basically defined on the same principle as the cell coverage, i.e. based on the "proximity". Voronoi diagrams are the solution of the closest-site equation (a Voronoi cell is the locus of points which are closest to the cell's site). In a mobile network, in the ideal situation where all BS's are identical in all parameters, a cell would also be defined by the proximity principle. This is, of course, not accurate in reality, as a terminal, seeks connection to a BS which provides the highest signal strength. That's why Voronoi diagrams can't give an accurate representation of the coverage.

In the same context, it should be highlighted that the term "coverage" being used in our paper refers to a conventional approach. This means that the coverage is calculated basing on 3D maps of the covered area taking into account the parameter settings of BSs and antennas used in reality in that area. It doesn't refer to a coverage which is measured in the real terrain, but it's based on simulation outputs. Notice that the cell shape and size in the coverage are practically determined according to the "dominance coverage". In a dominance coverage, a cell is defined as the area where the probability that a terminal will be connected to the BS of that cell is higher comparing to surrounding BSs. This criterion is behind the irregularity in shapes and sizes, as the dominancy is influenced, among others, by the topography of the coverage area, the built-up area and by BSs and antennas settings (i.e. the number of antennas, the tower height, the transmission power, the antenna radiation pattern, etc.).

The relative position of cells to each other can easily be checked in the coverage map. This is easy when it's a matter of just a few cells. But in practice, it's required to investigate the relative position of many cells whose number is too large to be checked visually. Therefore, it's indispensable to automate the whole process.

Technically, the analysis of the coverage data and finding its cells' interrelations cannot be done efficiently by processing all cells' vertices as an input, due to the high complexity of the coverage. Therefore, it would be efficient to limit the number of attributes per cell as low as possible. We propose to deal with just two attributes per cell (see Fig. 2):

1. BS' position of the cell (i.e. tower); denoted as BA for cell A.
2. Cell centroid position of the cell; denoted as CA for cell A.

Basing on these two attributes, we define another two ones: cell-vector and azimuth. The cell-vector of cell A is written as $\mathbf{r}_A$ (see Fig. 2). It is a vector quantity[2] and defined as the vector whose origin point is given by B_A and terminal point is C_A, i.e. $\mathbf{r}_A = \overrightarrow{B_A C_A}$. The azimuth of cell A, θ_A, is the azimuth of its1 cell-vector, i.e. the angle measured from the north vector to vector $\mathbf{r}_A$ (see Fig. 2).

---

[2] Please notice that vector quantities are written in bold (i.e. $\mathbf{r} = \overrightarrow{r}$).

The position of BSs and centroids are given by their GPS coordinates, or by other equivalent geodetic coordinate systems. The GPS coordinates of the BS and the centroid consists of two components: (i) latitude, and (ii) altitude. While the BS position is the position of the real physical tower on the ground, the centroid is a virtual point which represents the "center of gravity" of the polygon which represents the cell coverage. In addition to the positions of BSs and centroids, we also assume to know the number of sectors per given BS, let's call it N. (see Fig. 3). The parameter N typically ranges from 2 to 4 sectors. For example, the base station B_A in Fig. 3, ensures coverage to 3 sectors (cells), which are: cell A1, cell A2 and cell A3, whose centroids, resp. cell-vectors are C_{A1}, C_{A2} and C_{A3}, resp. $\mathbf{r}_{A1}$, $\mathbf{r}_{A2}$ and $\mathbf{r}_{A3}$.

Figure 4 shows a sample of a coverage area which contains BSs of different number of sectors, such as B_A, B_B and B_C which have 2, 3 and 4 sectors resp. The cell-vector representation of Fig. 4 is illustrated in Fig. 5.

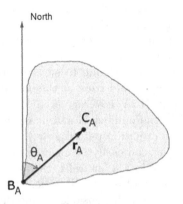

Fig. 2. Basic cell attributes.

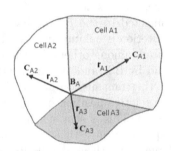

Fig. 3. Cell attributes notation.

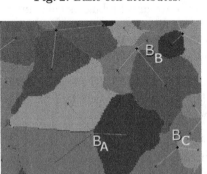

Fig. 4. Example of real covered area.

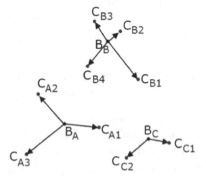

Fig. 5. Cell-vector representation of the real covered area.

The cell interrelation mainly depends on the relative position of the cell-vectors of cells being investigated, rather than the detailed shape of each cell. Thus, the coverage map becomes a map of "clusters of vectors", where each cluster is the group of vectors originating from the same BS, as shown in Fig. 5. It should be emphasized that for a certain cell, only cells in its proximity are investigated to know their relative position to the given cell. That's because farther cells have no impact on the cell of interest.

4   Relative Position Scenarios

Potential scenarios of the relative position of two proximity cells are listed in Table 1. For simplicity, the two cells being studied are referred to as cell A and cell B, whose cell-vectors are r_A and r_B. It should be stressed that the cell A and cell B could be either of the same technology (e.g. GSM) or of different technology or frequency (e.g. GSM and LTE). The classification of cell position is done according to the following two criteria: (i) the angle between the r_A and r_B. (ii) the relative position between the r_A and r_B (i.e. whether the vectors face each other, diver from each other, or head in the same direction).

In the table, we use notations by means of arrows to illustrate the relative position of cell-vectors. For example, the notation ↑↑ refers to cell-vectors which are parallel and in the same direction (scenario 1 in Table 1), while the notation ↑↓ refers to cell-vectors which are parallel but in the opposite direction (scenario 2 in Table 1). In contrast, the notation ╲╱, resp. ╲ ╱ refer to two diverging cell-vectors, where the cell vectors share (Fig. 6a), resp. do not share, (Fig. 6b and c), the same BS. Basing on the knowledge of which sub-scenario two cells match, it's possible to estimate the probability of being neighbors for the given pair of cells, which is indicated in Table 1 as well.

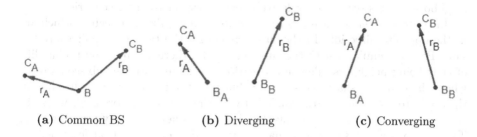

(a) Common BS   (b) Diverging   (c) Converging

Fig. 6. Non-parallel cell-vectors.

Scenario 1 refers to two cells with parallel cell-vectors which are in the same direction. This scenario include 3 sub-scenarios. In the sub-scenario 1_a, the two cells don't share the same BS and have a medium probability of being neighbors (as they have at least a common point, which is the BS itself). In the sub-scenario 1_b, the two cells share the same BS. This case occurs when there are two cells

Table 1. Possible scenarios of the 2-cell interrelation

Scenarios			sub-scenarios			
Notation	Description		Notation	Abbreviation	Description	Probability of being neighbors
1 $r_A \uparrow\uparrow r_B$	parallel vectors – same direction	a	$r_A \nearrow\nearrow r_B$	trapezoid/ parallelogram	parallel; same direction	medium
		b	$r_A \rightarrow\!\!\rightarrow r_B$	closed Joint	joined; linear; same direction	neighbors
		c	$r_A \rightarrow \rightarrow r_B$	queue	linear; same direction	low
2 $r_A \uparrow\downarrow r_B$	parallel vectors – opposite direction	a	$r_A \nearrow\swarrow r_B$	trapezoid/ parallelogram	parallel; opposite direction	medium
		b	$r_A \leftrightarrow r_B$	linear joint	joined; linear; opposite direction	neighbors
		c	$r_A \leftarrow \rightarrow r_B$	back-to-back	linear; opposite direction; diverging	low
		d	$r_A \rightarrow \leftarrow r_B$	face-to-face	linear; opposite direction; converging	high
3 $r_A \nwarrow\nearrow r_B$	non-parallel vectors	a	$r_A \vee r_B$	open Joint	Joined; diverging	neighbors
		b	$r_A \nwarrow\nearrow r_B$	divergence	non-parallel; diverging	low
		c	$r_A \nearrow\nwarrow r_B$	convergence	non-parallel; converging	medium

of different technologies sharing the same BS, and their antennas beams head in the same direction (with some tolerance δ). The two cells in this case are overlapping each other, and they are neighbors. Thus, they can be considered as linear, but their sizes may or may not be equal. As for the sub-scenario 1_c, the two cell-vectors don't share the same BS, and there's high probability to have at least one cell between them, and therefore the probability of cells being neighbors for this scenario is relatively low in respect to other scenarios.

In contrast, scenario 2 refers to two cells with parallel cell-vectors which are in the opposite direction. In the sub-scenario 2_a, the two cells don't share the same BS and, similarly with the sub-scenario 1_a, there is a medium probability of cells being neighbors. The sub-scenario 2_b is similar to the sub-scenario 1_b, where the two cells share the same BS, but here the two cell-vectors heading in the opposite direction, i.e. the angle between the two cell-vectors is ideally π. In this case, the two cells are not overlapping each other, and they are neighbors. Two cells of the sub-scenario 2c have distinct BSs, and there is at least one, or more, cells between them. Therefore, there's a very low probability that two cells of this scenario are neighbors. As for the sub-scenario 2_d, the two cell-vectors don't share the same BS, but they are heading toward each other. Since it's assumed that the two cells of interest have a certain proximity, there is a high probability that these two cells being neighbors.

As for the last subset of scenarios, scenario 3, the cell-vectors of the two cells of interest are not parallel. The sub-scenario 3_a 3a is very similar to the sub-scenario 2_b, except that the angle between the two cell-vectors is not π (nor 0).

Here too, the two cells are neighbors. The sub-scenarios 3_b and 3_c refer to cells having distinct BSs, with non-parallel cell-vectors. The two cell-vectors in the sub-scenario 3_b (3_c resp.) are diverging (converging resp.), i.e. the angle between them is obtuse, $\Delta\varphi < \pi/2$ (acute, $\Delta\varphi > \pi/2$ resp.). Therefore, the probability of being neighbors in sub-scenario 3b is higher than in sub-scenario 3_c.

In case of two or more cells sharing a common BS (tower), cell-vectors of these cells would be oriented in such a way that the relative angles between them, $\Delta\varphi_i$, are ideally equal; i.e. for N sectors, $\Delta\varphi_i = 2\pi/N$. For example, in the case of $N = 3$, the angles between $\mathbf{r}_{A1}, \mathbf{r}_{A2}$ and $\mathbf{r}_{A3}$ will ideally be $\Delta\varphi_1 = \Delta\varphi_2 = \Delta\varphi_3 = 2\pi/3$ (see Fig. 7), but practically, these cell-vectors maybe deviated within some margin as shown in the figure. This case specifically matches scenarios 2_b and 3_a.

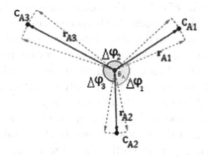

Fig. 7. Angles and sizes of cell-vectors belonging to one BS.

Notice that the arrow notations in Table 1 are symbolic, and they do not mean that cell-vectors are exactly of the same magnitude and/or angles (which is the ideal case). For example, cell-vectors which are classified as parallel (scenarios 1 and 2) may not be exactly parallel. That is, two cell-vectors should be considered parallel, if the relative angle between them doesn't exceed a certain tolerance or margin δ; i.e. $|\Delta\varphi \bmod \pi| < \delta$, where $\Delta\varphi = \varphi_1 - \varphi_2$; φ_1 and φ_2 are the azimuths of the two cell vectors being considered.

It should also be highlighted that cell-vectors discussed in all sub-scenarios are not necessarily of the same technology; they may be belonging to different layers.

Having several operating technologies in parallel, the coverage map becomes formed of several layers (see Fig. 8). These layers represent either different generations of mobile networks (e.g. 2G, 3G, 4G) or different frequencies of the same generation (e.g. LTE800, LTE1800). It's possible to have BSs in different layers at the same points, which happens when the antennas of different technologies are mounted on the same tower, see for example tower B in Fig. 7. Thus, the multi-layer coverage map is of high degree of complexity to be analyzed. As mentioned previously, the number and the positions of BSs in different layers are not necessarily identical; neither the cell-vectors are. For example, only the tower B is common for all three illustrated networks (2G, 3G, 4G) in Fig. 7.

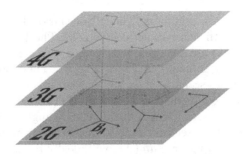

Fig. 8. A multi-layer coverage map.

The radii and azimuths of cells varies according to several factors. For example, the cell radius mainly depends on the population density and also on the network generation. Cells in highly-populated areas tend to be small to meet the growing needs of capacity in those areas. Therefore, microcells are common in these areas. On the other hand, cells in rural areas, and generally in areas of low population density, are larger, and typically classified as macrocells.

An example cell radii and azimuths of a coverage area, consisting of 230 cells, is illustrated in Fig. 9 (radius) and Fig. 10. (azimuth). The x-axis represents the radius (resp. azimuth), and the y-axis represents the number of cells.

As can be observed in Fig. 9, the cell radii follow the logarithmic distribution. The mean value of is 1528 m, and about 68% of cells have radius ranging between 500 m and 2000 m. This is because the coverage sample is taken from a suburban area with a middle population density.

Figure 10 illustrates the distribution of cells' azimuths of the same sample. As can be seen from the figure, the azimuth distribution is not linear. There are about 3 peak zones which are mainly due to the sectorization, where majority of BSs in the sample have 3 or 4 sectors.

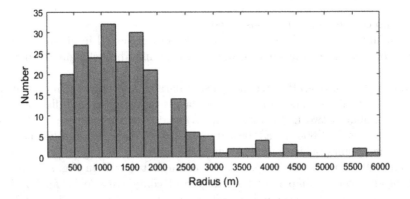

Fig. 9. Cell-radii distribution.

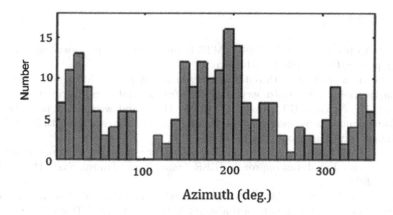

Azimuth (deg.)

Fig. 10. Cell-azimuths distribution.

5   Conclusions

The knowledge of the coverage topology is useful when analyzing the user mobility. The coverage in most areas is composite of several layers with a large number of cells, depending on the technologies deployed in each area. Therefore, disposing a practical tool to have this knowledge and look up the cells-interrelation is challenging. To be efficient, such tool needs to be firstly automated, and secondly based on a limited number of cell attributes, in order to make the process easy and time-effective. Addressing these challenges, we introduce a simplified graphical representation of the coverage, which helps investigating the cells interrelation and predicting whether a pair of cells are neighbors or not. Additionally, we identify and discussed possible scenarios of two proximity cells interrelations.

The proposed approach can be useful in several coverage-related applications. For example, when studying the user-mobility using the Cell-ID positioning technique, there's a challenge to determine whether a user is static or moving. This uncertainty occurs when the terminal generates successive signaling messages which indicate diverse cells, which may not reflect a real movement of the user. This mainly happens when the terminal is located in a bordering area between two cells or due to a temporal change in propagation conditions. Therefore, it's efficient to know cells-interrelations in order to distinguish the real movement and fake movement indication.

In the next phase, we plan to design and test an algorithm that would find out cells-interrelation. We will first apply it on simulated coverage data, and then test it on real date.

Acknowledgments. This research work was supported by the Grant Agency of the Czech Technical University in Prague, grant no. SGS18/181/OHK3/3T/13.

References

1. Magic Mobile Future 2010–2020, UMTS Forum (2005). http://www.3gpp.org/ftp/pcg/pcg_14/Docs/PDF/PCG14_17.pdf
2. Mobile data traffic growth outlook. Ericson report (2018). https://www.ericsson.com/en/mobility-report/reports/june-2018/mobile-data-traffic-growth-outlook
3. Dash, M., Nguyen, H.L., Hong, C., et al.: Home and work place prediction for urban planning using mobile network data. In: Proceedings of the 15th IEEE MDM (HumoComp Workshop), pp. 37–42 (2014)
4. Senaratne, H., Mueller, M., et al.: Urban mobility analysis with mobile network data: a visual analytics approach. IEEE Trans. Intell. Transp. Syst. **19**(5), 1537–1546 (2018)
5. Di Lorenzo, G., Sbodio, M., Calabrese, F., Berlingerio, M., et al.: Visual exploration of cellphone mobility data to optimise public transport. IEEE Trans. Vis. Comput. Graph. **22**(2), 1036–1050 (2016)
6. Liu, Zh., Qiao, Y., Tao, S., et al.: Analyzing human mobility and social relationships from cellular network data. In: 13th International Conference on Network and Service Management (CNSM) (2017)
7. Dufkova, K., Ficek, M., Kencl, L., Danihelka, J.: Active GSM cell-id tracking: where Did You Disappear? In: Proceedings of the ACM International Workshop on Mobile Entity Localization and Tracking in GPS-less Environments, San Francisco, California, USA, 19 September 2008
8. Trevisani, E., Vitaletti, A.: Cell-ID location technique, limits and benefits: an experimental study. In: Proceedings of the Sixth IEEE Workshop on Mobile Computing Systems and Applications (2004)
9. Molinari, M., Fida, M.R., et al.: Spatial interpolation based cellular coverage - prediction with crowdsourced measurements. In: Proceedings of the 2015 ACM SIGCOMM Workshop on Crowdsourcing and Crowdsharing of Big (Internet) Data (2016)
10. Sonntag, S., et al.: Mobile network measurements - it's not all about signal strength. In: IEEE Wireless Communications and Networking Conference (WCNC) (2013)
11. Portela, J., Alencar, M.: Cellular coverage map as a Voronoi diagram. J. Commun. Inf. Syst. **23**(1), 3–4 (2008)
12. O'Sullivan, D.: Graph-cellular automata (graph-CA) of urban and regional change. In: Centre for Advanced Spatial Analysis, University College London, 1–19 Torrington Place, London WC1E 6BT, England (2001)
13. Gondor, S., Uzun, A.: Predicting user mobility in mobile radio networks to proactively anticipate traffic hotspots. In: International Conference on MOBILe Wireless MiddleWARE, Operating Systems and Applications (2013)
14. Zhao, Zh., Zhang, P., et al.: User mobility modeling based on mobile traffic data collected in real cellular networks. In: 11th International Conference on Signal Processing and Communication Systems (ICSPCS) (2017)
15. Neidhardt, E., Uzun, A., et al.: Estimating locations and coverage areas of mobile network cells based on crowdsourced data. In: Çelebi, Ö.F., Zeydan, E., et al. (eds.) 6th Joint IFIP Wireless and Mobile Networking Conference (WMNC) (2013). On Use of Big Data for Enhancing Network Coverage Analysis, ICT (2013)
16. Baltzis, K.B.: Hexagonal vs circular cell shape: a comparative analysis and evaluation of the two popular modeling approximations. In: Cellular Networks - Positioning, Performance Analysis, Reliability (2011)

17. Ghosh, R.K.: Wireless Networking and Mobile Data Management. Springer, Singapore (2017). https://doi.org/10.1007/978-981-10-3941-6
18. Aurenhammer, F.: Voronoi diagrams-a survey of a fundamental geometric data structure. ACM Comput. Surv. **23**, 345–405 (1991)

Communication Model of Smart Substation for Cyber-Detection Systems

Radek Fujdiak[1], Petr Blazek[1], Petr Chmelar[2], Petr Dittrich[2],
Miroslav Voznak[3(✉)], Petr Mlynek[1], Jan Slacik[1], Petr Musil[1], Pavel Jurka[2],
and Jiri Misurec[1]

[1] Brno University of technology, Technicka 12, Brno 61600, Czech Republic
fujdiak@feec.vutbr.cz
[2] GreyCortex, Purkynova 649/127, 61200 Brno, Czech Republic
petr.chmelar@greycortex.com
[3] VSB - Technical University of Ostrava, 17. listopadu 2172/15,
708 00 Ostrava, Czech Republic
voznak@ieee.org

Abstract. Intrusion detection (prevention) systems (IDS/IPS) are already widely used in an information network, but their popularity is growing as well in the industrial environment due to the recent security incident, especially in energy sector. However, to build strong defense via IDS/IPS high-quality data are needed, but these often sensitive data are not easy to obtain. Therefore, the smart grid testbeds start to appear across the world to provide an experimental test environment and also provide sufficient data models. This paper focused on smart substation as a crucial part of the distribution network in the Smart Grid. The paper provides extensive analysis of Smart Grid protocols with close focus on promising protocol IEC 61850. The communication and the data model is provided and an inexpensive experimental environment is introduced.

Keywords: SCADA · ICS · Communication · Generator · IEC · 61850

1 Introduction

Over the past years, the Operational Technologies (OT) and Industrial Control Systems (ICS) have evolved into the main method of monitoring and controlling large processes over many different areas including industry or energy sector [13]. Nowadays, these systems are leading parts of new phenomena such as Industry 4.0 [45] or Smart Grid [40]. Before the Stuxnet computer virus, the ICS security was running in the background and most systems were running with major security imperfections [28]. Since then, there has been a big rise of attention in this

The research was financed by the National Sustainability Program under grant LO1401 and the Ministry of the Interior under grant no. VI20172019057 and partially by the Ministry of Education, Youth and Sports withins SGS project no. SP2019/41. For the research, the infrastructure of the SIX Centre was used.

© Springer Nature Switzerland AG 2019
P. Gaj et al. (Eds.): CN 2019, CCIS 1039, pp. 256–271, 2019.
https://doi.org/10.1007/978-3-030-21952-9_20

area and cybersecurity became a major issue for these systems. The importance of growth even more with including the ICS into the country's national critical infrastructure, which provides the control, monitoring, and distribution of resources [46]. Moreover, Ukraine's incident in 2015 raises many questions for the cybersecurity of ICS systems used in Smart Grid [43]. The attack started by malware via phishing mail and ended by denial of service together with large power outage [10]. The energy sector also remains today the most vulnerable ICS area, where the ICS components make more than half of the threats [27]). To provide sufficient defense, the cyber-detection systems such as Intrusion Detection System (IDS) or Anomaly Detection System (ADS) are used [6,38]. However, these systems are mostly based on artificial intelligence and machine learning, where various high-quality training/learning data are needed [26]. Unfortunately, due to the sensitiveness of the data and global-danger of testing these algorithms in a real environment, experimental laboratories are the only options for how to obtain required data. The expensiveness of real electrical equipment leads to simulation, emulation and virtualization methods. However, this approach needs an accurate architecture, communication and data model, which would ensure close-to-real experience. Therefore, this paper focuses on critical energy infrastructure represented by the distribution network with a deeper look at the smart substations as a crucial part. The closer attention is given to the ICS communication protocols (especially newer protocol IEC 61850), communication and data model for the substation. Also, the real cost-effective testbed is introduced.

2 Related Works

The crucial function of Smart Grid is without doubts communication. There are existing approx. 150–200 SCADA and ICS protocols used in Smart Grid [24]. However, most of the current surveys are focused on network/routing protocols [39], transmission/access protocols [47], or architecture of Smart Grid itself [44]. The ICS/SCADA protocols are mostly only mentioned or mixed with other protocols. To improve the current state of the art, Sect. 3 provides an extensive analysis of more than 50 ICS protocols widely used in Smart Grid, which should help to better orient in this complex environment. However, the communication is only one part of the test environment and just communication solution will be insufficient without accurate communication/data model. The crucial part of Smart Grid is the distribution network, which represents the "veins" holding together all the parts. Therefore, the failure of the distribution network is the main reason for power outages [29]. As the most crucial parts of the distribution network, the substation represents the key node, intelligence as well as the attack point. Thus, it is necessary to provide an accurate model of the key communication node in the distribution network for the test environment or laboratory. Cintuglu et al. [11] analyzes 37 laboratories across the world and discovers that most of the laboratories are focused on older protocols and very simplified substation without advanced functionalities or not dealing the substation model at all. The extensive communication model for substation automation together with infrastructure and control center logic is provided in Sect. 4. The real example along with a data model for selected protocol IEC 61850 and a brief overview of testbed buildings blocks are held in Sect. 5. Finally, the comparison with selected related solutions is provided in Sect. 6 followed by final summarization and conclusion in Sect. 7.

3   Analysis of Communication Protocols in ICS

As previously mentioned, the ICS/SCADA contains large group of protocols/standards. This section provide analysis of the most common protocols divided into the following groups: (i) Substation automation (Table 1), (ii) General purpose protocols (Table 2), and (iii) Advanced Metering Infrastructure (Table 3).

Table 1. Protocols used for substation automation [7, 12, 18, 20, 23, 30, 31, 33, 36, 41, 42].

Protocol	Description
Siemens S7	Proprietary group of communication protocols (S-200/S-300/S-400/S-1200 and S/1500) for programming logical components with four-layers, command-oriented and client-server architecture
Modbus	3–6 layer open non-standardized communication protocol with client-server architecture and request-response logic (three main message - request, response, and negative response), TCP/RTP version
Profibus	Three-layer fieldbus (upto 5 with IEC 60870-5-104 combination) communication protocol with bus or star topology and master-slave architecture (max. 32 stations in the network))
ProfiNet	3–4 layer communication protocol based on standard Ethernet and TCP/IP (compatible with IEEE 802.3), defines the whole network model, high performance, but not for time critical applications
IEC 61158, IEC 61784	Inexpensive three-layer fieldbus communication protocol based on the master-slave architecture for very short update times ($\leq 100\,\mu s$) with low jitter and precise synchronization ($\leq 1\,\mu s$)
IEC 62413	Known as Ethernet/IP, based on standard Ethernet and TCP/IP (compatible with IEEE 802.3), adapts the Common Industrial Protocol (CIP), based on predefined Ethernet nodes

(continued)

Table 1. (*continued*)

Protocol	Description
C37.118	Four-layer specialized and standardized protocol for synchrophasor measurements in power systems, defines synchrophasors, frequency, and rate of change of frequency for measurement under all conditions
DNP3	Group of protocols for data collection, master/slave oriented with response/request logic, based on Enhanced Performance Architecture model with four-layer structure
IEC 60870	Group of protocol for communication and control in SCADA systems, master/slave oriented with response/request logic, based on Enhanced Performance Architecture model with three-layer structure
IEC 61850	Group of standardized protocol for easier compatibility in ICS (main protocols MMS, GOOSE, SMV and SNTP), seven-layers structure, client-server oriented, very robust

Table 2. General or multi purpose Smart Grid protocols [5,9,14–16,19,20,35,37].

Protocol	Description
FIPA	Foundation for Intelligent Physical Agents (FIPA), body for developing and setting computer software standards for heterogeneous and interacting agent-based systems
TCNet	Standardized protocol for real-time Ethernet communication, small-to-medium size networks, cyclic transmission, priority implemented, high performance/reliability with no-delay
VNetIP	High-reliability 1GB/s real-time and stable communication protocol for control networks
Sercos III	Open protocol for communication between industrial control, I/O devices and Ethernet nodes, master-slave oriented (up to 511 slaves)
EPA	Open Network Control System for automation via Ethernet (EPA), RTC, oriented on publish/subscriber and client-server, message-based
P-Net	Very simple communication protocol over RS232 or IP for basic automation, client-server/master-slave oriented
Ether-S-Bus	Four-layer communication protocol with/between process control devices, client-server oriented
Ether-S-I/O	Protocol for data transfer between PLC and remote I/O devices, uses telegrams, one telegram used for multiple I/O stations, uses UDP:6060
HART IP	Global standard for communication between intelligent devices and control/monitoring systems, metallic medium, client-server oriented, TCP/UDP:5094
OPC DA, OPC UA	Plug-and-play protocol for monitoring real-time processes, client-server or server/server oriented, OPC DA - COM and OPC UA - Ethernet, only L7 defined

(*continued*)

Table 2. (*continued*)

Protocol	Description
HomePlug	Power line communication technology for smart appliance in HAN, where HomePlug Green PHY defines low power, cost-optimized PLC
U-SNAP	Universal Smart Network Access Port (U-SNAP), group of proprietary communication protocols for HAN and demand-response applications
OpenADR	Communication protocols for dynamic pricing and demand response application, defining whole infrastructure
EN50170	Factory Instrumentation Protocol (FIP), used for real-time monitoring and supervision of industrial devices, three-layers model, periodic message transmission
ISO 16484-5	Building Automation and Control Network protocol (BacNet), four-layer model, devices defined by objects with specific properties (more than 54 types)
ISA 100.11a	Frou-layer multi-purpose wireless communication protocol for industrial automation and process control applications, mesh oriented
SAE	Group of standards for electrical energy transfer (J2293), communication for electric vehicles (J2836), PEVs to Grid (J2847), hybrid electric vehicles (J2841), short-range communication (J2735) and others
IEC 60870-6	- Group of protocols for data exchange between utility control centers, utilities, power pools and regional control centers
IEC 61970 IEC 61969	Provides Common Information Model (CIM) for transmission domain (IEC 61970) and distribution domain (IEC 61969)
IEC 62351	Group of standards defining cybersecurity via different objectives for the communication protocols in smart grid networks

Table 3. Protocols used in Advanced Metering Infrastructure (AMI) [1–4, 17, 20, 21, 32].

Protocol	Description
EN 13757	The three-layer protocol M-Bus for remote readings (gas, electricity, etc.), master-slave architecture (with a max. of 250 stations in one bus) with client-server logic
ZigBee	Four-layer Wireless communication technology built on the IEEE 802.15.4 standard for PAN/HAN, for large networks (hundreds or thousands of nodes) with small data transmission (max. 250 kbps), possible star/tree/cluster/mesh topology
Z-Wave	Alternative to ZigBee, handles the interference with 802.11/b/g, open-standard solution (under ITU 9959), source-routed mesh, designed for low-latency and small data, up to 100 kbps

(*continued*)

Table 3. (*continued*)

Protocol	Description
OSGP	Three-layer Open Smart Grid Protocol (OSGP), data transmission from electro-meters, direct load control modules, solar panels, gateways, and other smart grid devices
ANSI C12.X	Group of protocols for bi-directional communication with the smart meters, via optical port (C12.18), from/to terminal (C12.19), via telephone modem (C12.21), with AES encryption (C12.22)
ITU G.9955 ITU G.9956	Standardized narrow-band orthogonal frequency division multiplexing power line communication - PHY Layer (G.9955) and Link Layer (G.9956)
PRIME	Specification for narrow band power line communication, open global standard for multi-vendor interoperability
G3-PLC	Specification for power line communication, provides interoperability, reliability, cybersecurity, and robustness
IEC 61107	Standardized communication protocol for AMI, half-duplex, based on ASCII data via serial port, protocol is suppressed by 62056 (but still widely used)
IEC 62056	Two-layer communication standards DLMS/COSEM for measuring electrical energy, data exchange for reading meter, tariff control and load control, client-server architecture

Focusing on the substation automation protocols, the physical devices are in most cases connected via a serial link (RS232/RS485) or Ethernet via older but widely used Modbus or Profibus protocol [8]. These protocols are already very well described and known. The ICS control communication is handled most often via three main protocols - DNP3, IEC 60870 and IEC 61850 [34]. This is also proven by Cintuglu et al. [11], who provides an extensive survey on smart grid testbeds with analyzing 37 laboratories across the world. However, IEC 61850 was included only in 27% of analyzed laboratories. Moreover, laboratories including IEC 61850 often used incomplete models. Table 4 illustrates the main differences between these three dominant protocols. It is visible that using only part of the IEC 61850 leads to degraded model and not full functionality. Due to the complexity, novelty and popularity, this paper will further focus on full-IEC 61850 data model.

Table 4. Comparison of DNP3 and IEC 61850 [12, 22, 25].

Description	DNP3	IEC 61850	IEC 60870
Read/Write	✓	✓	✓
Reporting	✓	✓	✓
Control (SBO and direct)	✓	✓	✓
Enhanced Control (with reports)	*	✓	-
Files	✓	✓	✓
Start/Stop	✓	*	-
Event logs	*	✓	✓
Substitution (Forcing)	✓	✓	✓
Object Discovery	✗	✓	✗
Substation Configuration Language	✗	✓	✗
Peer-to-peer messaging	✗	GOOSE/GSSE	✗
Sampled Measured Values	✗	SMV	✗

*Not native in the standard, but can be implemented later
Note: SBO ... Select Before Operate, GOOSE ... Generic Object Oriented Substation Events, GSSE ... Generic Substation State Events, and SMV ... Sample Measured Values

4 Established Smart Substation Model

As previously said, the substation is one of the crucial parts of the critical infrastructure and distribution network. Figure 1 introduces the component model of Smart Substation divided into five main blocks (A–E):

Block A includes parts, which ensure the control function, monitoring and data collection over the smart substation. Transmission between the substation and the SCADA system is via cellular network or optic fiber core. The blocks ensure: (A1) Data processing for high voltage, (A2) Data processing for low voltage, and (A3) Data storage for low voltage.

Block B is created by RTU (Remote Terminal Unit) and the communication interface (gateway/modem). This block ensures control function and data collection from connected sensors, concentrates data and ensure the gateway function between transmission technologies to Block A. The reliability is ensured via backup UPC (Uninterruptible power supply). The blocks provide: (B1) Voltage protection for the outage, over-/under- voltage, and (B2) Batteries.

Block C ensures controlling function and indication of failures occurred on HV (high-voltage) network. In this block, the basic traffic information are collected and abnormal states are signalized. In case of failure, there is a possibility for disconnecting the high-voltage part from the smart substation. The communication between Block C and RTU is ensured via serial link or Ethernet. Remote disconnection of HV is implemented via relays. The blocks ensure: (C1) State signalization and controlling, (C2) measurements of V (Voltage),

P (Electric Power), Q (Reactive power), I (Current) and fault signalization, and (C3) compact high voltage switchboard.

Block D deals with monitoring and controlling of low voltage (LV). This block saves the basic values of the main variables - V, P, Q, I. These values are saved to internal memory for future processing. Obtained data serve for a detailed overview of traffic in the smart substation. Based on knowledge gained from these data, it is possible to control and optimize the traffic of a specific network segment. In case of an abnormal state, it is possible to manually or automatically shut-down the main circuit breaker. The blocks ensure: (D1) state signalization and controlling, and (D2) measurements of V, P, Q, I, histograms, records of events on voltage and function of voltage protection.

Block E serves for ensuring operations of the smart substation. It provides overflow protection, an indication of short-circuit current and indication of fuses. Sensors are also part of Block E, mostly for simple measurements of temperature and humidity. These sensors are also able to indicate non-standard states (e.g., too high temperature around critical component). Communication is ensured via serial link or Ethernet. The blocks ensure: (E1) Protection against P overflow to high voltage, main switch and signalization, (E2) measurement of P and Q, (E3) signalization of fuses state, and (E4) main switch detection.

The substation model includes three main connections: (i) Ethernet or serial (RS485 or RS232), which serves to general/multi purpose, (ii) relay contacts via serial interface (switching the connected relays, which might be a motor,

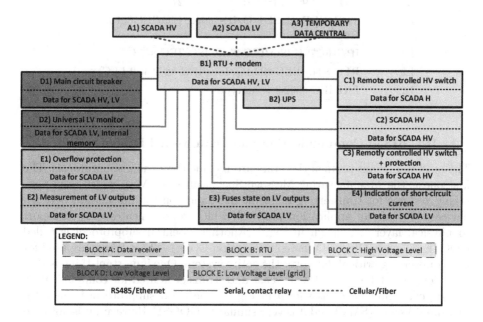

Fig. 1. Technological component blocks for smart substation.

transformer, or others), and high-speed cellular or fiber connection, which establish the connection between RTU/Modem and SCADA servers.

5   Cyberphysical Environment for Smart Substation with IEC 61850 Protocol

5.1   IEC 61850 in ISO/OSI Context

All protocols (open, proprietary, standardized) could not be simply combined, and cooperation between them is very complex and expensive. For this reason, IEC 61850 standard has been developed. This standard collects a total of 10 main parts. For communication, it is used all seven ISO/OSI layers (displayed in Table 5).

Table 5. Communication model over ISO/OSI layers for protocol IEC 61850 [25].

ISO/OSI	Description
L7	TimeSync SNTP, SV, GOOSE, GSSE, MMS ISO 9506, ISO/IEC 8649/8650/10035, ACSI/ACSE
L6	ANS.1 BER, ISO/IEC 8824.1, ISO/IEC 8649/10035, ISO/IEC 8822/8823
L5	ISO/IEC 9548, ISO/IEC 8326/8327
L4	TCP/UDP, GSSE ISO/IEC 8602, ISO/IEC 8073, RFC 1006
L3	IP/RFC791, ISO/IEC 9542, ISO/IEC 8473
L2	RFC 894, IEEE 802.1Q, ISO/IEC 8802.3(.2 LLC)
L1	ISO/IEC 8802.3 Ethertype

5.2   IEC 61850 Communication and Data Model Design

The most important parts of this standard include three main subprotocols. The first one is the Manufacturing Message Specification (MMS), which used messaging systems for transferring real-time data and control information between devices. MMS is an application protocol, which communicates over transport to physical layers as shown in Fig. 2. Another essential subprotocol is called GOOSE (Generic Object Oriented Substation Event). These events are used for fast transferring critical data over the entire system. One important thing is that time delay must be at most 4 ms. Fast data transfer is used for communication strictly over Link Layer as shown in Fig. 2. The third subprotocol is Sampled Measured Values (SMV) and it is very similar to GOOSE. However, it is not used for critical events. SMV protocol sends high-speed multi-cast messages, which contain user-defined values. The second layer (Link Layer) of the ISO/OSI model is used for communication same as in case of GOOSE (see Fig. 2).

The IEC 61850 is implemented in hardware platform Raspberry Pi 3B+ (OS Rasbian) via Library *libiec61850*[1]. Moreover, the library was also used for implementing the virtual environment framework. This provides the possibility to emulate, simulate and virtualizate end-nodes, RTUs, and more, which helps to obtain the closest possible environment to the real distribution network. The library *libiec61850* provides an implementation of all three previously mentioned subprotocols (MMS, GOOSE, SMV). Our approach allows mounting different communication modules (LTE modem, RS232, RS485, and others), which are used in ICS systems, as well as various end-devices, might be simulated via GPIO such as relays, switches, and others.

The cornerstone of each device communicating with the IEC 61850 is the data model, which contains a list of all important information about the device. The data model structure stores all input data, which can be processed. Access to variables is again mediated by querying the given data structure. Re-accessing variables are mediated by querying on the data structure. The data model of the devices is described in a markup language XML (extensible Markup Language). The structure of the data model is based on Fig. 3. According to IEC 61850-7-1, each device must have at least three logical nodes - LPHD (physical device information), LLN0 (logical node information, data sets, and others) and LN (protection functions). For our purposes, this was supplemented by a GPIO logical node, which is used for input/output (DO 1–4) on Raspberries.

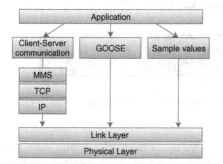

Fig. 2. Implemented IEC 61850 protocols

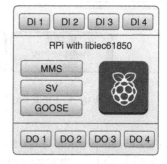

Fig. 3. Implemented data model

5.3   Experimental Environment and Results

The final implementation of our testbed is shown in Fig. 4. The network is based on previous Fig. 1. The environment contains three Double Ethernet Raspberry Pi 3B+ (with several sensors, actuators, relays and others connected via GPIO), two main standard routers (one with firewall), main server (Backup and NTP is included, HMI is based on OpenScada SW) and probe (ProfiShark 1 Gbps). The main message types of IEC 61850 - MMS, GOOSE and SV were implemented via

---

[1] http://libiec61850.com/libiec61850/.

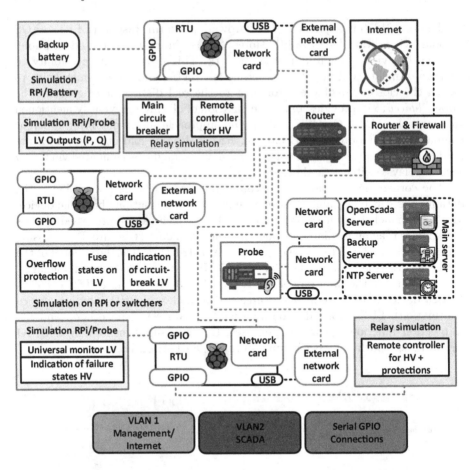

Fig. 4. Final implemented test-case environment.

model shown in previous Figs. 2 and 3. Two main network circuits were implemented - VLAN1 for management and VLAN2 for SCADA metering/control.

MMS implementation – MMS protocol is based on client-server communication. In *libiec61850* there are two library iec61850_server.h and iec-61850_client.h, which provide client-server communication. Figure 5 shows the recorded MMS message via Wireshark. This is a request from an HMI station to write a Boolean variable with a True value (boolean: True). The goal of the write variable is the digital output (itemId) on the Concentrator (domainId) which can e.g. control relay.

GOOSE implementation – Compared to MMS, GOOSE is based on multi-cast communications called publisher-subscriber. Communication is mediated through three libraries goose_publisher.h, goose_subscriber.h and goose_receiver.h. The publisher sends multicast messages, which are received by subscribers based on an identifier. The library goose_receiver.h is additional library for subscribers and it used to receive GOOSE messages. An example of realized GOOSE communication is in Fig. 6. The communication goes through the Data Link Layer with MAC addresses. For the AllData variable, four data structures are representing the digital outputs of RPi called Outstation A. Values are sent periodically to the HMI, where are graphically displayed by 1 (on) or 0 (off).

SV implementation – As GOOSE, Sampled Values uses the publisher-subscriber messaging architecture. In *libiec61850* are two main library sv_publisher.h and sv_subscriber.h, which are used to mediate communication. An example of a recorded SV message is shown in Fig. 7. For the transmission of messages is used Data Link Layer as with GOOSE. In the case of this message, two float values (svID) with the name svpub1 and svpub2 are transmitted. Values are randomly generated from Outstation A or Outstation B and sent to the Concentrator.

Fig. 5. Captured MSS message.   **Fig. 6.** Captured GOOSE message.   **Fig. 7.** Captured SMV message.

6   Functionality and Model Comparison

Table 6 provides comparison of our solution (#0) with selected relevant solutions [11]: FIU Testbed (#1), MSU SCADA Security Lab (#2), ISU Power-Cyber Testbed (#3), UNC at Charlotte Testbed (#4), Texas University A&M Cyber-Physical Testbed (#5), USF Smart Grid Power System Lab (#6), QUB Cybersecurity Testbed (#7), Queensland University testbed (#8), UCB Intrusion and defense testbed (#9), IIT Galvin Center PEV Testbed (#10), and NCSU Electric Vehicle Testbed (#11). The comparison was made for following parameters: (i) Heterogeneity (HET), (ii) Infrastructure type (Type) - simulation (SM), hardware (HW) or hybrid (HB), (iii) Infrastructure implemented (Inf), (iv) Substation model (Sub), (v) Security support (Sec), (vi) GOOSE (GSE), (vii) MMS, (viii) SMV, and (ix) Time synchronization (TS). From the results, our environment offers all crucial functionalities for smart substation automation testbed for data generation purposes. Moreover, backup server offers fast recovery in case of testing the outage scenarios or security incidents.

Table 6. Comparison of presented substation model with other models.

#	Type	Het	Inf	Sub	Sec	GSE	MMS	SMV	TS
0	HB	✓	✓	✓	✓	✓	✓	✓	✓
1	HW	✓	✓	✓	✓	✓	✗	✗	✓
2	SM	✓	✓	✓	✓	✓	✗	✗	✓
3	SM	✗	✓	✓	✓	✓	✓	✗	✗
4	SM	✗	✓	✗	✗	✓	✗	✓	✓
5	SM	✗	✓	✗	✓	✗	✗	✓	✗
6	SM	✗	✓	✗	✗	-	-	-	-
7	SM	✗	✗	✓	✓	✓	✓	✗	✓
8	SM	✗	✗	✓	✗	✓	✓	✓	✓
9	HB	✗	✗	✓	✗	✓	✓	✓	✓
10	HW	✓	✗	✗	✗	✓	✗	✗	✗
11	HW	✗	✗	✗	✗	✗	✗	✓	✗

7   Conclusion

The paper introduces the issue of insufficient data for learning algorithms of defense systems in ICS, especially in Smart Grid. The extensive ICS Smart Grid protocol analysis in different application areas is provided. However, the main contribution of this paper and novelty relies on the proposed Smart Substation model, which should serve as a baseground for building strong data generators in the area of the distribution network (Smart Grid). This ends in the selection of IEC 61850 as one of the most promising newest protocol. The brief context of ISO/OSI is given. Further, the data model together with realized network infrastructure is described. The environment is experimentally verified with captured IEC 61850 frames to show its functionality. Also, the comparison with 11 related testbeds and models is given, which shows the advances of our proposed solution. The following research should focused on another protocols such as TASE.2 (one of the novel and very promising Smart Grid protocol), AMI infrastructure, and P2P trading infrastructure.

References

1. Alliance, Z.: ZigBee specification (2015). Technical documentation
2. ANSI: ANSI C12.18-2005. Rosslyn (2006). https://www.scribd.com/doc/112623956/ANSI-C1218
3. ANSI: ANSI C12.21-2006: American National Standard Protocol Specification for Telephone Modem Communication. Rosslyn (2006). https://www.scribd.com/document/325940508/ANSI-C12-21-2006
4. ANSI: ANSI C12.19-2008: American National Standard For Utility Industry End Device Data Tables. Rosslyn (2009). https://www.scribd.com/document/320288707/ansi-c12-19-2008

5. ASHRAE: Proposed addendum bj to standard 135–2016, BACnet - a data communication protocol for building automation and control networks (2016). aNSI/ASHRAE (BSR(ASHRAE) Standard 135–2016
6. Bass, T.: Intrusion detection systems and multisensor data fusion. Commun. ACM **43**(4), 99–105 (2000). https://doi.org/10.1145/332051.332079
7. Beckhoff: Ethercat system documentation (2018). Technical documentation (v5.3)
8. Bhattacharyya, D.: The taxonomy of advanced scada communication protocols. J. Secur. Eng. Res. **5**(6), 517–526 (2008). https://www.earticle.net/Article/A119092
9. Bush, S.F.: Smart Grid: Communication-Enabled Intelligence for the Electric Power Grid. Wiley, Hoboken (2014)
10. Case, D.U.: Analysis of the cyber attack on the Ukrainian power grid. In: Electricity Information Sharing and Analysis Center (E-ISAC) (2016)
11. Cintuglu, M.H., Mohammed, O.A., Akkaya, K., Uluagac, A.S.: A survey on smart grid cyber-physical system testbeds. IEEE Commun. Surv. Tutor. **19**(1), 446–464 (2017). https://doi.org/10.1109/COMST.2016.2627399
12. Clarke, G.R., Reynders, D., Wright, E.: Practical modern SCADA protocols: DNP3, 60870.5 and related systems. Newnes (2004)
13. Colbert, E.J.M., Kott, A. (eds.): Cyber-security of SCADA and Other Industrial Control Systems. AIS, vol. 66. Springer, Cham (2016). https://doi.org/10.1007/978-3-319-32125-7
14. Control, S.B.: Ether-s-bus (sbus) documentation, technical documentation
15. Control, S.B.: Ether-s-i/o (esio) documentation, technical documentation
16. Demartini, C., Valenzano, A.: The en50170 standard for a european fieldbus. Comput. Stand. Interfaces **19**(5–6), 257–273 (1998). https://doi.org/10.1016/S0920-5489(98)00027-0
17. ETSI: Open smart grid protocol (OSGP): Smart metering/smart grid communication protocol (2016). Technical specification (version 2.1.1)
18. Felser, M.: Profibus manual (2017). https://www.felser.ch/profibus-manual/index.html. Technical documentation
19. FieldComm Group: Hart communication protocol specification (2013). Technical documentation (HCF SPEC-13)
20. Gungor, V.C.: Smart grid technologies: communication technologies and standards. IEEE Trans. Ind. Inform. **7**(4), 529–539 (2011). https://doi.org/10.1109/TII.2011.2166794
21. IEC: DLMS/COSEM architecture and protocols: companion specification for energy metering (2014). http://dlms.com/documents/Excerpt_GB8.pdf
22. IEEE: IEEE trial-use recommended practice for data communications between intelligent electronic devices and remote terminal units in a substation. IEEE Std 1379–1997 (1998). https://doi.org/10.1109/IEEESTD.1998.86094
23. IEEE: IEEE standard for electric power systems communications - distributed network protocol (DNP3). IEEE Std 1815–2010, pp. 1–775 (2010). https://doi.org/10.1109/IEEESTD.2010.5518537
24. Igure, V.M., Laughter, S.A., Williams, R.D.: Security issues in SCADA networks. Comput. Secur. **25**(7), 498–506 (2006). https://doi.org/10.1016/j.cose.2006.03.001
25. IntelliGrid: data management and exchange technologies (2004)
26. Jokar, P., Leung, V.C.: Intrusion detection and prevention for ZigBee-based home area networks in smart grids. IEEE Trans. Smart Grid **9**(3), 1800–1811 (2018). https://doi.org/10.1109/TSG.2016.2600585
27. Kaspersky: Threat landscape for industrial automation systems in H2 2017 (2018). Report (H2 2017)

28. van der Knijff, R.M.: Control systems/SCADA forensics, what's the difference? Digit. Investig. **11**(3), 160–174 (2014). https://doi.org/10.1016/j.diin.2014.06.007

29. Ma, S., Chen, B., Wang, Z.: Resilience enhancement strategy for distribution systems under extreme weather events. IEEE Trans. Smart Grid **9**(2), 1442–1451 (2018). https://doi.org/10.1109/TSG.2016.2591885

30. Mackiewicz, R.: Overview of IEC 61850 and benefits. In: 2006 IEEE PES Power Systems Conference and Exposition, pp. 623–630. IEEE (2006). https://doi.org/10.1109/PSCE.2006.296392

31. Martin, K., et al.: An overview of the IEEE standard C37. 118.2 - synchrophasor data transfer for power systems. IEEE Trans. Smart Grid **5**(4), 1980–1984 (2014). https://doi.org/10.1109/TSG.2014.2302016

32. Miehlisch, F.: The M-Bus: a documentation (1998). http://www.m-bus.com/files/MBDOC48.PDF

33. Modbus: Modbus application protocol specification (2012). Technical docummentaiton (version 1.1b)

34. Mohagheghi, S., Stoupis, J., Wang, Z.: Communication protocols and networks for power systems-current status and future trends. In: 2009 IEEE/PES Power Systems Conference and Exposition, pp. 1–9. IEEE (2009). https://doi.org/10.1109/PSCE.2009.4840174, http://ieeexplore.ieee.org/document/4840174/

35. O'Brien, P.D., Nicol, R.C.: FIPA - towards a standard for software agents. BT Technol. J. **16**(3), 51–59 (1998). https://doi.org/10.1023/A:1009621729979

36. ODNetVA: Quick start for vendors handbook: a guide for Ethernet/IP developers (2008). Technical documentation

37. OLE: Data Access Automation Specification: Data Access Automation Interface Standard (1999). oPC Data Access Automation Specification (version 2.02)

38. Pasqualetti, F., Dorfler, F., Bullo, F.: Attack detection and identification in cyber-physical systems. IEEE Trans. Autom. Control. **58**(11), 2715–2729 (2013). https://doi.org/10.1109/TAC.2013.2266831

39. Saputro, N., Akkaya, K., Uludag, S.: A survey of routing protocols for smart grid communications. Comput. Netw. **56**(11), 2742–2771 (2012). https://doi.org/10.1016/j.comnet.2012.03.027

40. Sharma, K., Saini, L.M.: Performance analysis of smart metering for smart grid: an overview. Renew. Sustain. Energy Rev. **49**, 720–735 (2015). https://doi.org/10.1016/j.rser.2015.04.170

41. Siemens: The Siemens S7 communication - part 1 general structure (2016). Technical documentation

42. Siemens: Snap7: Overview (2018). Technical documentation

43. Sun, C.C., Hahn, A., Liu, C.C.: Cyber security of a power grid: state-of-the-art. Int. J. Electr. Power Energy Syst. **99**, 45–56 (2018). https://doi.org/10.1016/j.ijepes.2017.12.020

44. Wang, W., Xu, Y., Khanna, M.: A survey on the communication architectures in smart grid. Comput. Netw. **55**(15), 3604–3629 (2011). https://doi.org/10.1016/j.comnet.2011.07.010

45. Wollschlaeger, M., Sauter, T., Jasperneite, J.: The future of industrial communication: automation networks in the era of the internet of things and industry 4.0. IEEE Ind. Electron. Mag. **11**(2), 17–27 (2017). https://doi.org/10.1109/MIE.2017.2649104

46. Yadav, D., Mahajan, A.R., Thomas, A.: Security risk analysis approach for smart grid. Int. J. Smart Grid Green Commun. **1**(3), 206–215 (2018). https://doi.org/10.1504/IJSGGC.2018.091349

47. Zaballos, A., Vallejo, A., Selga, J.M.: Heterogeneous communication architecture for the smart grid. IEEE Netw. **25**(5), 30–37 (2011). https://doi.org/10.1109/MNET.2011.6033033

Virtual Oversampling of Sparse Channel Impulse Response

Marcin Kucharczyk[1(✉)], Grzegorz Dziwoki[1], Jacek Izydorczyk[1],
Wojciech Sułek[1], and Bartłomiej Ulfik[1,2]

[1] Institute of Electronics, Faculty of Automatic Control,
Electronics and Computer Science, Silesian University of Technology, Gliwice, Poland
{marcin.kucharczyk,grzegorz.dziwoki,jacek.izydorczyk,
wojciech.sulek,bartlomiej.ulfik}@polsl.pl
[2] WB Electronics S.A., Ożarów Mazowiecki, Poland

Abstract. The paper presents the problem of determining the impulse response of a sparse channel, in which the propagation path delays are not restricted to the sampling grid of the signal converters. The OMP (Orthogonal Matching Pursuit) algorithm performs very well, if the nonzero channel impulse response coefficients are in synchrony with the sampling grid. We propose in this paper to virtually increase the resolution of the channel sampling model in order to increase the estimation effectiveness. We describe the importance of the interpolation filter proper selection. Then the slightly modified OMP algorithm is proposed to find coefficients that estimate the transmission channel characteristics close to the ideal one. The equalizer quality has been increased in comparison with the one determined by the basic OMP.

Keywords: OFDM · Channel estimation · Channel equalization · Sparse channels · Compressed sensing

1 Introduction

Today, the OFDM (Orthogonal Frequency Division Multiplexing) modulation is widely used in the digital data transmission systems. The reception of data with low error rate in an OFDM wireless multipath propagation scenario is possible if the transmission channel parameters are properly estimated in the equalizer. The channel equalization problem in an OFDM transmission system was widely discussed [1–3]. The simplest solution is an independent one tap equalizer for each OFDM subchannel calculated with Zero Forcing algorithm, using the received pilot frame signal [4]. This is practically useful solution especially when combined with SNR estimation by the MMSE (Minimum Mean Square Error) algorithm [4,5].

This MMSE algorithm has been implemented in the wireless transmission system model, which is a basis of the research presented in this paper. However, the estimation quality of MMSE itself is not outstanding and it can be improved

© Springer Nature Switzerland AG 2019
P. Gaj et al. (Eds.): CN 2019, CCIS 1039, pp. 272–286, 2019.
https://doi.org/10.1007/978-3-030-21952-9_21

significantly. The research goal was to improve the equalizer without redesign of the whole system. Analysis of the channel impulse response calculated from the received pilot signals, recorded in a real multipath radio propagation environment, revealed that the estimation is often inaccurate in the case of channel with just 2 close propagation paths. Such channel is characterized by a frequency response characteristic with a single deep fade. The sparsity of the observed channel impulse responses redirected the research to compressed sensing algorithms [6–8] designed to estimate the signal based on its simple physical form.

Theoretically, the two paths channel has only two nonzero coefficients in the channel impulse response. In the band limited systems, this kind of impulse response is observed if the distance between paths is equal to multiplicity of sampling period and sampling moments of the signal in the transmitter and the receiver are perfectly synchronized. Shortly: the delays of the paths are in the grid of the signal sampling. In this case, iterative compressive sensing algorithms like OMP [7] and CoSaMP [8] perform well in channel impulse response recovery. But if the synchronization is not perfect, the channel impulse response is observed as the convolution of the real channel response (with the number of coefficients limited to the number of propagation paths) and the impulse response (theoretically infinite) of the low pass filter representing the model of transmitter-receiver system in the baseband [9,10].

The idea presented in this paper is based on separation of the real channel impulse response from the system filter response and making a precise recovery of the delays of the channel propagation paths with the compressive sensing algorithm. We propose to use oversampling [11] of the channel impulse response in the equalizer and execute a path search with OMP algorithm on the signal with time resolution increased over the sampling frequency. Estimated impulse response of the channel is then filtered by a low pass filter, which models the overall filtering implemented in the transmission system, giving more coefficients for equalizer estimation.

The resolution of the oversampled channel impulse response is higher than original, but it is still limited. Therefore we propose to use in the estimation stage of the OMP algorithm not only the delay values obtained in the search stage, but also the nearest neighbors, assuming that the real path delays are in that range. This idea was partially used in [10], but without signal oversampling. The simulation results confirmed the validity of the proposed approach.

In the next section, the description of the channel impulse response estimation problem will be presented in more detail. Section 3 deals with the problem of using the compressive algorithm with the oversampled signal. In Sect. 4 we consider the errors caused by inaccurate delay estimation and mismatching the filter parameters used in the transmission system and in the equalizer coefficients calculations. Then, this analysis leads to the proposed estimation algorithm improvement. The simulations results of this algorithm are presented in Sect. 5. Some conclusions are collected in the last section.

2   Problem Statement

The research concerns a model of an OFDM transmission system with pilot signals transmitted in each subchannel in the first OFDM frame, in which the parameters of the channel equalizer are calculated completely in the frequency domain. The OFDM system involves K subchannels for transmission, thus the vector $\mathbf{P}_x(k)$ of signal samples (in the frequency domain) contains K pilot signals in the first OFDM frame. After IFFT transform, the (time domain) sampled signal $\mathbf{p}_x(n)$ is extended with a cyclic prefix and after D/A conversion transmitted through the channel with impulse response $h(t)$. The signal is additionally distorted by the noise $n_0(t)$ and then it is sampled and A/D conversion is made on the receiver side. After removing the cyclic prefix, the $\mathbf{p}_y(n)$ signal is obtained and transformed using FFT to $\mathbf{P}_y(k)$ – a vector, which is used for estimation of K one-tap equalizers using Zero Forcing algorithm [4]:

$$\mathbf{E}(k) = \frac{1}{\mathbf{H}(k)} = \frac{\mathbf{P}_x(k)}{\mathbf{P}_y(k)} \tag{1}$$

The above formula is very simple and it does not exploit any information about physical form of the channel. It is known [12,13] that real channels can be effectively modeled as a short FIR filters with coefficients that represent signal propagation paths. In most cases the considered number of paths does not exceed six [14]. The closer investigation of our transmission system led us to notice that in many cases the signal has one main path, but the serious problem in channel estimation occurs, when non-neglected second path with similar delay is present. This type of channel, with only two close paths, leads to very deep fade in channel frequency response and makes a number of subchannels almost useless for transmission. However, if the result of channel estimation is precise and similar to the real shape, the chance of proper error correction in the FEC (Forward Error Correction) decoder grows significantly [15].

The discrete impulse response of the channel $\mathbf{h}(n)$ can be calculated from (1) using IFFT, therefore the received training data defined by pilots in frequency domain is enough to create an equation, which includes impulse response of the channel [6]:

$$\mathbf{P}_y(k) = \mathbf{P}_x(k)FFT(\mathbf{h}(n)) = \mathbf{P}_x(k)\mathbf{F}(k,n)\mathbf{h}(n), \text{ or} \tag{2}$$

$$\mathbf{H}(k) = FFT(\mathbf{h}(n)) = \mathbf{F}(k,n)\mathbf{h}(n), \tag{3}$$

where k and n are data indices respectively in the frequency domain and in the time domain. The sizes of matrices are limited to the number of pilots in the frequency domain K and the number of coefficients of the channel impulse response N in the time domain. Both values determine the size of the Fourier transformation matrix $\mathbf{F}$. If the pilots are transmitted in all subchannels, the matrix sizes are equal ($K = N$) and are equal to the size of the FFT used in the OFDM system.

The unknown in the above equations is the channel impulse response $\mathbf{h}(n)$, which essentially involves also the noise added to the transmitted signal. The result is a vector of length N, but as it was mentioned, the real channel impulse response is much shorter. Actually, it should be shorter than cyclic prefix length P to avoid inter-symbol interferences.

The noise is omitted in the above formulas, but it exists in the real transmission system. After taking into account the noise, the equation describing the operation of the system takes the form (4), where $\mathbf{E}_N(n)$ represents an error value caused by the noise:

$$\mathbf{H}(k) = \mathbf{F}(k,n)\mathbf{h}(n) + \mathbf{E}_N(n). \tag{4}$$

2.1 FIR Filter Design

Solution of the estimation according to Eq. (3) can be perceived as a design of a FIR filter $\mathbf{h}(n)$ of the known frequency response $\mathbf{H}(k)$. In this point of view, the matrix $\mathbf{F}(kn)$ represents the base signals for subsequent frequencies. Typically, it is assumed that the length of the designed FIR filter is limited to the length of the cyclic prefix P. The cyclic prefix is added to the signal to avoid inter-symbol interference (ISI), therefore its time-length should be greater than the length of the channel impulse response $h(t)$. Assuming that the receiver is well synchronized, non-zero coefficients of the channel impulse response are expected in the first P samples. Therefore, according to this assumption, it is possible to decrease dimensions of the matrix $\mathbf{F}(k,n)$ and the result $\mathbf{h}(n)$.

After decreasing the size, the $\mathbf{F}(k,n)$ matrix has more rows than columns, thus the set of Eq. (2) is overdetermined and the solution $\mathbf{h}(n)$ is an approximation. Using LS algorithm [16], the calculated impulse response is the nearest to the optimal in terms of ℓ_2-norm between received and calculated pilot signal. The solution does not make any assumption about the real channel impulse response $h(t)$. However, it is expected $h(t)$ is sparse.

2.2 Compressive Sensing

The real channel impulse response $h(t)$ has limited number of paths [9,14], so it is a sparse signal. A solution that minimizes approximation error of the sparse response $h(t)$, with a predetermined number of coefficients, can be achieved with a compressive sensing algorithm. It is known that the FFT matrix $\mathbf{F}(k,n)$ can be used as a sensing matrix in such algorithms [6], so the Eq. (4) describes the sparse system. Considering the noise, the optimization problem can be defined as:

$$\min \|\mathbf{h}(n)\|_{\ell_0} \text{ subject to } \|\mathbf{F}(n,k)\mathbf{h}(n) - \mathbf{H}(k)\|_{\ell_2} \leq \epsilon, \tag{5}$$

where ϵ bounds the estimation error to the noise level in the received signal. The minimization of $\mathbf{h}(n)$ with ℓ_0-norm can be done with algorithms like Orthogonal Matching Pursuit (OMP) and Compressive Sampling Matched Pursuit (CoSaMP). Both algorithms search for the best position of $\mathbf{h}(n)$ filter coefficients and then calculate the best approximation of them using LS-algorithm, like in the filter design case described above. If the path delays of the signal are properly identified, the approximated impulse response is close to the real one with the minimum variance.

Unfortunately, neither OMP nor CoSaMP does not suggest how many coefficients need to be found. The CoSaMP algorithm searches always for a defined number of paths [8], which number is actually not known in advance. The simple OMP can also be implemented to search for a defined number of paths, but other stop conditions were proposed in the literature as well [10]. These conditions are based on the power of residue value (internal data of the mentioned greedy algorithms) in the subsequent iterations. The good stop condition is the power of residue just lower than the power of noise [10, 13], but the noise level is unknown to the receiver. However, it can be more or less accurately estimated [17, 18]. A different approach is to compare residue power between iterations and stop, when its value changes slower comparing to previous iterations [10].

Remark some inaccuracy in the above description: it was mentioned that the $h(t)$ is sparse, but the compressive sensing algorithm searches for the best approximation of $\mathbf{h}(n)$. If the distance between signal reflections is equal to the sampling period and the sampling moments of the signal in the transmitter and in the receiver are perfectly synchronized, the discrete channel impulse response $\mathbf{h}(n)$ is exactly equivalent to the continuous response $h(t)$. If those conditions are not satisfied, signal paths are out of the sampling grid, and the number of coefficients in the digital impulse response of the channel $\mathbf{h}(n)$ is greater than the number of paths. It is because the signal is not only propagated by channel $h(t)$ with limited number of paths but, in the real system, it is also filtered to the desired transmission band in both the transmitter and the receiver.

3   Filtered Impulse Response of the Channel

The signal in the transmission system needs to be filtered to the desired band to fulfill regulations and avoid interferences with other systems. The processing path of the signal contains both analog and digital filters. Influence of these filters to equalizer design is not negligible, but it is somehow automatically taken into account in the basic zero forcing algorithm (1).

Even if an ideal band-pass filter is used, which does not disturb the signal in frequency domain, the shape of the signal in the time domain changes. Impulse response of the channel $\mathbf{h}(n)$ in the (2) might be no longer sparse and compressive sensing algorithms do not work as well as expected. The main problem is the number of coefficients to find [10].

It is possible to include these band filters to the equalizer design equations, similarly to the model presented in literature [19], and it leads to the following computation problem:

$$\mathbf{P}_y(k) = \mathbf{P}_x(k) \cdot \mathbf{F}(k, n) \cdot \mathbf{S}(n, t) \cdot h(t), \qquad (6)$$

where $\mathbf{P}_x(k)$ and $\mathbf{P}_y(k)$ are still the same transmitted and received pilot signals in frequency domain, hence the system does not have any additional measurements of the signal. The number of elements in the matrix $\mathbf{F}(k, n)$ is also unchanged, as the rows of the matrix correspond to the channel frequencies used in the OFDM transmission, while the columns correspond to the number of coefficients in the digital impulse response of the channel.

The vector $\mathbf{h}(t)$ corresponds to the real analog sparse impulse response of the channel. But the system is digital, so it is a sampled signal. The sampling frequency needs to be higher than the one used in the D/A and A/D converters in the transmitter and the receiver. The oversampling here is only a virtual, mathematical operation, because the real sampling frequency of the signal remains the same. The last element of the Eq. (6), the matrix $\mathbf{S}(n, t)$, describes the digital FIR filter, which represents signal filtering by the processing path of the transmission system.

The main advantage of this proposed model is that the impulse response $h(t)$ in the equation is really sparse. Some of the problems that needs more investigation are: oversampling factor and the required density of the impulse response of the channel $h(t)$, impulse response of the FIR filter $\mathbf{S}(n, t)$ and system properties in view of compressive sensing algorithms - properties of the modified sensing matrix, which is now multiplication of two matrices $\mathbf{F}(k, n)$ and $\mathbf{S}(n, t)$.

3.1   The Shaping Filter

Basic resampling procedure of a signal consists of 3 stages [11]: interpolation, filtering and decimation. The last 2 stages are implemented in the Eq. (6). The oversampled signal $h(t)$ requires filtering to become band-limited impulse response of the channel and decimation to match the number of samples with the number of pilot signals.

The filter $\mathbf{S}(n, t)$ primarily models the behavior of the real system, among others antialiasing filtering. The filter parameters can be measured for the specific transmitter and receiver pair, but generally a real shape of this filter is unknown. The known required parameter is a frequency response of the filter, therefore assuming that the system is equipped with ideal band pass filter, the *sinc* function for the shaping of the channel impulse response may be used. Another function that can be used in the general case is a raised cosine function [19].

If the oversampling factor will be set to 1, which means no oversampling, then the shaping function transforms to the Kronecker delta. It has zero value for all elements except the main one and the matrix $\mathbf{S}(n, t)$ becomes identity matrix, so the equation (6) takes back to the base form (2).

3.2   The Sensing Matrix

Compressive sensing algorithms give useful results, if the sensing matrix fulfills restricted isometry property [6,7]. If the matrix and the sparse signal satisfies RIP of order m, then m-sparse vector is enough to reconstruct the signal. In practice it is difficult to compute the RIP for a given sensing matrix $\mathbf{A}$, as a search must be performed over all combinations of the matrix columns. The frequently used parameter for the practical measurement of the sensing matrix for sparse recovery is its coherence [6]. For a given matrix $\mathbf{A}$ it is defined as follows:

$$\mu = \max_{i \neq j} \frac{|a_i^H a_j|}{\|a_i\|_{\ell_2} \|a_j\|_{\ell_2}}, \tag{7}$$

where a_i, a_j are the columns of $\mathbf{A}$. The lower the μ value is, the better the sensing matrix $\mathbf{A}$ for distinguishing elements of the sparse signal in compressive sensing algorithm is. The low value of μ means that columns of sensing matrix are highly incoherent and significant values in sparse result will be better distinguished by OMP or CoSaMP greedy algorithms. There is no sharp boundary binding μ value with signal sparsity, but the following formula is sometimes used [6,20]:

$$M < \frac{1}{2} \left(\frac{1}{\mu} + 1 \right), \tag{8}$$

which allows to estimate maximum sparsity of the signal properly calculated with the given sensing matrix. After searching for M results, the next index values could be not so well selected from the remaining ones by matching pursuit algorithms. But in general, the lower value of μ is calculated, the better the matrix for compressive sensing algorithms is.

The base sensing matrix of the presented algorithm is an FFT matrix, which is known to behave well in algorithms searching for sparse signals [6]. The columns of FFT matrix are highly incoherent as the value of μ is zero, so such matrix can be used for searching signal with any sparsity. Virtual oversampling and the filtering of the impulse response by the matrix $\mathbf{S}$ changes parameters of sensing matrix. For the base FFT matrix with size of 128, the calculated μ value is 0.64 and 0.90 for the 2× and 4× oversampling respectively with ideal $\sin(x)/x$ filtering.

The μ value close to 1 means that in the sensing matrix, there exist two columns with similar selective properties. But closer look at the distribution of correlation values from the right side of Eq. (7) shows that the high ones are only near to central, autocorrelated columns. An example of a distribution of μ value near to the center for 4× oversampling are presented in Fig. 1.

The distance between values in Fig. 1 is a quarter of sampling period and it depends on oversampling factor. There are two conclusions from the data in the figure: (1) do not expect that the algorithm with oversampling will be able to distinguish paths with the distance smaller than the system's sampling period and (2) if the sparse response search algorithm finds paths closer than the sampling period, they are the result of "blurring" of the same path.

4   Estimation Errors

The Eq. (6) includes continuous time $h(t)$ response. It means that the delay of the propagation path in the real system can be observed with any resolution. But in fact, the discrete time values are calculated here with the accuracy depending on the oversampling factor. Higher overampling factor leads to possible higher

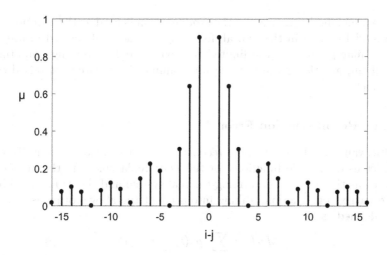

Fig. 1. The values of μ parameter from (7) around the central point for 4× oversampling.

precision of determining the delay of the each signal path. On the other hand, with increasing the number of samples of the channel impulse response, computation complexity grows. And the last but not least: as it was shown in the previous section, selective properties of the sensing matrix decrease when the oversampling factor increases.

The precision of determining the delay limited by the oversampling factor is the first source of the errors in the process of channel estimation in presented method. The second source of the estimation error is the difference between characteristics of the low pass filter used in the system and the one included in the sensing matrix in the Eq. (6). We expect that both errors decreases, when the calculated delay is closer to the real one and the filter used in the calculation better matches the real ones.

4.1   The Discrete Channel

Based on the [9], the bandlimited discrete channel can be defined as:

$$g_n(t) = \sum_{k=1}^{K} a_k(t)\alpha(k, n), -N \leq n \leq N, \tag{9}$$

where $\alpha(k, n)$ defines the shaping function sampled with the period T:

$$\alpha(k, n) = func\left[\frac{\tau_k}{T} - n\right]. \tag{10}$$

Parameters a_k and τ_k define the amplitudes and the delays of the K paths in the channel model. The length of the model is limited to $2N + 1$, where N is the number of columns in the sensing matrix in (6).

For channel modeling, usually the *sinc* function is used as the function *func*. Such a model is used in the Matlab following the model described in literature [9,21]. Similar properties, the limited bandwidth and decreasing amplitude in time domain, are also characterizing for example the square root raised cosine (*srrc*) function.

4.2   The Reconstruction Error

The MSE value can be used as the reconstruction error measurement. To calculate it, let us define two functions for the *func* replacement in the model (10): the *func1* as the function describing the real system and the *func2* used in the sensing matrix construction. Taking into account the above formulas, the MSE error is defined as:

$$MSE = \sum_{n=1}^{N} (g_n(t) - h_n(t))^2,$$ (11)

where:

$$g_n(t) = \sum_{k=1}^{K} a_k(t) func1 \left[\frac{T_k}{T} - n \right] \text{ and}$$ (12)

$$h_n(t) = \sum_{l=1}^{L} b_l(t) func2 \left[\frac{T_l}{T} - n \right].$$ (13)

The first source of the error seems to be obvious: the *func1* is different than *func2*. The second source of the error is truncating of the impulse response which takes the compressive form: 1-tap filter defined by delay goes to infinite *sinc* like function which is truncated in model definition.

4.3   Summary

The above analysis leads to the conclusion that a non-zero estimation error is expected for a virtual oversampling algorithm: the oversampling factor must be limited in order to reduce computational complexity, and additional measurements are required to ensure that the filter used in the calculation algorithm matches the real one. In the [19], the authors suggest that the function *func* used in the algorithm is not very important. Moreover, they proposed to skip close neighbors of the coefficients found in the search stage of greedy algorithm in the estimation step. Based on the above analysis and the numerical results, some modifications have been made in the algorithm. They will be described in the next section.

5   Numerical Simulation

The analysis presented earlier focuses on determining the impulse response of the channel with the resolution higher than sampling period. The next stage in the receiver is creation of the standard 1-tap equalizer for the each channel of the OFDM system. It is made out of reversal of the product of elements from the right side of the Eq. (6):

$$\mathbf{E}(k) = \frac{1}{\mathbf{F}(k,n) \cdot \mathbf{S}(n,t) \cdot h(t)}. \tag{14}$$

Number of elements in the vector $\mathbf{E}(k)$ is equal to the number of pilots. And the Eq. (14) consists of 2 operations: filtering of the channel impulse response by the shaping filter $\mathbf{S}(n,t)$ and FFT operation on that filtered value. The filtering here is an operation known from creating the model of the channel [9]. Taking into account the results of the analysis in the previous chapter, better results are expected, if the same filter is used in both cases: modeling and estimating the channel. In the real system, it means that the filter parameters need to be measured in the transmission devices.

5.1   The Algorithm

The basic algorithm based on the above analysis is as follows:

1. Create a sensing matrix $\mathbf{H} = \mathbf{PFS}$ as in the (6).
2. Use the OMP algorithm [7] to estimate channel impulse response $\mathbf{h}$ from (6).
3. Calculate the equalizer coefficients using formula (14).

Such algorithm works fine, if the delays are in the grid of oversampled impulse response $\mathbf{h}$ and were correctly found in the search stage of the OMP algorithm. But when the delays are off the grid and because of poor distinguishing properties of the sensing matrix $\mathbf{H}$ (see Fig. 1), the searching stage was slightly extended. The matrix used in the least square (LS) algorithm in OMP consists of the maximum found in each iteration of the search phase and the neighboring columns from the sensing matrix. Similar procedure is used in the block OMP algorithm [22], which assumes that the result is block-sparse.

Summarizing, the step 2, the OMP operation, from the above algorithm consists the following operations:

(a) find position n_i of maximum value in the residue part of channel impulse response;
(b) with the LS algorithm calculate the temporary $\mathbf{h}(n)$ using previously found positions and additional 3 ones: $[n_i - 1, n_i, n_i + 1]$;
(c) if the stop criterion is not met go back to the step (a).

5.2   Simulation Parameters

Presented algorithm was tested in the environment that simulates the transmission system. The OFDM modulation was used with QPSK constellation in each of 128 channels. The size of cyclic prefix was set to 32 samples, sufficient value compared to the reflection paths observed in the measured channels. Most of the

simulations were made with the channel with close 2 paths with similar attenu-
ation. The distance between paths was in a range 0.5 to 2.5 of sampling period
and the sampling moment was changed in the time of one sampling period. This
type of channel does not look complex in the time domain but is very difficult
to equalize due to the existence of deep fades in the frequency domain.

The second path in the model was rotated between simulations to fulfill the
whole $2pi$ period. It results in moving of the fades in the whole frequency band
of the transmission system. To get results statistically useful, 30 similar channels
were simulated and more than 30000 QPSK symbols (1 frame with pilots and
250 frames with random data) were sent by each of them. The channel model
is created according to the literature [9], particularly the paths were filtered by
sinc function to get values off the grid.

The channel estimated using OMP with oversampling algorithm presented in
this paper was compared with the standard OMP algorithm. For the reference it
was also compared to the result obtained using the simple Zero Forcing equalizer
with noise estimation (*ZF MMSE* in figures) and with the data obtained when
the simulated channel state information was used for equalization (*known CSI*
in figures). So the *known CSI* result is the limit to reach with the equalizer based
on estimated channel.

5.3   The Results

The different equalizers were compared in terms of symbol error rate (SER) in
relation to the signal to noise ratio (SNR).

The first result presented in Fig. 2 is a comparison of the basic OMP algorithm and the algorithm with oversampling (*OS F=X* in the figure, where X is
the oversampling factor) for the channels with two paths in the 1 or 2 sample
distance. So the channels are in the range defined earlier, but the delays are
always in the sampling grid. As it can bee seen in the figure, the basic OMP
works almost perfectly and the oversampling does not provide an improvement.
All versions of the OMP algorithm searches for the 2 paths (simple stop criterion). The closer look on the results shows that the distance between estimated
and known channel response is about 0.3 dB at $SNR = 25$ dB.

The next results, in the Fig. 3 are obtained in more realistic model, when
the paths are not in the sampling grid. When searching 2 paths the standard
OMP algorithm does not give useful result. Based on our previous work [10] the
additional stop criterion was implemented in the OMP search algorithm which
analyzes the power of residual error. So the number of coefficients in channels
impulse response differs between simulations. The result (*OMP stop*) is much
better than basic OMP, but still not satisfying.

The oversampling algorithm performs well in this model. The estimated channel is in distance about 0.7 dB at $SNR = 25$ dB to the *known CSI* result, even
closer for lower SNR values. The poor properties of the sensing matrix results
in the almost negligible gain for higher values of oversampling factor.

The results in the Fig. 4 presents influence of the mismatching of the shaping
filter in the transmission system and in the equalizer. For the low SNR values, the

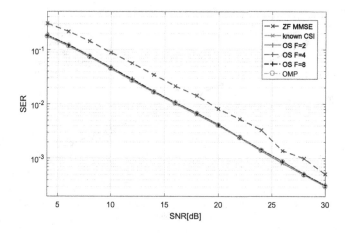

Fig. 2. SER = f(SNR) for the channel with 2 paths in the grid

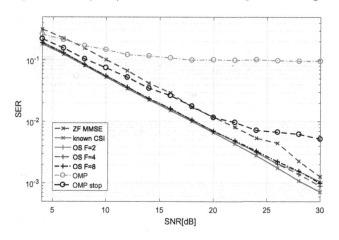

Fig. 3. SER = f(SNR) for the channel with 2 paths off the grid

filter used in the equalizer can differ from the one used in the system, but when the SNR grows matching is required. The factor 4 was used in the oversampled models here. Channel was generated with the *sinc* function and the equalizer uses *srrc* with various α values.

In the last presented result the channel type was changed. Based on the COST-207 model [14] a number of channels with 6 propagation paths were created. Gain values from the COST-207 were used but the absolute delay values from that model was compressed to the range of cyclic prefix in the simulation environment. Nevertheless, the relationships between the delay values have been preserved. The number of paths to search in the OMP algorithm was increased to 6.

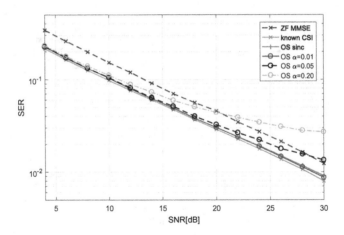

Fig. 4. SER = f(SNR) for the channel with 2 paths in the grid for different shaping filters

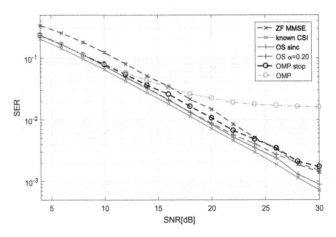

Fig. 5. SER = f(SNR) for the 6 paths COST-207 channel model.

The results for the COST-207 model presented in the Fig. 5 are similar to the ones obtained for channel with 2 paths only. Significant differences are observed for higher SNR values. Basic OMP again does not perform well and the importance of the right shaping filter selection is visible. These results shows that the proposed algorithm performs well not only for simple 2 paths channels, which initiated the research presented in this paper.

6   Conclusion

Orthogonal greedy algorithms like the OMP allow for inclusion of physical channel properties in channel estimation for equalization. The results are very satisfying when delays of the propagation paths of the channel are in the sampling grid. Otherwise the algorithm needs some modifications to get the required performance.

The paper presents some simple modifications to the base OMP algorithm that allows for improvement of the greedy algorithm results for channels with delays beyond the resolution of the sampling system. The sampling frequency of the channel impulse response is increased during calculations, so the delay resolution at search stage of the OMP algorithm grows. Transmission system uses the same sampling frequency, but it is virtually increased before the search stage of the OMP algorithm. Delay values detection is made with better resolution, so results are closer to real path delays. After additional modifications of the OMP algorithm, the final equalizer performs well in comparison to the one using basic OMP. For the entire SNR range used in simulations, the results achieved with equalizer estimated using proposed algorithm is in distance about 0.7–0.8 dB to the results got with equalizer calculated from known CSI value.

The main disadvantage of the algorithm is growing computational complexity in comparison to the base OMP algorithm. In addition, we realize that some additional theoretical analysis is required to determine best oversampling factor values. From the simulations results it can be concluded that the factor of 4, or at most 8 is sufficient.

References

1. Liu, Y.S., Tan, Z.H., Hu, H.J., Cimini, L.J., Li, G.Y.: Channel estimation for OFDM. IEEE Commun. Surv. Tutorials **16**(4), 1891–1908 (2014)
2. Ozdemir, M.K., Arslan, H.: Channel estimation for wireless OFDM systems. IEEE Commun. Surv. Tutorials **2**(4), 18–48 (2007)
3. Hwang, T., Yang, C., Wu, G., Li, S., Li, G.Y.: OFDM and its wireless applications: a survey. IEEE Trans. Veh. Technol. **58**(4), 1673–1694 (2009)
4. Mark, J.W., Zhuang, W.: Receiver techniques for fading dispersive channels. In: Wireless Communications and Networking, Prentice Hall, New Jersey (2003)
5. Schniter, P.: Low-complexity equalization of ofdm in doubly selective channels. IEEE Trans. Signal Process. **54**(4), 1002–1011 (2004)
6. Candes, E.J., Wakin, M.B.: An introduction to compressive sampling. IEEE Signal Process. Mag. **25**(2), 21–30 (2008)
7. Tropp, J., Gilbert, A.: Signal recovery from random measurements via orthogonal matching pursuit. IEEE Trans. Inf. Theory **53**(12), 4655–4666 (2007)
8. Needell, D., Tropp, J.A.: CoSaMP: iterative signal recovery from incomplete and inaccurate samples. Appl. Comput. Harmonic Anal. **26**(3), 301–321 (2009)
9. Jeruchim, M.C., Balaban, P., Shanmugan, K.S.: Simulation of Communication Systems, 2nd edn. Kluwer Academic/Plenum, New York (2000)
10. Dziwoki, G., Kucharczyk, M., Izydorczyk, J.: Modified OMP algorithm for compressible channel impulse response estimation. In: Gaj, P., Sawicki, M., Suchacka, G., Kwiecień, A. (eds.) CN 2018. CCIS, vol. 860, pp. 161–170. Springer, Cham (2018). https://doi.org/10.1007/978-3-319-92459-5_13
11. Eldar, Y.C.: Sampling Theory: Beyond Bandlimited Systems. Cambridge University Press, Cambridge (2015)
12. Berger, C.R., Wang, Z.H., Huang, J.Z., Zhou, S.L.: Application of compressive sensing to sparse channel estimation. IEEE Commun. Mag. **48**(11), 164–174 (2010)

13. Dziwoki, G., Izydorczyk, J.: Iterative identification of sparse mobile channels for TDS-OFDM systems. IEEE Trans. Broadcast. **62**(2), 384–397 (2016)
14. Failli, M.: Digital Land Mobile Radio Communications COST 207. Technical report, European Commission (1989)
15. Sułek, W.: Protograph based low-density parity-check codes design with mixed integer linear programming. IEEE Access **7**, 1424–1438 (2018)
16. Press, W.H., Teukolsky, S.A., Vetterling, W.T., Flannery, B.P.: Numerical Recipes: The Art of Scientific Computing, 3rd edn. Cambridge University Press, New York (2007)
17. Khan, A.M., Jeoti, V., Rehma, M.Z.U., Jilani, M.T.: In: 30th IEEE Canadian Conference on Electrical and Computer Engineering (CCECE), Windsor, ON, Canada (2017)
18. Hendriks, R.C., Jensen, J., Heusdens, R.: Noise tracking using DFT domain subspace decompositions. IEEE Trans. Audio Speech Lang. Process. **16**(3), 541–553 (2008)
19. Chen, C., Zoltowski, M.D.: A modified compressed sampling matching pursuit algorithm on redundant dictionary and its application to sparse channel estimation on OFDM. In: 45th Asilomar IEEE Conference on Signals, Systems, and Computers, pp. 1929–1934. IEEE, Pacific Grove, CA (2011)
20. Donoho, D.L.: Compressed sensing. IEEE Trans. Inf. Theory **52**(4), 1289–1306 (2006)
21. MathWorks, Fading Channels. https://www.mathworks.com/help/comm/ug/fading-channels.html
22. Eldar, Y.C., Kuppinger, P., Bolcskei, H.: Block-sparse signals: uncertainty relations and efficient recovery. IEEE Trans. Signal Process. **58**(6), 3042–3054 (2010)

Recreation of Containers for High Availability Architecture and Container-Based Applications

Rafal Pawlik and Jan Werewka[(✉)]

Department of Applied Computer Science,
AGH University of Science and Technology, Kraków, Poland
{rpawlik,werewka}@agh.edu.pl

Abstract. Over the past few years containers have become a very popular solution for virtualization purposes. They have all major advantages of virtual machines and, additionally, reduced hardware and time requirements and built-in support for detecting faults in containers and recreating the ones which are not working. Docker is an open platform which enables running applications in a loosely connected environment called a container. A Docker Swarm framework may be used on a Docker platform to cluster and schedule containers. The time needed for detecting and recreating failed containers was tested in this paper. The time needed to recreate a given container depends on the size of the container and the reason of its failure. Docker Swarm needs from 8 to 31 s to recreate a failed container, however, for some purposes, e.g. e-commerce systems, it is too long. In such systems the administrators attempt to minimize the time of website unavailability. To mitigate this problem, we propose the architecture of a lightweight system for fast monitoring of containers and scheduling their recreation. The prototype based on the proposed architecture is insensitive to a container image size, and recreation times depend only on the reason of a container failure. The recreation times range from 3 to 10 s. If we omit the time needed to create containers, which depends solely on internal Docker Engine implementation, our solution detects a container failure and schedules the recreation of a container instance almost instantaneously. This paper presents a part of the holistic solution of fast and lightweight container creation, monitoring and responding to failures.

Keywords: Resilience · Containers · Docker Swarm

1 Introduction

Container environments are a desirable solution for microservice-based systems. It is very important to maintain high availability in these systems, which is achieved mainly through the use of immune solutions (resilience). This means

© Springer Nature Switzerland AG 2019
P. Gaj et al. (Eds.): CN 2019, CCIS 1039, pp. 287–298, 2019.
https://doi.org/10.1007/978-3-030-21952-9_22

that we want to achieve high availability by proposing solutions for quick recovery from emergency states. Due to the complexity of the distributed environment, the considered issue is also a complex one.

The paper proposes an architecture model of the system which monitors a cluster and containers operating on this cluster. This system will detect and respond to failures, which will result in an improvement in its resilience.

Container solutions are important because they allow for reducing hardware and time requirements - fewer hardware requirements for scaling and a faster launch of a new service instance. Docker Swarm is a good solution for the future because it allows us to quickly replicate services with low consumption of hardware resources. It uses one image to run multiple instances, which leads to the same start time, regardless of the size of the image.

The analysis of Docker Swarm and its effect on resilience can help to determine how quickly it can detect a container failure and play it back and the limitations in which Docker Swarm scenarios work and in which they do not.

In the proposed cluster monitoring architecture, the system will detect failures of both containers and cluster nodes and will react to them. The terms recovery and recreation will be used as follows. Recovery is an operation of returning a system to normal health. Recreation is a part of the process of creating something in the system from scratch.

2   Characteristics of the Container Environment

Virtualization technologies have become very popular in recent years. They come with a few advantages, like the easiness of moving a virtual machine between physical nodes, restoring the machine to a stable form or resource isolation. Containers offer the same benefits as virtual machines and bring more. Virtualization technologies require the Operating System to be installed on every virtual machine, which increases hardware resources usage. Containers are light-weight, they do not require Guest Operating System and an additional start-up time. Containers are more efficient and portable. Multiple containers can share the same OS kernel, where each of them is run as an isolated process. Because containers use fever hardware resources, they can handle more applications on the same machine than VMs [1].

Figure 1 compares VM and container architectures. There are several differences between them. VM includes everything that is required to run an application, like the operating system, all 3rd party binaries and application files. Container includes only executables with the required dependencies. In addition, different containers deployed on the same machine share the operating system, and containers based on the same container image share 3rd party binaries.

A container image is a standalone, lightweight package that has everything software requires to run, like a code, libraries, tools, settings and runtime [2]. When a container image is run on Docker Engine, it becomes a container. Docker is the most popular container engine. It is available for both Windows and Linux. One advantage of containers is that, regardless of hosting OS and infrastructure,

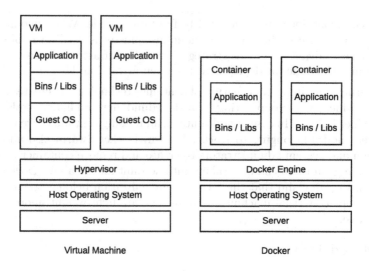

Fig. 1. A comparison of virtual machine and Docker architectures.

the containerized software will always run the same way. Like virtual machines, containers isolate applications and software from their environment. Using containers, we can quickly deploy and scale applications into any environment.

The first release of Docker technology took place in 2013. It is still unique today because it focuses on the separation of application dependencies from infrastructure. Its success in the Linux world led to a partnership with Microsoft and soon Docker Engine became available for Windows Servers [2]. Over the last years Docker was leveraged by all major cloud providers and data center vendors. It is also used by leading open source serverless frameworks, like Apache OpenWhisk or Iron.io.

Docker Engine 1.12 introduces new cluster management and orchestration feature called Docker Swarm. With Swarm, application and system administrators can manage Docker cluster as a single virtual system. Docker nodes within a cluster run in a swarm mode. Some of them act as managers (they manage membership and delegation) and others act as workers (running swarm services). It is also possible for one node to perform both roles [3].

Docker Swarm decides where a newly requested container should be deployed. It balances containerized application workloads to ensure that every container was deployed and run on a system with adequate resources. Swarm uses three strategies to achieve that:

– Spread – it is a default setting, deploying containers across all nodes in clusters based on CPU and RAM availability as well as a number of already running containers. The biggest advantage of this strategy is that when one node fails, only several containers will have to be recreated.

- BinPack – containers are scheduled to fully use a node. When one node has fully used resources, then the next node in a cluster is used for deployment. The biggest advantage of this strategy is smaller infrastructure usage.
- Random – the node for deployment is chosen at random [4].

Clustering is an important feature of container technology. It creates a group of systems which, while cooperating, provide redundancy and handle node outage and container failures. In case of node outage, all missing containers are recreated on different nodes belonging to Docker Swarm. Docker Swarm also allows the application and system administrators to create and remove container instances when demands on them change. Taking into account all the above Docker characteristics, we can claim that the future of virtualization belongs to containers. They will replace virtual machines because they are light-weight, more efficient and portable.

3   Related Works

There are papers investigating efficiency of using containers in different environments, and some of them consider specific situations aimed at improving given quality attributes.

Paper [5] investigates the sizes of virtual machines which host containers, with the purpose of efficient resource usage. The obtained results point out up to 7.55% decrease of average energy consumption in comparison to the scenario with fixed sizes of virtual machines.

In [6] a checkpoint-and-restore technique is proposed to increase the availability of Docker Swarm. A remote storage server saves images at check points. The proposed method can save 50% of storage space with a low overhead on executing processes.

Containers enable using microservices architectures thanks to their being lightweight, delivering fast start-up times, and possessing a low overhead. The performance of microservices architectures using containers is demonstrated for two models in [7]: master-slave and nested-container. The goal of the paper is to compare the performance of CPU and network for these models and to present the results to system designers.

In [8] the performance of Docker-based MQTT server implementation on the internet of things device is evaluated. The obtained results show that Docker containers can be an effective platform for implementation of IoT components. Other papers are more general and provide tools or frameworks which may be used to improve some parameters concerning containers usage.

Paper [9] presents an approach used for modeling Docker containers. The proposed solution is demonstrated on the example of an event processing application, for which a significantly better compromise between performance and development costs is obtained in comparison to the basic Docker container solution.

Paper [10] illustrates work-in-progress development of a framework for container resiliency for cloud applications. The proposed framework uses Linux containers to deliver resiliency to services, and its aim is to orchestrate and manage the container lifecycle.

Docker has become the de facto containerization platform and Docker Swarm delivers general-purpose solutions. In [11] different problems for Docker Swarm are defined and solutions are given in the form of a design pattern. The analysis of the related works reveals that the topic is of growing importance and can be investigated from different points of view.

4 Case Study

We take into account an e-commerce application like eBay but the one that is designed for microservices. The following parts can be distinguished in this application: (1) database engine, (2) search engine, (3) mail server, (4) payment service, (5) account creation page, (6) account management page, (7) products list and details pages. Each of these parts will be deployed as a separate container. Some of them are crucial for the correct working of the whole system (database), others are vulnerable for overloading (website).

In today's world fast page loading is crucial for e-commerce portals. Research shows that after 2 s of waiting for page loading, the internet users lose interest and close the web browser [12], which, for an e-commerce portal, means a loss of a customer. To mitigate this consequence, the containers recreation time is crucial.

The whole e-commerce application is divided into smaller parts and a failure of one container does not have to mean the breakdown of the whole website. If a container hosting a registration module fails, users who already have accounts can search for and purchase products. Unfortunately, this issue has an impact on the number of new customers. To mitigate this problem, fast container recreation needs to be ensured. In the following experiments the time needed for recreating containers deployed on Docker Swarm will be verified.

5 Experiments

To verify Docker Swarm recreation times we must know why containers fail. We can distinguish two reasons of their failure: container internal exceptions and hardware failures. Container internal exceptions may occur when memory consumption increases over time and will be used up at the point. Hardware failures can be related to an electric power failure, network issues or server components damages.

In this set of experiments, we considered two failure scenarios: (1) a failure of a container and its recreation time, (2) a failure of a node and recreation time of a container initially deployed on this node. Tasks carried out by the damaged containers are not handled by Docker Swarm and handling them will not be addressed in this paper.

The tests were performed on the Docker Swarm cluster with 4 nodes (Core i7, 2 GB RAM, Ubuntu) and for default parameters of Docker. To measure the efficiency of Docker Swarm, we implemented a monitoring application that uses

Docker REST API to gather information about running containers and to calculate how much time it takes to recreate failed ones. The following Docker images with different sizes were used in the tests: alpine (5 MB), nginx (109 MB), mongo (382 MB), WebLogic (3.7 GB). Over 100 samples were taken for every container. The creation time of a single container is about 3 s for the small ones and about 5 s for the big ones. The rest is the time of detecting a container failure and scheduling recreation.

During these tests, a new Docker Service for every tested Docker image was created. The monitoring application detects on which node the container is running, kills it and monitors all nodes to detect where and after what time the container will be recreated. To simulate node failure, the monitoring application disables the network card of a node hosting container. To minimize the impact of external factors, only the container under test was deployed in Docker Swarm.

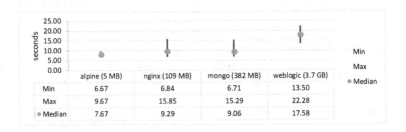

Fig. 2. Container recreation time for a container failure.

In case of a container failure, the results show (Fig. 2) that the recreation time is about 8 s for small containers ranging from several to several hundreds of megabytes. The difference is significant for very big containers, like WebLogic (3.7 GB). In this case, the average recreation time is over 17 s. When we take into account the time of container creation, in the tested environment Docker Swarm required 5 s for small containers and 12 s for big containers to detect and schedule recreation. In case of a cluster node failure, the recreation times are noticeably higher (Fig. 3): about 21 s for small containers and 31 s for big ones. The creation time for new containers remains the same as in case of container failures, which means that about 18 s for small containers and 26 s for big containers are needed to detect and schedule recreation.

In both cases the recreation time was also verified for overloaded nodes (high CPU usage, other containers in Swarm), and the impact on the recreation time was negligible.

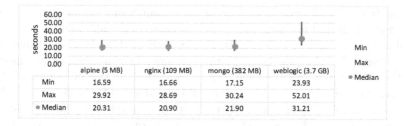

	alpine (5 MB)	nginx (109 MB)	mongo (382 MB)	weblogic (3.7 GB)
Min	16.59	16.66	17.15	23.93
Max	29.92	28.69	30.24	52.01
Median	20.31	20.90	21.90	31.21

Fig. 3. Container recreation time for a node failure

6 Improvement of Container Resilience

To improve container resilience we should consider scenarios for a container failure and a node failure separately. Docker Swarm detects failures and recreates missing containers but the problem is that the required time for detecting a failure and scheduling recreation is too long. In our study, the improved solution will be presented and a prototype tested to compare it with standard Docker Swarm solution. Detecting failures is the first step to improve the application resilience. To achieve it, we need to create a monitoring system. The presented architecture (Fig. 4) consists of at least one manager node and N worker nodes. Manager nodes are responsible for deciding where new containers will be deployed and, in case of a failure, where the missing containers will be recreated. Worker nodes will be responsible for running application containers. The Manager container will be deployed on each manager node. In turn, the Agent container will be deployed on each worker node. The Agent containers will be responsible for gathering containers health data, recreating failed containers, if possible, and sending heartbeat to the Manager container. The Manager container will be

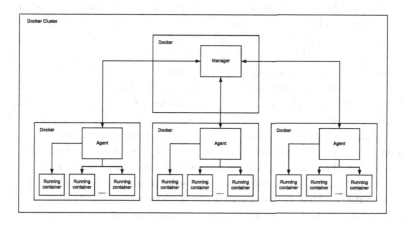

Fig. 4. Architecture of a container health monitoring system

responsible for deciding on which worker new containers will be deployed and for recreating all containers on other nodes in case of a node failure.

Failure detecting will be performed on two-levels. The Agent container will be responsible for the first level. If any container fails because of an internal exception, the Agent container will try to recreate it on the same node. If this operation fails, the Manager container will be notified. The Manager container will be responsible for the second level of failure detection. In case of the first level failure, it will schedule container recreation on a different node. Also, it will detect if one of the worker nodes fails and then will request the recreation of missing containers on different nodes.

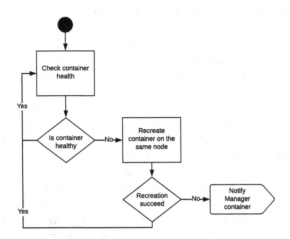

Fig. 5. 1st level failure detection and recreation workflow

The 1st level failure detection (Fig. 5) is performed on every individual node within the Docker cluster. It is responsible for health checking every 50 ms. The monitoring system for this level is deployed on every node within the cluster as a separate container. When a container failure is detected, it recreates it on the same node. If this operation fails, the container is removed from this node and the Manager container is notified about recreation failures.

The 2nd level (Fig. 6) is responsible for checking two sources for information about failures. The first one is listening for notifications from the 1st level. When such a notification appears, the Manager decides on which node to deploy a container, then updates the internal database with container deployment details and sends a deployment request to the Agent container located on a chosen node. The second source of information about failures is a process inside the Manager container that checks heartbeats to detect hardware failures. When a failure is detected, the Manager gathers data about containers deployed on a corrupted node and passes this information to a module responsible for deciding and deploying containers on other nodes within the cluster.

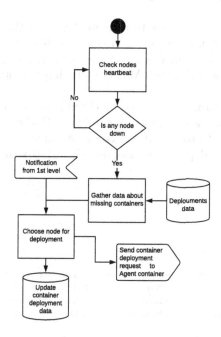

Fig. 6. 2nd level failure detection and recreation workflow

7   Proposal of Suitable Architecture Patterns and Prototype Solution

Analyzing other cases similar to the ones presented in this paper, we can consider three modules of the system:

1. Failure detection - for fast detection of a container or node failure we are going to use a heartbeat pattern. It will notify a monitoring application if containers and nodes are up.
2. A system returning to normal health - when we take into account the characteristics of containers, there is no reason to analyze the causes of a failure and to try to restore a container to normal health. It is cheaper and faster to recreate the container.
3. Proactivity - detecting potential failures and handling them. We can predict memory exhaustion and move a container to a different node within the cluster. This part is not addressed in this paper.

The prototype implements the workflow from a failure to recovery presented in this paper. It consists of two applications representing two levels of failure detection and recreation presented on workflows. The first one is an Agent application deployed on each node within the cluster and the second one is a Manager application deployed on one node. Both parts were implemented as .Net Core applications using Docker REST API to monitor and manage Docker Engine. In addition, both applications are deployed as Docker containers. When Agent fails, Manager detects it and tries to recreate Agent on the same node. In case of a node failure, all containers are recreated on different nodes. A failure of a node with Manager container is not a part of this paper. This case can be handled by duplicating nodes with Manager and implementing synchronization between them.

The tests were performed in the same environment as before, the same application for measurements was used and for the same Docker images. To minimize the impact of external factors, only the container under test and Manager with Agent containers were deployed in Docker. Over 100 samples were taken for every container.

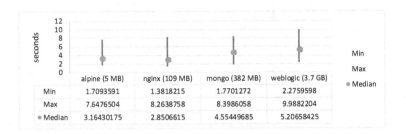

Fig. 7. Container recreation time for a container failure

In case of a container failure, the results show that the average recreation time takes from 3 to 5 s (Fig. 7). Also, it is worth mentioning that for this solution the Docker image size does not impact the recreation time. The creation time of a new container stays the same as in the previous experiment. This time is strictly connected to internal Docker Engine implementation. When we take into account the container creation time, the detection of a failure and scheduling recreation using the proposed solution is almost immediate. In case of a cluster node failure, the recreation times are higher (Fig. 8). It takes around 9–10 s to recreate a failed container. Also for this part, the size of the container image does not impact the recreation time. If we take into account the container creation time, the proposed prototype needs about 6 s to detect and schedule container recreation.

8   Conclusions and Further Work

Using containers and Docker Swarm brings a lot of advantages. They provide portability, isolation and allow for reducing hardware and time requirements. Every year containers become increasingly popular and replace virtual machines more and more frequently.

To provide high availability of containers, the creators of Docker Engine have implemented Docker Swarm. It monitors the Docker cluster, detects container failures and recreates the missing ones.

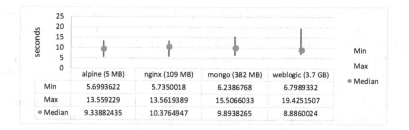

	alpine (5 MB)	nginx (109 MB)	mongo (382 MB)	weblogic (3.7 GB)
Min	5.6993622	5.7350018	6.2386768	6.7989332
Max	13.559229	13.5619389	15.5066033	19.4251507
● Median	9.33882435	10.3764947	9.8938265	8.8860024

Fig. 8. Container recreation time for a node failure

This papers describes the experiment we conducted to verify how much time Docker Swarm needs to recreate a failed container. The results differ depending on the size of a container image and the reason of a failure. The quickest recreation is for small images and a container failure - about 8 s, while the longest - for big container images and a node failure - about 31 s.

Next, we propose the architecture of a lightweight container monitoring system that improves the detection of failed containers and scheduling recreation. The results are promising. For container and node failures, we have received a solution which is not sensitive to a container image size. The recreation times are also better: about 9–10 s for a node failure, which takes only 32% of Docker Swarm recreation time. In case of the container recreation time, it is about 3 to 5 s (37% of Docker Swarm recreation time).

Further work should focus on the container creation time. For Docker Engine, it takes from 3 to 5 s to create a new container and for the solution presented in this paper the container creation time makes a significant impact on the recreation time. Creating a lightweight and isolated Linux process to run a container can be a step in the right direction.

References

1. Docker Overview. https://docs.docker.com/engine/docker-overview/. Accessed 5 Jan 2019
2. What is a Container. https://www.docker.com/resources/what-container. Accessed 6 Jan 2019
3. Swarm Mode Key Concepts. https://docs.docker.com/engine/swarm/key-concepts/. Accessed 6 Jan 2019
4. Create a Swarm Manager. https://docs.docker.com/swarm/reference/manage/#options. Accessed 6 Jan 2019
5. Piraghaj, S.F., Dastjerdi, A.V., Calheiros, R.N., Buyya, R.: Efficient virtual machine sizing for hosting containers as a service (SERVICES 2015). In: 2015 IEEE World Congress on Services, pp. 31–38. IEEE, New York City (2015)
6. Huang, C.-H., Lee, C.-R.: Enhancing the availability of Docker Swarm using checkpoint-and-restore. In: 2017 14th International Symposium on Pervasive Systems, Algorithms and Networks & 2017 11th International Conference on Frontier of Computer Science and Technology & 2017 Third International Symposium of Creative Computing (ISPAN-FCST-ISCC), pp. 357–362. IEEE, Exeter (2017)

7. Amaral, M., Polo, J., Carrera, D., et al.: Performance evaluation of microservices architectures using containers. In: 2015 IEEE 14th International Symposium on Net-work Computing and Applications, pp. 27–34. IEEE, Cambridge (2015)
8. Przyłucki, S., Czerwiński, D., Sierszeń, A.: A performance evaluation of Docker-based MQTT server implementation on internet of things device. Stud. Inform. **38**, 89–99 (2017)
9. Paraiso, F., Challita, S., Al-Dhuraibi, Y., Merle, P.: Model-driven management of Docker containers. In: 2016 IEEE 9th International Conference on Cloud Computing (CLOUD), pp. 718–725. IEEE, San Francisco (2016)
10. Merino Aguilera, X., Otero, C., Ridley, M., Elliott, D.: Managed containers: a framework for resilient containerized mission critical systems. In: 2018 IEEE 11th International Conference on Cloud Computing (CLOUD), pp. 946–949. IEEE, San Francisco (2018)
11. Vohra, D.: Docker Management Design Patterns. Apress, Berkeley (2017)
12. Nah, F.: A study on tolerable waiting time: how long are Web users willing to wait? Behav. Inf. Technol. **23**, 153–163 (2004)

Queueing Theory and Queuing Networks

Analysis of Energy Consumption in Cloud Center with Tasks Migrations

Youssef Ait El Mahjoub[1,2]([⊠]), Jean-Michel Fourneau[1]([⊠]), and Hind Castel-Taleb[2]([⊠])

[1] DAVID - UVSQ, Univ Paris-Saclay, Versailles, France
youssef.ait-el-mahjoub2@uvsq.fr, Jean-Michel.Fourneau@uvsq.fr
[2] SAMOVAR - CNRS UMR 5157, Télécom Sud Paris, Évry, France
hind.castel@telecom-sudparis.eu

Abstract. Reducing the energy consumption of a data center is a major goal in today's digital world. We consider a cloud system represented by a set of physical servers hosting several Virtual Machines, and each one running tasks. We define a resource management policy that implements task migrations between overused servers to unused servers. The advantage of the policy is to balance the load and reduce the energy consumption. We model the system by a multi-server Jackson network, where each station represents a physical server, and the Virtual Machines are the servers. We derive an analytic formula for the energy consumption of the physical servers. We provide an upper bound of the migration energy for task migrations to reduce energy consumption. Moreover, we optimize the energy consumption, and we compute the migration rate that minimizes energy consumption.

Keywords: Energy consumption · Data center · Jackson network · Optimization

1 Introduction

In order to reduce high operating costs, as well as CO_2 emissions, one of the greatest future challenges is the improvement of energy efficiency in data centers. In [1], the authors analyze using traces some shutdown policies in order to study their impact on energy saving. Shutdown policies are often combined with consolidation algorithms that gather the load on a few servers to favor the shutdown of the others. The resources (servers, Virtual Machines (VMs)) can be in one of the following states: OFF, ON, and the ON state is divided into two sub-states: Idle or Run. The power consumption for OFF state is low (sometimes considered zero), but in the Idle state as the resource is powered On, then the consumption could reach half of its maximum consumption when it is in activity. In the Run state, the consumption increases with the resource utilization. Moreover, energy consumption for turning ON, and turning OFF is not negligible and has to be taken into account for the shutdown policies.

© Springer Nature Switzerland AG 2019
P. Gaj et al. (Eds.): CN 2019, CCIS 1039, pp. 301–315, 2019.
https://doi.org/10.1007/978-3-030-21952-9_23

So turning OFF resources could be energy saving only if the idle periods are long enough to make the transitions ON/OFF and OFF/ON. Migration of tasks or VMs can also be used in order to balance the load between overused servers to unused servers [2]. This helps to preserve the nodes from overheating and to prevent overused nodes to have technical problems prematurely. It is linked to the principle that it is better to have a working node than a turned ON node which is doing nothing. Several papers have addressed the problem of load balancing in distributed networks to avoid server overload, and to improve the response time [3–5]. Another approach, is to use consolidation [6] in order to overload some servers and to shut down others. It could be efficient for long shut down periods. In [7] this approach is used with PIKA simulation framework aiming at reducing the brown energy consumption (i.e. from non-renewable energy sources), and improving the usage of renewable energy for small mono-site data centers. Simulators as "CloudSim, SimGrid, ns-2 and ns-3, ecofen (for wired networks) [8] ..." have been widely used in the last decade to evaluate cloud performance and energy consumption. Virtualization in cloud data centers allows one to create several VMs on a physical server, and therefore, reduce the amount of hardware in use and improve the utilization of resources. Many studies as consolidation algorithms, load balancing algorithms, power saving under performance/power budget/CO_2 emission constraints are discussed in [6,7].

In our study, we use the load balancing as a resource management, in order to see the impact on energy consumption. We use Jackson multi-server network to model the servers of the data center. Queueing theory and Mean Field analysis have already been used for performance analysis of clouds. Models include M/M/C queues to study the minimum of resources required for the performance [9] or the reliability of the cloud [10]. IaaS clouds have been modeled by M/G/m/m+k queues in [11]. Mean Field approximation have been proposed for the asymptotic analysis of large scale load balancing network with general distribution [12]. Dealing with general arrivals and services processes, one can also use a diffusion approximation [13,14]. Such an approach was published recently to build a model which explicitly represent energy and data in sensor networks with energy harvesting [15].

In this paper we study how to optimaly minimize the energy consumption using an exact analysis of the queueing network with customer migration. We use the multi-server Jackson network to represent the behavior of the data center. Each physical server is represented by a multi-server station, and the VM (Virtual machines) are the servers of the stations. The routing between the stations represents the task migrations. The exact distribution provided by Jakson model allows to compute the optimal migration rate. An extension of this work consists in using the new product form result published by Gelenbe and his colleagues [16,17]. This model explicitly represents the energy needed by data transfer and computation.

The paper is organized as follows: next, we give the energy consumption equations computed in a data center, then in Sect. 3.3 we derive the mathematical formula of the energy consumption and we give the upper bound of the migration energy which reduces the global energy consumption. We give the migration rate which minimizes the energy consumption. In Sect. 4, we show with a numerical example how to choose the parameters (energy migration, migration rates) in order to minimize energy consumption.

2   Energy Consumption Model

In a data center, the energy consumption can be divided into static and dynamic parts. The static parts are the base costs of running the data center when being idle and the dynamic costs depend on the current usage [18]. For the data center with n servers, we denote s_i the energy consumption of server i, for $i = 1, \ldots, n$. Each server hosts a set of (VMs) Virtual Machines that perform tasks, and m represents the number of VMs hosted by a server.

Let Eng_{total} be the total energy consumed in the data center, and $Eng(s_i)$ the energy consumed by a physical server. From [19] we have:

$$Eng_{total} = \sum_{i=1}^{n} Eng(s_i).$$

The energy consumed by a server is the sum of static energy (δ) when the server is powered ON (and idle) and the dynamic part (β and α) depends on the CPU resource utilization of the VMs hosted in the server when they are performing tasks. As the CPU is one of the highest power consumers on a node, in many studies this is the most considered (compared to the RAM or disk). In [19], the consumption of a server is defined as a polynomial function of degree two of the CPU load of this server. It is therefore an extension of the linear model, which is a polynomial of degree one.

$$Eng(s_i) = \delta + CPU(s_i) * \beta + CPU^2(s_i) * \alpha, \tag{1}$$

We consider that $CPU(s_i)$ is the CPU utilization ratio of the server i, which is divided between each VM of the server:

$$CPU(s_i) = \sum_{j=1}^{m} CPU(vm_j),$$

where $CPU(vm_j)$ is the CPU ratio allocated to the vm_j. We suppose that all the virtual machines are homogeneous in terms of CPU utilization [11,20]. That means that the physical resources allocated to each VM are of the same capacity and type. We consider a single-task policy, each job or task is served by one virtual machine, so the mean number of activated VMs corresponds to the mean number of tasks in service. As the virtual machines are homogeneous, then $CPU(vm_j) = CPU(vm)$, for all j. And if $E(n_{s,i})$ is the mean number of tasks in service (or activated VMs), then:

$$CPU(s_i) = E(n_{s,i}) * CPU(vm).$$

As an example, if we consider a data center with 2 physical servers, then Equation (1) gives, for $i \in \{1, 2\}$:

$$Eng(s_i) = \delta + E(n_{s,i}) * CPU(vm) * \beta + (E(n_{s,i}) * CPU(vm))^2 * \alpha.$$

The consumption of a physical server is upper bounded by $\delta + \beta + \alpha$: let's m be the number of VMs in a physical server, $CPU(vm) = \frac{1}{m}$ since the VMs are all homogeneous and receive the same utilisation ratio as mentioned above. As $E(n_{s,i}) \leq m$ then we deduce that $\delta + \beta + \alpha$ is an upper bound of $Eng(s_i)$.

We note that in this section, we have not considered migrations of tasks and therefore energy consumption due to migrations, but it will be added in the following. Next, we compute analytically the energy consumption in a data center using multi-server Jackson network with task migrations for load balancing.

3   Jackson Network Model for Task Migrations

We consider a data center system with two physical servers. We suppose now that the system performs migrations from overloaded to unloaded server in order to avoid to shutdown the unloaded servers. We model the system with a Jackson network. Each station represents a physical server, where tasks are either waiting in the queue, or in activity (in service or in migration). This limitation to system with two servers is due to the optimization process detailed in Sect. 3.3. Of course the queueing model can accommodate larger systems.

A task can migrate from one station to another one. We assume that for each station i ($i = 1, 2$), the external arrivals of tasks follow independent Poisson processes with rate λ_i. We also assume that the services and the migration durations are independent and distributed according to exponential distributions with rate μ_i ($i = 1, 2$) and $\gamma_{i,j}$ ($i, j = 1, 2,$ and $i \neq j$) for the task migration from station i to station j.

We suppose that each physical server is associated to m logical servers (VMs) as in [9]. If the number of tasks is larger than m, then they wait in the queue, and only m among them are in service or involved in a migration. We also assume that the queueing capacity is infinite. No task can be rejected at its arrival.

The logical service rate in a queue could be a real service or a migration. As both activities are independent and exponentially distributed, their parallel composition can be seen as an exponential duration with rate $\mu_i + \gamma_{i,j}$ ($i, j = 1, 2$ and $j \neq i$). After this logical service, a task can leave the system (being stopped) or stochastically route to the other queue (migrate). This is typically a race condition between two events as in a stochastic Petri net. The probabilities of these events are given in Lemma 1.

3.1   Analytic Results for the Queues

The state of station i is x_i, which represents the number of tasks waiting or in activity. The task scheduling discipline is FCFS (First Come First Serve).

We assume that both stations contain m logical servers. Therefore $min(m, x_i)$ task are in service or in a migration activity while $(x_i - m)^+$ are queued. Each queue in isolation can be seen as a M/M/m queue, using Kendall notations. Under the classical assumptions we mention earlier, $X_t = (x_1, x_2)_t$ is a Markov chain, and the transitions are as follows:

$$
\begin{aligned}
(x_1, x_2) &\rightarrow (x_1 + 1, x_2) \text{ with rate } \lambda_1, \\
&\rightarrow (x_1, x_2 + 1\}), \text{ with rate } \lambda_2, \\
&\rightarrow (\max\{0, x_1 - 1\}, x_2), \text{ with rate } min\{x_1, m\} \cdot \mu, \\
&\rightarrow (\max\{0, x_1 - 1\}, x_2 + 1), \text{ with rate } min\{x_1, m\} \cdot \gamma_{1,2}, \\
&\rightarrow (x_1, \max\{0, x_2 - 1\}) \text{ with rate } min\{x_2, m\} \cdot \mu, \\
&\rightarrow (x_1 + 1, \max\{0, x_2 - 1\}) \text{ with rate } min\{x_2, m\} \cdot \gamma_{2,1}.
\end{aligned}
$$

We know that under ergodicity conditions, Jackson networks have a steady-state distribution $\Pi(x_1, x_2) = \Pi_1(x_1)\Pi_2(x_2)$ which has a product form [21]. $\Pi_1(x_1)$ and $\Pi_2(x_2)$ are steady-state distributions of two $M/M/m$ queues the parameters of which are given by the flow equations between traffic intensities ρ_1 and ρ_2.

Theorem 1. *Let $(x)_t$ be the Markov chain that represents an M/M/m queue. x is the number of tasks in the station. Let $\rho < 1$ be the traffic intensity and $\Pi(x)$ the steady-state distribution. Then from [21], we have:*

$$
\Pi(x) = \begin{cases} \Pi(0)\dfrac{(m\rho)^x}{x!}, & x < m \\ \Pi(0)\dfrac{\rho^x m^m}{m!}, & x \geq m \end{cases} \tag{2}
$$

and

$$
\Pi(0) = \left[1 + \frac{(m\rho)^m}{m!(1-\rho)} + \sum_{x=1}^{m-1} \frac{(m\rho)^x}{x!} \right]^{-1}. \tag{3}
$$

It remains to write the flow equation for our model.

Lemma 1. *In the networks we consider, the fresh arrivals at station i $(i = 1, 2)$ follow Poisson processes with rate λ_i. The parallel composition of service and migration results in an activity with an exponential duration with rate $\mu_i + \gamma_{i,j}$ $(i, j = 1, 2$ and $i \neq j)$, followed by a departure with probability $\frac{\mu_i}{\mu_i + \gamma_{i,j}}$ $(i, j = 1, 2, i \neq j)$ or a migration with probability $\frac{\gamma_{i,j}}{\mu_i + \gamma_{i,j}}$ $(i, j = 1, 2, i \neq j)$. Using the flow equations for Jackson networks, we get:*

$$
\rho_1 = \frac{\lambda_1 + m(\mu 2 + \gamma_{2,1})\rho_2 \frac{\gamma_{2,1}}{(\mu 2 + \gamma_{2,1})}}{m(\mu 1 + \gamma_{1,2})}, \qquad \rho_2 = \frac{\lambda_2 + m(\mu 1 + \gamma_{1,2})\rho_1 \frac{\gamma_{1,2}}{(\mu 1 + \gamma_{1,2})}}{m(\mu 2 + \gamma_{2,1})}.
$$

And after simplifications we obtain the flow equation of the model with migration:

$$
\rho_1 = \frac{\lambda_1 + m\rho_2\gamma_{2,1}}{m(\mu 1 + \gamma_{1,2})} \quad and \quad \rho_2 = \frac{\lambda_2 + m\rho_1\gamma_{1,2}}{m(\mu 2 + \gamma_{2,1})}. \tag{4}
$$

We now define some notations we will use in the following of the paper. Consider a task in a station, it is either served or implied in a migration to another station. Let P_1 (resp. P_2) be the probability of service and $1\text{-}P_1$ (resp. $1\text{-}P_2$) the probability of migration:

$$P_1 = \frac{\mu_1}{\mu_1 + \gamma_{1,2}}, P_2 = \frac{\mu_2}{\mu_2 + \gamma_{2,1}}. \tag{5}$$

In order to compute energy consumption of the system using the relations mentioned in the previous section, we need the mean number of tasks in service which correspond to the mean number of active VMs. Since our system has a product form solution, the rewards will be calculated separately for each station. Let $E(n_{s,1})$ (resp. $E(n_{s,2})$) be the mean number of tasks in service in station 1 (resp. station 2). It is well-known that (see for instance [21] for a proof):

$$E(n_{s,1}) = m\rho_1, \quad and \quad E(n_{s,2}) = m\rho_2. \tag{6}$$

Lemma 2. *The total energy is the summation of the energy consumed by each station plus the energy Em needed for the migrations:*

$$Eng_{Jackson} = Eng(s_1) + Eng(s_2) + Em.$$

Combining the energy model (in Eq. 1), the value of the average queue size (Eq. 6) and the probability of serving (resp. migrating) (Eq. 5), we obtain for physical server $i \in \{1,2\}$:

$$Eng(s_i) = \delta + E(n_{s,i}) * CPU(vm) * P_1 * \beta + (E(n_{s,i}) * CPU(vm) * P_1)^2 \alpha, \tag{7}$$

while for the energy consumed by the migration:

$$Em = e_m E(n_{s,1})(1 - P_1) + e_m E(n_{s,2})(1 - P_2), \tag{8}$$

where e_m is the migration energy for a task.
Combining Eqs. 7, and 8, we finally obtained after summation:

$$Eng_{Jackson} = 2\delta + [P_1 E(n_{s,1}) + P_2 E(n_{s,2})] CPU(vm)\beta + \big[E(n_{s,1})^2 P_1^2 + E(n_{s,2})^2 P_2^2 \big] CPU(vm)^2 \alpha + Em.$$

Remember that, $CPU(vm) = \frac{1}{m}$ since the VMs are supposed homogeneous as already mentioned in Sect. 2. Then, we get:

$$Eng_{Jackson} = 2\delta + (\rho_1 P_1 + \rho_2 P_2)\beta + \big[(\rho_1 P_1)^2 + (\rho_2 P_2)^2 \big] \alpha \\ + m e_m \rho_1 (1 - P_1) + m e_m \rho_2 (1 - P_2). \tag{9}$$

Corollary 1. *The system without migration is modeled with migration rates equal to 0:*

$$\gamma_{1,2} = \gamma_{2,1} = 0.$$

Each physical server is an independent M/M/m queue. Let r_i be the traffic intensity of station i ($i \in \{1,2\}$) in the system without migration, then:

$$r_i = \frac{\lambda_i}{m\mu_i}, \tag{10}$$

And the mean number of tasks in service are $E(n_{s,i}) = mr_i$. Since $\gamma_{1,2} = \gamma_{2,1} = 0$, we get $P_1 = P_2 = 1$ (see Equation (5)) and the energy consumption of this system without migration is:

$$Eng_{MMm} = 2\delta + (r_1 + r_2)\beta + \left[r_1^2 + r_2^2\right]\alpha. \tag{11}$$

3.2 Comparing Both Systems

In this section we compare the energy consumption in both models. Clearly if migration cost is very high, there is no benefit to perform migration.

Assumption 1. *We consider that the two physical servers have different traffic intensities. More precisely and without loss of generality, we assume a heavy traffic in station 2 and a low traffic in server 1: $r_1 < r_2 < 1$. To minimize the energy consumption, we assume that the migrations only occur from the heavy traffic station (station 2) to the low traffic one (station 1). Thus, $\gamma_{1,2} = 0$.*

Lemma 3. *Under this assumption, we have $P_1 = 1$ and $P_2 = \frac{\rho_2}{r_2}$.*

Proof. As $P_1 = \frac{\mu_1}{\mu_1 + \gamma_{1,2}}$ the assumptions clearly implies that $P_1 = 1$. Furthermore we have the following relations:

$$P_2 = \frac{\mu_2}{\mu_2 + \gamma_{2,1}}, \quad r_2 = \frac{\lambda_2}{m\mu_2}, \quad \rho_2 = \frac{\lambda_2}{m(\mu_2 + \gamma_{2,1})}, \quad \rho_1 = \frac{\lambda_1 + m\rho_2\gamma_{2,1}}{m(\mu_1)}. \tag{12}$$

Clearly $P_2 = \frac{\rho_2}{r_2}$ holds.

Lemma 4. *Assume that VMs are homogeneous and the servers are equivalent (i.e. $\mu_1 = \mu_2 = \mu$), then the sum of the mean number of activated VMs in each model is a constant value.*

$$m\rho_1 + m\rho_2 = mr_1 + mr_2 = mK, \quad where \quad K \in [0, 2]. \tag{13}$$

Proof. By solving the equation system in Eq. (4) and using $\gamma_{1,2} = 0$, we get:

$$\rho_1 = \frac{\lambda_2\gamma_{2,1} + \lambda_1(\mu + \gamma_{2,1})}{m\mu(\mu + \gamma_{2,1})}, \quad and \quad \rho_2 = \frac{\lambda_2}{m(\mu + \gamma_{2,1})}. \tag{14}$$

So:

$$\Rightarrow \rho_1 + \rho_2 = \frac{\lambda_2\gamma_{2,1} + \lambda_1(\mu + \gamma_{2,1}) + \lambda_2\mu}{m\mu(\mu + \gamma_{2,1})} = \frac{\lambda_2(\mu + \gamma_{2,1}) + \lambda_1(\mu + \gamma_{2,1})}{m\mu(\mu + \gamma_{2,1})}.$$

$$\Rightarrow \rho_1 + \rho_2 = \frac{\lambda_2(\mu + \gamma_{2,1}) + \lambda_1(\mu + \gamma_{2,1})}{m\mu(\mu + \gamma_{2,1})}.$$

$$\Rightarrow \rho_1 + \rho_2 = \frac{\lambda_2 + \lambda_1}{m\mu} = \frac{\lambda_1}{m\mu} + \frac{\lambda_2}{m\mu}.$$

$$\Rightarrow \rho_1 + \rho_2 = r_1 + r_2.$$

$\square$

Lemma 5. *Assuming that $\gamma_{2,1} \geq 0$ then the following relations hold:*

$$\rho_2 \leq r_2 \quad and \quad \rho_1 \geq r_1.$$

Proof. From Eqs. (10) and (14) we have:

$$\rho_1 - r_1 = \frac{\lambda_2 \gamma_{2,1} + \lambda_1(\mu + \gamma_{2,1})}{m\mu(\mu + \gamma_{2,1})} - \frac{\lambda_1}{m\mu} = \frac{\lambda_2 \gamma_{2,1}}{m\mu(\mu + \gamma_{2,1})} \geq 0,$$

$$\rho_2 - r_2 = \frac{\lambda_2}{m(\mu + \gamma_{2,1})} - \frac{\lambda_2}{m\mu} = \frac{-\lambda_2 \gamma_{2,1}}{m\mu(\mu + \gamma_{2,1})} \leq 0.$$

$\square$

Lemma 6. *Under Assumption 1, if the following relation on the energy for a migration holds,*

$$0 \leq e_m \leq \left[\frac{(r_2 + r_2 P_2)(2\alpha + \alpha P_2^2) - 2K\alpha + P_2\beta}{m P_2} \right], \tag{15}$$

then $f(\rho_1, \rho_2) - g(r_1, r_2) \leq 0$.
This gives an upper bound on the energy needed for a migration which reduces the global energy consumption.

Proof. Let $f(\rho_1, \rho_2)$ (resp. $g(r_1, r_2)$) be the energy consumption function in the task migration model (resp. no task migration model). From Eqs. (9) and (11) and Assumption 1 we obtain:

$$\begin{bmatrix} g(r_1, r_2) = 2\delta + (r_1 + r_2)\beta + [r_1^2 + r_2^2]\alpha, \\ f(\rho_1, \rho_2) = 2\delta + (\rho_1 + \rho_2 P_2)\beta + [\rho_1^2 + (\rho_2 P_2)^2]\alpha + m e_m \rho_2(1 - P_2). \end{bmatrix} \tag{16}$$

Using $P_2 = \frac{\rho_2}{r_2}$ and $r_1 + r_2 = \rho_1 + \rho_2 = K$ (see Lemma 4) then:

$$f(\rho_1, \rho_2) - g(r_1, r_2) = P_2(\rho_2 - r_2)\beta + \left(\frac{\rho_2^4 - r_2^4}{r_2^2} \right) \alpha$$
$$+ (r_2 - \rho_2)\left[(2K - \rho_2 - r_2)\alpha + m e_m P_2 \right].$$

After factorization:

$$f(\rho_1, \rho_2) - g(r_1, r_2) = (r_2 - \rho_2)$$
$$* \left[m e_m P_2 + (2K - \rho_2 - r_2)\alpha - P_2\beta - \frac{(\rho_2 + r_2)(\rho_2^2 + r_2^2)\alpha}{r_2^2} \right].$$

Remember that $\rho_2 \leq r_2$ (see Lemma 5) and after some algebraic manipulations, the proof is complete. $\square$

Remark that the best case is when $e_m = 0$ but it's not realistic, as migration of tasks costs energy.

3.3   Optimization of Energy Consumption

Next, we study the optimal solution of the equation describing energy consumption for the model with tasks migrations (see Eq. (9)). We know that $\rho_1 = K - \rho_2$ (see Lemma 4), $P_2 = \frac{\rho_2}{r_2}$ and $P_1 = 1$ (see Lemma 3). Then we can express Eq. (9) as a function of one variable ρ_2. Let $f()$ be this function:

$$f(\rho_2) = 2\delta + (K - \rho_2 + \rho_2 P_2)\,\beta + \left((K - \rho_2)^2 + (\rho_2 P_2)^2\right)\alpha + mem\rho_2(1 - P_2).$$

After some algebraic manipulations, we get the expression of $f()$ as a degree 4 polynomial:

$$f(\rho_2) = \left(\frac{\alpha}{r_2^2}\right)\rho_2^4 + \left(\frac{\alpha r_2 - mem + \beta}{r_2}\right)\rho_2^2 + (mem - \beta - 2\alpha K)\rho_2 + K^2\alpha + K\beta + 2\delta.$$

$$(17)$$

Let us now study the domain of definition (the possible values of ρ_2). The set of possible values for (ρ_1, ρ_2) comes from all the constraints we got:

$$\begin{bmatrix} \rho 1 & \leq 1, \\ \rho 1 & \geq r_1 \geq 0, \\ \rho 2 & \geq 0, \\ \rho 2 & \leq r_2 \leq 1, \\ \rho 1 + \rho 2 = K = r_1 + r_2. \end{bmatrix}$$

Let $\mathcal{S}$ be this set. It is a compact set of R^2. Let $\mathcal{D}$ be its projection for the second component. By construction $\mathcal{D}$ is a compact of R. Depending of the value of K, D can be:

$$\begin{bmatrix} [0, K] & \quad if\ 0 < K < 1, \\ [0, 1] & \quad if\ K = 1, \\ [K - 1, 1] & \quad if\ K > 1. \end{bmatrix}$$

Theorem 2. *As a polynomial function, f is continuous and it takes value from a compact set $\mathcal{D}$ to the real numbers. According to Weierstrass extreme value theorem, it has a minimal and a maximal values on the compact set. Therefore it exists a value of ρ_2 in the domain of definition which minimizes the energy consumption.*

Lemma 7. *If the following condition on the parameters of the energy function holds:*

$$mem - \beta - \alpha r_2 \leq \frac{6\alpha}{r_2}((K - 1)^+)^2,$$

$$(18)$$

then, f is convex on $\mathcal{D}$.

Proof. The first and second derivative of f are:

$$f'(\rho_2) = \left(\frac{4\alpha}{r_2^2}\right)\rho_2^3 + \left(\frac{2\alpha r_2 - 2mem + 2\beta}{r_2}\right)\rho_2 + (mem - \beta - 2\alpha K),$$

and, $$f''(\rho_2) = \left(\frac{12\alpha}{r_2^2}\right)\rho_2^2 + \left(\frac{2\alpha r_2 - 2me_m + 2\beta}{r_2}\right),$$

while $$f'''(\rho_2) = \left(\frac{24\alpha}{r_2^2}\right)\rho_2.$$

Function $f''()$ is a degree 2 polynomial which is increasing on $\mathcal{D}$ as $f'''()$ is positive on R^+ and $\mathcal{D} \subset R^+$. To prove that $f''(\rho_2) \geq 0$ on $\mathcal{D}$, we just have to compute $f''(\rho_2)$ for the smallest element of $\mathcal{D}$. We have two cases:

– if $K \leq 1$, the minimal element of $\mathcal{D}$ is 0. We obtain the following condition

$$me_m - \beta - \alpha r_2 \leq 0,$$

which is consistent with our claim, as $(K-1)^+ = 0$ in that condition.
– if $K > 1$, the minimal element is $K - 1$ and we get:

$$me_m - \beta - \alpha r_2 \leq \frac{6\alpha}{r_2}(K-1)^2,$$

which is equivalent to our claim as $(K-1)^+ = K - 1$ when $K > 1$.

And the proof is complete.   □

Let us state how we can obtain the optimal solution for the energy consumption by looking at function $f'()$ and its zero. We look for the value of ρ_2 which satisfies the following equation:

$$\rho_2^3 + \left(\frac{(\alpha r_2 - me_m + \beta)r_2}{2\alpha}\right)\rho_2 + \left(\frac{(me_m - \beta - 2\alpha K)r_2^2}{4\alpha}\right) = 0, \quad (19)$$

Let x (resp. y) be the minimal (resp. maximal) value on $\mathcal{D}$: $if(k \leq 1)$ then x = 0 and y = K, else x = K − 1 and y = 1.

– If this equation has no solution in $\mathcal{D}$ then:

$$\rho_{2,opt} = \rho_2' \quad where \quad f(\rho_2') = min(f(x), f(y)).$$

– Otherwise, Let $\rho_{2,j}$ with $j \in \{0,1,2\}$ (at most the three solutions are in $\mathcal{D}$) be solutions that satisfies $\rho_{2,j} \in \mathcal{D}$ and let define $\rho_{2,t}$ such as:

$$f(\rho_{2,t}) = min(f(\rho_{2,j})) \quad \forall j \in \{0,1,2\},$$

then: $$\rho_{2,opt} = \rho_2' \quad where \quad f(\rho_2') = min(f(x), f(y), f(\rho_{2,t})).$$

Lemma 8. *When $K \leq 1$ and f is convex, then the global minimum of the energy function (Eq. (17)) is unique.*

Proof. Using Cardan formula, to solve Eq. (19), we first compute Δ:

$$\Delta = -\left[4\left(\frac{(\alpha r_2 - me_m + \beta)r_2}{2\alpha}\right)^3 + 27\left(\frac{(me_m - \beta - 2\alpha K)r_2^2}{4\alpha}\right)^2\right].$$

Theorem 2 assures that f admits at least one global minimum on $\mathcal{D}$. Condition in Lemma 7 states that f is convex and that $\Delta \leq 0$. f is convex guaranties that global minimum is not on the reachable bounds of $\mathcal{D}$ and $\Delta < 0$ implies that f' admits one real solution. We conclude that the unique solution of f' is the global minimum of f.   □

Now we can deduce $\gamma_{2,opt}$ the migrations rate that optimizes energy consumption. So we have:

$$\gamma_{2,opt} = max\left(0, \frac{\lambda_2}{m\rho_{2,opt}} - \mu\right) \text{ with } \rho_{2,opt} \in [0, r_2], \tag{20}$$

is the optimal solution of:

$$f(\gamma_{2,1}) = 2\delta + \left(K - \frac{\lambda_2}{m(\mu + \gamma_{2,1})} + \left(\frac{\lambda_2}{m(\mu + \gamma_{2,1})}\right)^2 r_2^{-1}\right)\beta$$

$$+ \left(\left(K - \frac{\lambda_2}{m(\mu + \gamma_{2,1})}\right)^2 + \left(\frac{\lambda_2}{m(\mu + \gamma_{2,1})}\right)^4 r_2^{-2}\right)\alpha$$

$$+ me_m\left(\frac{\lambda_2}{m(\mu + \gamma_{2,1})} - \left(\frac{\lambda_2}{m(\mu + \gamma_{2,1})}\right)^2 r_2^{-1}\right). \tag{21}$$

Notice that $\rho_{2,opt} = 0$ could be possible (in theory), since $\gamma_{2,1} \in [0, +\infty[$.

Corollary 2. *Equation* (20) *shows that if* $\frac{\lambda_2}{m\rho_{2,opt}} < \mu$ *then the migration of tasks will only increase energy consumption of the data center. Therefore the optimal solution is to not perform the migration* ($\gamma_{2,opt} = 0$) *under the current energy parameters. Otherwise, it means that an optimal solution that acutely reduce energy consumption exists and its value is* $\frac{\lambda_2}{m\rho_{2,opt}} - \mu$.

4   Numerical Results

Let's consider the following energy parameters: $E_{min} = \delta = 95$, $\beta = 25$, $\alpha = 100$ so the maximal energy that can be consumed by a physical server without migration is $E_{max} = 220\,\text{W}$. Parameters E_{min} and E_{max} are inspired from Taurus server [18] consumption.

We use the total energy given by Eq. (11) for the system without migration, while for the system with migrations we have also to consider the migration energy Em given by Eq. (8), so it results Eq. (9). We suppose that $m = 20$ VMs on each physical server so the CPU ratio of each VM is $CPU(vm) = \frac{1}{20} = 0.05$. Arrival rate in server1 (resp. server2) is $\lambda_1 = 2$ (resp. $\lambda_2 = 17$), and service rate is $\mu = 1$. So traffic intensity is $r_2 = 0.85$ and $r_1 = 0.1$. Migration of tasks will only be performed from server2 to server1 (see Assumption 1) so $\gamma_{1,2} = 0$. When $\gamma_{2,1} = 0$, then no migrations are performed in the system (see Corollary 1).

Taking $e_m = 3$ and varying $\gamma_{2,1}$, we can see in Fig. 1 that data center energy consumption (green curve, circle points) is a convex function since numerical parameters satisfies condition in Lemma 7. The behavior of black curve (rectangular points) and red curve (star points) clearly shows the effect of tasks migrations from server 2 to server 1, also shown in Fig. 2. Migration rate should be cleverly chosen in order to reduce data center consumption. Let's analyze optimal case: using Cardan formula to solve Eq. (19) that becomes

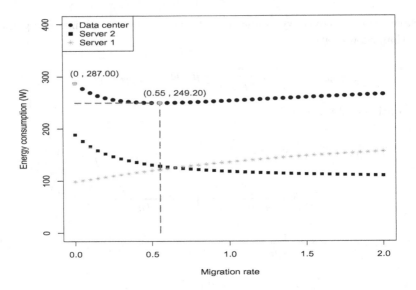

Fig. 1. Energy consumption under $\gamma_{2,1}$ variation

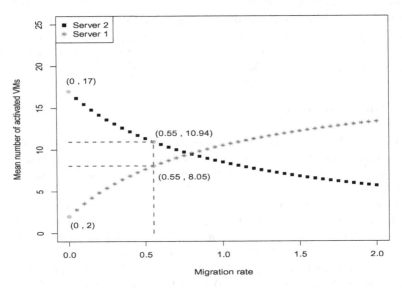

Fig. 2. Activated VMs under $\gamma_{2,1}$ variation

$\rho_2^3 + 0.2125\rho_2 - 0.2799 = 0$, we have $\Delta \leq 0$, then we obtain one solution $\rho_{2,opt} = 0.5305$ in $[0 , 0.95]$ so $\gamma_{2,opt} = 0.553$ which corresponds to an energy consumption of 249.20 W instead of an energy consumption of 287 W in the model without migrations. Then energy gain is 13.16%. Notice that, in Fig. 2, the load of expected number of activated VMs in server2 (10.94) is not equal to

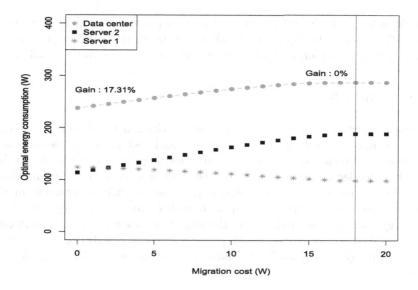

Fig. 3. Energy consumption in optimal case, under e_m variation

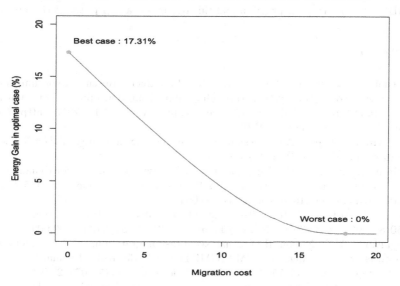

Fig. 4. Energy gain in optimal case, under e_m variation

the expected number of activated VMs in server1 (8.05), when the total energy consumption is minimized.

In Fig. 3, we have calculated optimal energy consumption for several value of e_m. We can see that the data center consumption increases with migration cost until some point ($e_m = 18$) where the migration is no more efficient, then $\forall \ e_m \geq 18$ optimal solution of Eq. (21) is $\gamma_{2,opt} = 0$. The best case, but not realistic, is when the migration does not cost energy then we achieve an energy gain of 17.31% and the worst case is when the energy is costly but in that case no migrations are performed so the energy gain is 0% as shown in Figs. 3 and 4.

5   Conclusion

We model a data center with a multi-server Jackson network in order to represent the resources (physical servers, VMs), and task activities (services and migrations) between the servers. We study task migrations between overloaded server to unloaded server in order to see the effect on energy consumption. Using closed form of the steady-state probability, we derive analytic formulas for the energy consumption. So we can compute bounds on migration energy for reducing energy consumption, and we give the migration rate minimizing energy consumption. We have considered in this paper the case of two stations, as a future work, we already have generalized Lemma 4 to a data center with $n > 2$ physical servers.

Acknowledgment. Youssef Ait el mahjoub is supported by Labex Digiscosme (ANR 11-LABX-0045) PHD grant program and this research is a part of the Perfeco project.

References

1. Benoit, A., Lefevre, L., Orgerie, A.C., Rais, I.: Reducing the energy consumption of large scale computing systems through combined shutdown policies with multiple constraints. Int. J. High Performance Comput. Appl. **32**(1) (2017). https://doi. org/10.1177/1094342017714530
2. Lefevre, L., Orgerie, A.C.: Designing and evaluating an energy efficient cloud. J. Supercomputing **51**(3), 352–373 (2010)
3. Marin, A., Balsamo, S., Fourneau, J.M.: LB-networks: a model for dynamic load balancing in queueing networks. In: Performance Evaluation, vol. 115 (2017). https://doi.org/10.1016/j.peva.2017.06.004
4. Leino, J., Virtamo, J.: Insensitive load balancing in data networks. Comput. Netw. **50**(8), 1059–1068 (2006). https://doi.org/10.1016/j.comnet.2005.09.009
5. Squillante, M.S., Nelson, R.D.: Analysis of task migration in shared-memory multiprocessor scheduling. In: ACM SIGMETRICS Performance Evaluation Review, vol. 19, no. 1, pp. 143–155 (1991). https://doi.org/10.1145/107972.107987
6. Beloglazov, A.: Energy-efficient management of virtual machines in data centers for cloud computing. Department of Computing and Information Systems. The University of Melbourne, Ph.D. thesis (2013)
7. Yunbo, L., Orgerie, A.C., Menaud, J.M.: Opportunistic scheduling in clouds partially powered by green energy. In: IEEE International Conference on Green Computing and Communications (GreenCom) (2015). https://doi.org/10.1109/ DSDIS.2015.80
8. Orgerie, A.C., Amersho, B.L., Haudebourg, T., Quinson, M., Rifai, M. : Simulation toolbox for studying energy consumption in wired networks. In: CNSM International Conference on Network and Service Management. hal-01630226 (2017)

9. Huang, G., et al.: Auto scaling virtual machines for web applications with queueing theory. In: ICSAI The 3rd International Conference on Systems and Informatics (2016)

10. Yang, B., Tan, F., Dai, Y.-S., Guo, S.: Performance evaluation of cloud service considering fault recovery. In: Jaatun, M.G., Zhao, G., Rong, C. (eds.) CloudCom 2009. LNCS, vol. 5931, pp. 571–576. Springer, Heidelberg (2009). https://doi.org/10.1007/978-3-642-10665-1_54

11. Chang, X., Wang, B., Muppala, J.K., Liu, J.: Modeling active virtual machines on IaaS clouds using an M/G/m/m+k queue. IEEE Trans. Serv. Comput. **9**(3), 408–420 (2016) https://doi.org/10.1109/TSC.2014.2376563

12. Aghajani, R., Xingjie, L., Ramanan; K.: Mean-field dynamics of load-balancing networks with general service distributions (2015)

13. Whitt, W.: A diffusion approximation for the G/GI/n/m queue. Oper. Res. **52**(6), 922–941 (2004). https://doi.org/10.1287/opre.1040.0136

14. Czachórski, T., Fourneau, J.M., Nycz, T., Pekergin, F.: Diffusion approximation model of multi-server stations with losses. Electr. Notes Theor. Comput. Sci **232**, 125–143 (2009). https://doi.org/10.1016/j.entcs.2009.02.054

15. Omer, H.A., Gelenbe, E.: A Diffusion model for energy harvesting sensor nodes. In: 24th IEEE International Symposium on Modeling, Analysis and Simulation of Computer and Telecommunication Systems, MASCOTS, pp. 154–158 (2016)

16. Gelenbe, E., Ceran, E.T.: Central or distributed energy storage for processors with energy harvesting. In: Sustainable Internet and ICT for Sustainability, SustainIT, pp. 1–3 (2015) https://doi.org/10.1109/SustainIT.2015.7101380

17. Gelenbe, E., Ceran, E.T.: Energy packet networks with energy harvesting. IEEE Access **4**, 1321–1331 (2016). https://doi.org/10.1109/ACCESS.2016.2545340

18. Kurpiez, M., Orgerie, A.C., Sobe, A.: How much does a VM cost? Energy-proportional accounting in VM-based environments. In: PDP Euromicro International Conference on Parallel, Distributed, and Network-Based Processing, p. 8 (2016). https://doi.org/10.1109/PDP.2016.70

19. Le Louët, G.: Maîtrise énergétique des centres de données virtualisés: D'un scénario de charge à l'optimisation du placement des calculs. École nationale supérieure des mines de Nantes, Ph.D. thesis (2014)

20. Ghosh, R., Longo, F., Naik, V.K., Trivedi, K.S.: Modeling and performance analysis of large scale IaaS clouds. Future Gen. Comput. Syst. **29**(5), 1216–1234 (2013). https://doi.org/10.1016/j.future.2012.06.005

21. Jain, R.: The Art of Computer Systems Performance Analysis, Techniques for Experimental Design Measurement, Simulation and Modeling. Wiley Professional Computing (1992). ISBN 0471503361

Queueing Systems
with Non-homogeneous Customers
and Infinite Sectorized Memory Space

Oleg Tikhonenko[1] and Marcin Ziółkowski[2(✉)]

[1] Faculty of Mathematics and Natural Sciences, College of Sciences,
Cardinal Stefan Wyszyński University in Warsaw,
Ul. Wóycickiego 1/3, 01-938 Warsaw, Poland
`o.tikhonenko@uksw.edu.pl`
[2] Faculty of Applied Informatics and Mathematics,
Warsaw University of Life Sciences, Ul. Nowoursynowska 159,
02-787 Warsaw, Poland
`marcin_ziolkowski@sggw.pl`

Abstract. In the paper, we investigate queueing systems with non-homogeneous customers. As non-homogenity, we mean that each customer is characterized by some random l-dimensional volume vector. The arriving customers appear according to a stationary Poisson process. Service time of a customer generally depends on his volume vector. Memory space is composed of l parts of infinity capacities in accordance with customers volume vectors components. As an example of such system, we consider Erlang-type $M/G/n/0$ system, for which we determine the joint distribution of l-dimensional random vector of total customers volumes and its marginal and mixed moments. An analysis of some special cases and some numerical examples are attached as well.

Keywords: Queueing systems with non-homogeneous customers ·
Queueing systems with sectorized memory · Laplace-Stieltjes transform

1 Introduction

In the theory of queueing systems with non-homogeneous customers, we investigate queueing models of telecommunication or computer systems taking into account both possible limitation of the summary volume of the customers and the character of dependency between customer's volume and his service time. In general, single customer volume can be understood as a random vector of l components containing different type of data that are stored separately (e.g. text parts, sound parts, video parts and so on) in buffers intended for them (see [4,5,19]). Service time of a customer usually depends on his volume vector. Analysed models are different in comparison to their classical analogies because we may have here additional losses of customers (in the case when summary volume of customers volume components is limited). Even if there are free servers

© Springer Nature Switzerland AG 2019
P. Gaj et al. (Eds.): CN 2019, CCIS 1039, pp. 316–329, 2019.
https://doi.org/10.1007/978-3-030-21952-9_24

or free places in the queue, arriving customer can be lost if his volume component is too big to be accepted to the proper limited buffer (that stores the components of the other customers until they finish their service). Therefore, the waiting queue is limited by the number of customers that can be present in the system at the same time and fixed buffer is limited by the summary volume of the proper components of all customers present in the system.

Total volume of all customers present in the system may be limited or unlimited. These facts cause that analyzed models belong to one of the four classes [10]: (1) total volume of customers is unlimited and customer's service time is independent on his volume vector, (2) total volume of customers is limited and customer's service time is independent on his volume vector, (3) total volume of customers is unlimited and customer's service time is dependent on his volume vector, (4) total volume of customers is limited and customer's service time is dependent on his volume vector.

Models of the first class are the simplest and can be investigated using the results of classical queueing theory [3,7,8]. Models belonging to the classes 2–4 require more advanced methods that have to take into account possible limitation of the total volume and the dependency between customer's volume and his service time (see [2,6,9,11,13,14]). The most interesting (from the practical point of view) are models belonging to the second and fourth classes because, in real systems, total volume is usually limited. For such models, we try to obtain customers number distribution and loss probability formulae (at least in steady state). For the models of the second class, it is not so complicated. If we have a formula defining steady-state customers number distribution for analogous classical queueing model, we may easily obtain these characteristics. Unfortunately, for the models of the fourth class, this problem becomes more complicated and exact relations are difficult to obtain – it is possible only for systems without waiting positions (e.g. the Erlang system $M/G/n/(0, V)$ and processor sharing one [12,15]).

Models of the third class seem to be impractical, because they assume unlimited total volume. We do not have additional losses of customers so we obtain here the same classical characteristics connected with the customers number distribution and loss probability. But in such situation we also want to obtain the characteristics of the summary volume that show the amount of information stored in buffers and depend strongly on the joint distribution of the customer volume vector components and his service time (even on the level of the first moment of the summary volume). On the other hand, their analysis, in many cases, let estimate loss characteristics in analogous models with limited total volume that cannot be precisely investigated (especially in the case when random volume vector of a customer is one-dimensional). Moreover, the results of such systems analysis can be useful during the process of designing of real systems with limited total volume. For example, we can estimate the value V of system memory size in the system with limited total volume, if we know the moments of the total volume for analogous system with unlimited buffer. So, investigation of such models is also important from the practical point of view (see [10,16,17]).

The most important general theoretical results connected with the analysis of the queueing models with non–homogeneous customers can be found in [10].

The purpose of our research is to present the basic theoretical investigations connected with analysis of the queueing systems with non–homogeneous customers and unlimited l-dimensional buffer, in which service time of a customer depends on his volume vector. As an example of such system, we consider Erlang-type $M/G/n/0$ one. For this system, we obtain steady-state distribution of the total volume vector in terms of l-dimensional Laplace–Stieltjes transform and calculate its mixed moments. We also present the analysis of some special cases together with numerical examples.

The rest of the paper is organized as follows: Sect. 2 contains basic notations. We also define here functions describing the system behavior and present most important general theoretical results. In Sect. 3, we present exact relations for the total volume characteristics for the Erlang multiserver loss system $M/G/n/0$, together with analysis of some special cases and numerical results. The last Sect. 4 presents conclusions and final remarks.

2   Basic Notations and Mathematical Background

We assume that each customer arriving to a queueing system is characterized by some random vector $\boldsymbol{\zeta} = (\zeta_1, \ldots, \zeta_l)$, where $l = 1, 2 \ldots$. The components ζ_i, $i = \overline{1, l}$, are non-negative random variables (RVs). Let $\eta(t)$ be the number of customers present in the system at time instant t, $\sigma_i(t)$, $i = \overline{1, l}$, be the sum of ith components of all these customers. Our purpose is the determination of the vector $\boldsymbol{\sigma}(t) = (\sigma_1(t), \ldots, \sigma_l(t))$ characteristics. We also assume that customer's service time ξ generally depends on his indication vector $\boldsymbol{\zeta}$. This dependence is determined by the following joint distribution function (DF):

$$F(\mathbf{x}, t) = F(x_1, \ldots, x_l, t) = \mathsf{P}\{\zeta_1 < x_1, \ldots, \zeta_l < x_l, \xi < t\} = \mathsf{P}\{\boldsymbol{\zeta} < \mathbf{x}, \xi < t\},$$

where $\mathbf{x} = (x_1, \ldots, x_l)$. Let $L(\mathbf{x}) = F(\mathbf{x}, \infty)$ be the joint DF of customer's indications, $B(t) = F(\infty, t)$ be the service time distribution function (here $\infty = \underbrace{(\infty, \ldots, \infty)}_{l}$).

Note that we can consider marginal DF of separate indication: $L_i(x) = L(\infty, \ldots, x, \ldots, \infty)$ or the joint DF of some separate indication and service time: $F_i(x, t) = F(\infty, \ldots, x, \ldots, \infty, t)$. We have evidently $B(t) = F_i(\infty, t)$ for all $i = \overline{1, l}$.

Let

$$\alpha(\mathbf{s}, q) = \int_{x_1=0}^{\infty} \cdots \int_{x_l=0}^{\infty} \int_{u=0}^{\infty} e^{-(\mathbf{s}, \mathbf{x}) - qu} \mathrm{d}F(\mathbf{x}, u)$$

$$= \int_{(\mathbf{0}; \infty)^l} \int_0^{\infty} e^{-(\mathbf{s}, \mathbf{x}) - qu} \mathrm{d}F(\mathbf{x}, u)$$

be Laplace-Stieltjes transform (LST) of the function $F(\mathbf{x}, t)$, where $\mathbf{s} = (s_1, \ldots, s_l)$, $(\mathbf{s}, \mathbf{x}) = s_1 x_1 + \cdots + s_l x_l$. It is clear that $dF(\mathbf{x}, u)$ has the following probability sense:

$$dF(\mathbf{x}, u) = \mathsf{P}\{\zeta_1 \in [x_1; x_1 + dx_1), \ldots, \zeta_l \in [x_l; x_l + dx_l), \xi \in [u; u + du)\}$$
$$= \mathsf{P}\{\boldsymbol{\zeta} \in [\mathbf{x}; \mathbf{x} + d\mathbf{x}), \xi \in [u; u + du)\}.$$

Denote by $\varphi(\mathbf{s}) = \alpha(\mathbf{s}, 0)$ LST (with respect to $\mathbf{x}$) of DF $L(\mathbf{x})$, $\beta(q) = \alpha(\mathbf{0}, q)$ LST (with respect to t) of the function $B(t)$, where $\mathbf{0} = \underbrace{(0, \ldots, 0)}_{l}$.

Let us introduce the l-dimensional vector $\mathbf{i} = (i_1, \ldots, i_l)$, where $i_m = 1, 2, \ldots$, $m = \overline{1, l}$. Denote by $\Delta(\mathbf{i}, j)$ the differential operator

$$\Delta(\mathbf{i}, j) = (-1)^{i_1 + \cdots + i_l + j} \frac{\partial^{i_1 + \cdots + i_l + j}}{\partial s_1^{i_1} \ldots \partial s_l^{i_l} \partial q^j}$$

and, analogously,

$$\Delta(\mathbf{i}) = (-1)^{i_1 + \cdots + i_l} \frac{\partial^{i_1 + \cdots + i_l}}{\partial s_1^{i_1} \ldots \partial s_l^{i_l}}.$$

Let $\alpha_{\mathbf{i}, j} = \Delta(\mathbf{i}, j)\alpha(\mathbf{s}, q)|_{\mathbf{s}=0,\, q=0}$ be the mixed $(i_1 + \cdots + i_l + j)$th moment of DF $F(\mathbf{x}, t)$ (if exists). Denote by $\varphi_{\mathbf{i}}$ the mixed $(i_1 + \cdots + i_l)$th moment of DF $L(\mathbf{x})$ and by β_i – the ith moment of DF $B(t)$.

We assume that the entrance flow is a stationary Poisson process with parameter a. Assume that service discipline does not depend on the indication vector $\boldsymbol{\zeta}$. Let $D(\mathbf{x}, t) = \mathsf{P}\{\boldsymbol{\sigma}(t) < \mathbf{x}\} = \mathsf{P}\{\sigma_1(t) < x_1, \ldots, \sigma_l(t) < x_l\}$ be DF of the vector $\boldsymbol{\sigma}(t)$. If steady state exists for a system under consideration, we have $\boldsymbol{\sigma}(t) \Rightarrow \boldsymbol{\sigma}$ in the sense of a weak convergence. Then, we have

$$\lim_{t \to \infty} D(\mathbf{x}, t) = D(\mathbf{x}) = \mathsf{P}\{\boldsymbol{\sigma} < \mathbf{x}\},$$

where $D(\mathbf{x})$ is DF of steady-state vector $\boldsymbol{\sigma}$ of the total volume.

Denote by

$$\delta(\mathbf{s}) = \mathsf{E}e^{-(\mathbf{s}, \boldsymbol{\sigma})} = \int_{(0;\infty)^l} e^{-(\mathbf{s}, \mathbf{x})} dD(\mathbf{x})$$

LST of DF $D(\mathbf{x})$.

Let $\boldsymbol{\chi}(t) = (\chi_1(t), \ldots, \chi_l(t))$ be the indication vector of a customer that is served at time instant t. Let $\xi^*(t)$ be the time from the service beginning to the moment t.

Lemma 1. *Let $E_y(\mathbf{x}) = \mathsf{P}\{\boldsymbol{\chi}(t) < \mathbf{x} \,|\, \xi^*(t) = y\}$ be the conditional DF of the random vector $\boldsymbol{\chi}(t)$ under condition $\xi^*(t) = y$. Then*

$$dE_y(\mathbf{x}) = [1 - B(y)]^{-1} \int_{u=y}^{\infty} dF(\mathbf{x}, u).$$

Proof.

$$dE_y(\mathbf{x}) = \mathsf{P}\{\chi(t) \in [\mathbf{x}; \mathbf{x} + d\mathbf{x}) \,|\, \xi^*(t) = y\} = \mathsf{P}\{\zeta \in [\mathbf{x}; \mathbf{x} + d\mathbf{x}) \,|\, \xi \geq y\}$$
$$= \frac{\mathsf{P}\{\zeta \in [\mathbf{x}; \mathbf{x} + d\mathbf{x}), \xi \geq y\}}{\mathsf{P}\{\xi \geq y\}} = [1 - B(y)]^{-1} \int_{u=y}^{\infty} dF(\mathbf{x}, u). \qquad \square$$

Hence, the function $E_y(\mathbf{x})$ takes the form:

$$E_y(\mathbf{x}) = \int_{(0;\mathbf{x})^l} dE_y(\mathbf{u}) = \mathsf{P}\{\zeta < \mathbf{x} \,|\, \xi \geq y\}$$
$$= \frac{\mathsf{P}\{\zeta < \mathbf{x}, \xi \geq y\}}{\mathsf{P}\{\xi \geq y\}} = \frac{L(\mathbf{x}) - F(\mathbf{x}, y)}{1 - B(y)},$$

where $\mathbf{x} = (x_1, \ldots, x_l)$.

Corollary. *LST of the function $E_y(\mathbf{x})$ has the form:*

$$e_y(\mathbf{s}) = \int_{(0;\infty)^l} e^{-(\mathbf{s},\mathbf{x})} dE_y(\mathbf{x}) = [1 - B(y)]^{-1} \int_{(0;\infty)^l} e^{-(\mathbf{s},\mathbf{x})} \int_{u=y}^{\infty} dF(\mathbf{x}, u).$$

3   System $M/G/n/0$ Analysis

Consider a queueing system $M/G/n/0$ with Poisson stationary entrance flow with parameter a. Introduce the notation $\rho = a\beta_1$. Assume that $\rho < \infty$. We shall analyze this system functioning in steady state (when $t \to \infty$). Let η be a number of customers present in it ($\eta(t) \Rightarrow \eta$ in the sense of a weak convergence).
    Let

$$P_i(t, y_1, \ldots, y_i) dy_1 \ldots dy_i$$
$$= \mathsf{P}\{\eta(t) = i, \xi_1^*(t) \in [y_1; y_1 + dy_1), \ldots, \xi_i^*(t) \in [y_i; y_i + dy_i)\},$$

where $\xi_j^*(t)$ is the time from service beginning of jth customer present in the system at time instant t to the moment t, $j = \overline{1, i}$. Denote by $P_0(t) = \mathsf{P}\{\eta(t) = 0\}$. When $t \to \infty$, we obtain

$$p_0 = \lim_{t \to \infty} P_0(t), \quad p_i(y_1, \ldots, y_i) = \lim_{t \to \infty} P_i(t, y_1, \ldots, y_i), \quad i = \overline{1, n}.$$

From the classical queueing theory [10], we have:

$$p_i(y_1, \ldots, y_i) = \frac{a^i}{i!} p_0 \prod_{j=1}^{i} [1 - B(y_j)], \quad i = \overline{1, n}, \qquad (1)$$

where $p_0 = \left[\sum_{i=0}^{n} \rho^i / i! \right]^{-1}$.
    It follows from the steady state existence (when $\rho < \infty$) that $\boldsymbol{\sigma}(t) \Rightarrow \boldsymbol{\sigma}$ in the sense of a weak convergence, where the distribution of $\boldsymbol{\sigma}$ does not depend on $\boldsymbol{\sigma}(0)$ one.

Theorem 1. *For the steady-state system M/G/n/0, LST of the joint DF of vector $\boldsymbol{\sigma}$ components has the form:*

$$\delta(\mathbf{s}) = \frac{\sum_{i=0}^{n} \left[-a\alpha_q'(\mathbf{s}, q)|_{q=0} \right]^i / i!}{\sum_{i=0}^{n} \rho^i / i!}. \tag{2}$$

Proof. Introduce the notation

$$\mathrm{d}D_i(\mathbf{x}, y_1, \ldots, y_i) = \mathsf{P}\{\boldsymbol{\sigma} \in [\mathbf{x}; \mathbf{x} + \mathrm{d}\mathbf{x}) \mid \eta = i, \xi_1^* = y_1, \ldots, \xi_i^* = y_i\}.$$

This is the conditional probability that the kth component of the vector $\boldsymbol{\sigma}$ lies in the interval $[x_k; x_k + \mathrm{d}x_k)$, $k = \overline{1, i}$, under condition that there are i customers in the system and the times from their service beginning equal $y_1, \ldots, y_i$, respectively. Note that, for $i \geq 1$, the components of the volume of jth customer in the system $\boldsymbol{\sigma}^j = (\sigma_1^j, \ldots, \sigma_l^j)$ depend on ξ_j^*, $j = \overline{1, i}$, only. Then, we have from Lemma 1 that

$$\mathsf{P}\{\boldsymbol{\sigma}^j \in [\mathbf{x}; \mathbf{x} + \mathrm{d}\mathbf{x}) \mid \xi_j^* = y_j\} = [1 - B(y_j)]^{-1} \int_{u=y_j}^{\infty} \mathrm{d}F(\mathbf{x}, u).$$

LST of the random vector $\boldsymbol{\sigma}^j$ under condition $\xi_j^* = y_j$ has the form (see corollary of Lemma 1):

$$e_{y_j}(\mathbf{s}) = [1 - B(y_j)]^{-1} \int_{(0;\infty)^l} e^{-(\mathbf{s}, \mathbf{x})} \int_{u=y_j}^{\infty} \mathrm{d}F(\mathbf{x}, u).$$

It is clear that, for $\eta = i$, we have $\sigma_m = \sum_{j=1}^{i} \sigma_m^j$, $m = \overline{1, l}$, and random vectors $\boldsymbol{\sigma}^j$ are independent under condition $\xi_1^* = y_1, \ldots, \xi_i^* = y_i$. Then LST $\delta(\mathbf{s}, y_1, \ldots, y_i)$ of the function $D_i(\mathbf{x}, y_1, \ldots, y_i)$ has the form of product:

$$\delta(\mathbf{s}, y_1, \ldots, y_i) = \int_{(0;\infty)^l} e^{-(\mathbf{s}, \mathbf{x})} \mathrm{d}D_i(\mathbf{x}, y_1, \ldots, y_i) = \prod_{j=1}^{i} e_{y_j}(\mathbf{s})$$

$$= \prod_{j=1}^{i} [1 - B(y_j)]^{-1} \int_{(0;\infty)^l} e^{-(\mathbf{s}, \mathbf{x})} \int_{u=y_j}^{\infty} \mathrm{d}F(\mathbf{x}, u).$$

Using Lemma 1 again and the relation (1), we obtain

$$\delta(\mathbf{s}) = p_0 + \sum_{i=1}^{n} \int_0^{\infty} \cdots \int_0^{\infty} \delta(\mathbf{s}, y_1, \ldots, y_i) p_i(y_1, \ldots, y_i) \mathrm{d}y_1 \ldots \mathrm{d}y_i$$

$$= p_0 \sum_{i=1}^{n} \frac{a^i}{i!} \prod_{j=1}^{i} \int_{(0;\infty)^l} e^{-(\mathbf{s}, \mathbf{x})} \int_{y_j=0}^{\infty} \mathrm{d}y_j \int_{u=y_j}^{\infty} \mathrm{d}F(\mathbf{x}, u),$$

where

$$\int_{(0;\infty)^l} e^{-(\mathbf{s}, \mathbf{x})} \int_{z=0}^{\infty} \mathrm{d}z \int_{u=z}^{\infty} \mathrm{d}F(\mathbf{x}, u) = \int_{(0;\infty)^l} \int_{u=0}^{\infty} e^{-(\mathbf{s}, \mathbf{x})} \mathrm{d}F(\mathbf{x}, u) \int_{z=0}^{u} \mathrm{d}z$$

$$= \int_{(0;\infty)^l} \int_{u=0}^{\infty} u e^{-(\mathbf{s}, \mathbf{x})} \mathrm{d}F(\mathbf{x}, u) = -\alpha_q'(\mathbf{s}, q)|_{q=0},$$

whereas, taking into consideration that $p_0 = \left(\sum_{i=0}^{n} \rho^i/i!\right)^{-1}$, we obtain the statement of the theorem. $\qquad\square$

Using the relation (2), we can determine mixed moments (if they exist) of the random vector $\boldsymbol{\sigma}$:

$$\delta(\mathbf{i}) = \mathsf{E}(\sigma_1^{i_1} \ldots \sigma_l^{i_l}) = \Delta(\mathbf{i})\delta(\mathbf{s})|_{\mathbf{s}=0}, \tag{3}$$

where $\mathbf{i} = (i_1, \ldots, i_l)$; $i_j, j = \overline{1,l}$, determines an order of the mixed moment with respect to jth component of the random vector $\boldsymbol{\sigma}$. Formula (3) follows from the properties of the Laplace–Stieltjes transform [18].

If we are interested in a mixed moment with respect to some (not all) components of $\boldsymbol{\sigma}$, we have to take the value 0 for all "unnecessary" components of it. So, we obtain a new function $\delta(\mathbf{s}')$, where the vector $\mathbf{s}'$ consists of components of our interest. Further determination of the moment is carried out analogously using the relation (3).

Note that, if $n \to \infty$, we obtain from (2) the relation for the function $\delta(\mathbf{s})$ characterizing a steady-state system $M/G/\infty$:

$$\delta(\mathbf{s}) = \exp\left[-\rho - a\alpha_q'(\mathbf{s}, q)|_{q=0}\right]. \tag{4}$$

Example 1. Let us consider the system $M/G/n/0$, where a is an arrival rate. Each customer is characterized by two dimensional random volume vector $\zeta = (\zeta_1, \zeta_2)$, where RVs ζ_1 and ζ_2 are independent and their DFs we denote by $L_1(x)$ and $L_2(x)$. The RV $\zeta = \zeta_1 + \zeta_2$ will be called a customer length. Assume that service time of the customer is proportional to his length: $\xi = c(\zeta_1 + \zeta_2)$, $c > 0$.

Let us determine

$$\alpha(\mathbf{s}, q) = \alpha(s_1, s_2, q) = \int_{x_1=0}^{\infty} \int_{x_2=0}^{\infty} \int_{t=0}^{\infty} e^{-s_1 x_1 - s_2 x_2 - qt} dF(x_1, x_2, t),$$

where $F(x_1, x_2, t) = \mathsf{P}\{\zeta_1 < x_1, \zeta_2 < x_2, \xi < t\}$. In this case, we can write out

$$\alpha(s_1, s_2, q) = \int_0^{\infty} \int_0^{\infty} e^{-s_1 x_1 - s_2 x_2} dL(x_1, x_2) \int_0^{\infty} e^{-qt} dB(t \,|\, \zeta_1 = x_1, \zeta_2 = x_2),$$

where $B(t \,|\, \zeta_1 = x_1, \zeta_2 = x_2) = \mathsf{P}\{\xi < t \,|\, \zeta_1 = x_1, \zeta_2 = x_2\}$ and $L(x_1, x_2) = \mathsf{P}\{\zeta_1 < x_1, \zeta_2 < x_2\} = L_1(x_1)L_2(x_2)$.

It is clear that, in our case, we have:

$$B(t \,|\, \zeta_1 = x_1, \zeta_2 = x_2) = \begin{cases} 1, & \text{if } t > c(x_1 + x_2); \\ 0, & \text{if } t \leq c(x_1 + x_2). \end{cases}$$

Hence, we obtain using delta-function:

$$\int_0^{\infty} e^{-qt} dB(t \,|\, \zeta_1 = x_1, \zeta_2 = x_2) = \int_0^{\infty} e^{-qt} \delta(t - c(x_1 + x_2)) dt = e^{-cq(x_1 + x_2)}.$$

Then, we have:

$$\alpha(s_1, s_2, q) = \int_0^\infty e^{-s_1 x_1 - cqx_1} dL_1(x_1) \int_0^\infty e^{-s_2 x_2 - cqx_2} dL_2(x_2) \tag{5}$$
$$= \varphi^1(s_1 + cq)\varphi^2(s_2 + cq),$$

where $\varphi^1(s)$, $\varphi^2(s)$ are LSTs of the functions $L_1(x)$, $L_2(x)$, respectively.
If we substitute $\alpha(s_1, s_2, q)$ from (5) to the relation (2), we obtain:

$$\delta(\mathbf{s}) = \delta(s_1, s_2) = \frac{\sum_{i=0}^n (-1)^i \left\{ ac \left[\varphi^{1'}(s_1)\varphi^2(s_2) + \varphi^1(s_1)\varphi^{2'}(s_2) \right] \right\}^i /i!}{\sum_{i=0}^n \rho^i/i!}, \tag{6}$$

where $\rho = ac(\varphi_1^1 + \varphi_1^2)$; φ_1^1 and φ_1^2 are the first moments of DF $L_1(x)$ and $L_2(x)$.

Assume now that the buffer memory is divided onto two sectors of infinite volume. The initial part of customer volume ζ_1 is placed to the first sector and the rest ζ_2 – to the second one. Let us determine the mean total customers volumes δ_1^1 and δ_1^2 present in the first and second sector. To determine δ_1^1, we first obtain:

$$\delta^1(s) = \delta(s, 0) = \frac{\sum_{i=0}^n (-1)^i \left\{ ac \left[\varphi^{1'}(s) - \varphi_1^2 \varphi^1(s) \right] \right\}^i /i!}{\sum_{i=0}^n \rho^i/i!},$$

Finally, we have:

$$\delta_1^1 = -\delta^{1'}(0) = \frac{\sum_{i=1}^n ac\rho^{i-1}(\varphi_2^1 + \varphi_1^1\varphi_1^2)/(i-1)!}{\sum_{i=0}^n \rho^i/i!}. \tag{7}$$

Analogously, we obtain:

$$\delta^2(s) = \delta(0, s) = \frac{\sum_{i=0}^n (-1)^i \left\{ ac \left[\varphi^{2'}(s) - \varphi_1^1 \varphi^2(s) \right] \right\}^i /i!}{\sum_{i=0}^n \rho^i/i!},$$

and

$$\delta_1^2 = -\delta^{2'}(0) = \frac{\sum_{i=1}^n ac\rho^{i-1}(\varphi_2^2 + \varphi_1^1\varphi_1^2)/(i-1)!}{\sum_{i=0}^n \rho^i/i!}. \tag{8}$$

Now, let us determine the mixed moment δ_{11} of the order $1+1$ for the random vector (σ_1, σ_2): $\delta_{11} = \left. \frac{\partial^2 \delta(s_1, s_2)}{\partial s_1 \partial s_2} \right|_{s_1=0, s_2=0}$, whereas we have:

$$\delta_{11} = p_0 \sum_{i=1}^n \frac{ac\rho^{i-2} \left[(i-1)ac(\varphi_2^1 + \varphi_1^1\varphi_1^2)(\varphi_2^2 + \varphi_1^1\varphi_1^2) + \rho(\varphi_1^2\varphi_2^1 + \varphi_1^1\varphi_2^2) \right]}{(i-1)!}, \tag{9}$$

where $p_0 = \left(\sum_{i=0}^n \rho^i/i! \right)^{-1}$.

Assume additionally that customer's volume vector components are exponentially distributed with parameters f and g, respectively. Then formulae (6)–(9) take the form:

$$\delta(\mathbf{s}) = \delta(s_1, s_2) = p_0 \sum_{i=0}^{n} \left[\frac{acfg(f + g + s_1 + s_2)}{(f + s_1)^2 (g + s_2)^2} \right]^i /i!;$$

$$\delta_1^1 = -\delta^{1\prime}(0) = \frac{ac}{f} \left(\frac{2}{f} + \frac{1}{g} \right) \left(1 - \frac{\rho^n p_0}{n!} \right);$$

$$\delta_1^2 = -\delta^{2\prime}(0) = \frac{ac}{g} \left(\frac{2}{g} + \frac{1}{f} \right) \left(1 - \frac{\rho^n p_0}{n!} \right);$$

$$\delta_{11} = \frac{acp_0}{fg} \sum_{i=1}^{n} \frac{\rho^{i-2} \left[(i-1)ac \left(\frac{2}{f} + \frac{1}{g} \right) \left(\frac{2}{g} + \frac{1}{f} \right) + 2\rho \left(\frac{1}{f} + \frac{1}{g} \right) \right]}{(i-1)!}, \tag{10}$$

where $p_0 = \left(\sum_{i=0}^{n} \rho^i / i! \right)^{-1}$ and $\rho = ac(1/f + 1/g)$.

Now we shall present some numerical results. Consider $M/G/3/0$ queueing system. Assume that $a = 1$ and $c = 1$. Computations for δ_1^1, δ_1^2 and δ_{11} are presented in Tables 1, 2 and 3. Numerical calculations were obtained with the help of *Mathematica* environment [1] using standard representation of float numbers. Note that the complexity of calculations is linear $(o(n))$ because we have only one iteration loop in every part of the formulae (10).

Table 1. Numerical values of δ_1^1 for special case of $M/G/3/0$ system - customer's service time is proportional to his length

δ_1^1	$f = 1$	$f = 2$	$f = 3$	$f = 4$
g = 1	2.36842	0.86567	0.49488	0.33864
g = 2	2.16418	0.70313	0.37242	0.24164
g = 3	2.07850	0.63843	0.32484	0.20448
g = 4	2.03181	0.60409	0.29990	0.18513

Table 2. Numerical values of δ_1^2 for special case of $M/G/3/0$ system - customer's service time is proportional to his length

δ_1^2	$f = 1$	$f = 2$	$f = 3$	$f = 4$
g = 1	2.36842	2.16418	2.07850	2.03181
g = 2	0.86567	0.70313	0.63843	0.60409
g = 3	0.49488	0.37242	0.32484	0.29990
g = 4	0.33864	0.24164	0.20448	0.18513

Table 3. Numerical values of δ_{11} for special case of $M/G/3/0$ system - customer's service time is proportional to his length

δ_{11}	$f = 1$	$f = 2$	$f = 3$	$f = 4$
$g = 1$	7.42105	2.79104	1.62799	1.12995
$g = 2$	2.79104	0.89063	0.47475	0.31134
$g = 3$	1.62799	0.47475	0.23992	0.15184
$g = 4$	1.12995	0.31134	0.15184	0.09375

Example 2. Consider again the system $M/G/n/0$, in which a is an arrival rate and each customer is characterized by two dimensional random volume vector $\zeta = (\zeta_1, \zeta_2)$ having independent components that are characterized by DF $L_1(x)$ and $L_2(x)$. Let $B(t)$ be the DF of customer's service time and β_1 be its first moment. This time we additionally assume that customers's service time and his volume vector are also independent. It means that joint DF $F(\mathbf{x}, t)$ has the following product form:

$$F(\mathbf{x}, t) = F(x_1, x_2, t) = L_1(x_1)L_2(x_2)B(t).$$

In this case, LST of the DF $F(\mathbf{x}, t)$ is determined by the following formula:

$$\alpha(s_1, s_2, q) = \varphi^1(s_1)\varphi^2(s_2)\beta(q),$$

where $\varphi^1(s), \varphi^2(s)$ and $\beta(q)$ are the LST of the functions $L_1(x), L_2(x)$ and $B(t)$, respectively. It leads to the following results:

$$\delta(\mathbf{s}) = \delta(s_1, s_2) = p_0 \sum_{i=0}^{n} \left[\rho\varphi^1(s_1)\varphi^2(s_2)\right]^i / i!; \tag{11}$$

$$\delta^1(s) = \delta(s, 0) = p_0 \sum_{i=0}^{n} \left[\rho\varphi^1(s)\right]^i / i!; \quad \delta^2(s) = \delta(0, s) = p_0 \sum_{i=0}^{n} \left[\rho\varphi^2(s)\right]^i / i!;$$

$$\delta_1^1 = -\delta^{1\prime}(0) = \rho\varphi_1^1 \left(1 - \frac{\rho^n p_0}{n!}\right); \tag{12}$$

$$\delta_1^2 = -\delta^{2\prime}(0) = \rho\varphi_1^2 \left(1 - \frac{\rho^n p_0}{n!}\right); \tag{13}$$

$$\delta_{11} = \frac{\partial^2 \delta(s_1, s_2)}{\partial s_1 \partial s_2}\bigg|_{s_1=0, s_2=0} = \rho\varphi_1^1\varphi_1^2 \left(1 - \frac{\rho^n p_0}{n!}\right); \tag{14}$$

where $p_0 = \left(\sum_{i=0}^{n} \rho^i / i!\right)^{-1}$ and $\rho = a\beta_1$.

Assume additionally that customer's volume vector components are exponentially distributed with parameters f and g and his service time consists of two independent phases having exponential distribution with parameters μ_1 and μ_2. Then we obtain the following results:

$$\delta(\mathbf{s}) = \delta(s_1, s_2) = p_0 \sum_{i=0}^{n} \left[\frac{\rho f g}{(f + s_1)(g + s_2)} \right]^i / i!;$$

$$\delta_1^1 = -\delta^{1\prime}(0) = \frac{\rho}{f} \left(1 - \frac{\rho^n p_0}{n!} \right); \ \delta_1^2 = -\delta^{2\prime}(0) = \frac{\rho}{g} \left(1 - \frac{\rho^n p_0}{n!} \right);$$

$$\delta_{11} = \frac{\partial^2 \delta(s_1, s_2)}{\partial s_1 \partial s_2} \bigg|_{s_1=0, s_2=0} = \frac{\rho}{fg} \left(1 - \frac{\rho^n p_0}{n!} \right), \tag{15}$$

where $\rho = a(1/\mu_1 + 1/\mu_2)$.

Consider now $M/G/3/0$ queueing system, assuming that $a = 1$, $\mu_1 = f$ and $\mu_2 = g$. Numerical results are presented in Tables 4, 5 and 6. Here we also used *Mathematica* environment and standard representation of float numbers to calculate needed characteristics.

Table 4. Numerical values of δ_1^1 for special case of $M/G/3/0$ system - customers' volume vector and his service time are independent

δ_1^1	$f = 1$	$f = 2$	$f = 3$	$f = 4$
$g = 1$	1.57895	0.64925	0.39590	0.28220
$g = 2$	1.29851	0.46875	0.26601	0.18123
$g = 3$	1.18771	0.39902	0.21656	0.14313
$g = 4$	1.12878	0.36245	0.19084	0.12342

Table 5. Numerical values of δ_1^2 for special case of $M/G/3/0$ system - customers' volume vector and his service time are independent

δ_1^2	$f = 1$	$f = 2$	$f = 3$	$f = 4$
$g = 1$	1.57895	1.29851	1.18771	1.12878
$g = 2$	0.64925	0.46875	0.39902	0.36245
$g = 3$	0.39590	0.26601	0.21656	0.19084
$g = 4$	0.28220	0.18123	0.14313	0.12342

If we compare numerical results from the Tables 1, 2, 3, 4, 5 and 6 (Examples 1 and 2), we simply notice that they differ, although, from the classical queueing theory point of view, investigated systems are equivalent i.e. they have the same

Table 6. Numerical values of δ_{11} for special case of $M/G/3/0$ system - customers' volume vector and his service time are independent

δ_{11}	$f=1$	$f=2$	$f=3$	$f=4$
$g=1$	1.57895	0.64925	0.39590	0.28220
$g=2$	0.64925	0.23438	0.13301	0.09061
$g=3$	0.39590	0.13301	0.07219	0.04771
$g=4$	0.28220	0.09061	0.04771	0.03085

parameter of Poisson entrance flow ($a=1$) and the same DF of service time that is determined (in the case when $f \neq g$) by the following formula [18]:

$$B(t) = (f-g)^{-1} \left[f \left(1 - e^{-gt} \right) - g \left(1 - e^{-ft} \right) \right].$$

It means that character of dependency between customer's volume vector and his service time has a substantial influence on the characteristics of the total volume of all customers present in the system, especially on their moments.

Example 3. Consider the system $M/G/\infty$ with all notations the same as in Example 1. For this system, we obtain analogously:

$$\delta_1^1 = ac(\varphi_2^1 + \varphi_1^1 \varphi_1^2), \ \delta_1^2 = ac(\varphi_2^2 + \varphi_1^1 \varphi_1^2),$$

$$\delta_{11} = ac(\varphi_2^1 \varphi_1^2 + \varphi_1^1 \varphi_2^2) + a^2 c^2 (\varphi_1^1 \varphi_1^2 + \varphi_2^2)(\varphi_1^1 \varphi_1^2 + \varphi_2^1).$$

Example 4. Consider the system $M/G/\infty$ with all notations the same as in Example 2. For this system, we obtain analogously:

$$\delta_1^1 = \rho \varphi_1^1, \ \delta_1^2 = \rho \varphi_1^2, \ \delta_{11} = \rho \varphi_1^1 \varphi_1^2.$$

4   Conclusions and Final Remarks

In the paper, we present basic results connected with the theory of queueing systems with non-homogeneous customers and infinite sectorized memory space. As an example, we investigate multiserver $M/G/n/0$ system. For this system, we prove formulae for the Laplace-Stieltjes transforms of the summary volume vector distribution function and present formulae for chosen numerical characteristics. We also investigate some special cases illustrating them with numerical computations. Our research shows that the character of dependency between customer's volume vector and his service time plays an important role and has an influence on the total volume vector characteristics.

References

1. Abell, M.L., Braselton, J.P.: The Mathematica Handbook. Elsevier, Amsterdam (1992)
2. Alexandrov, A.M., Katz, B.A.: Non-homogeneous demands flows service. Izvestiya AN SSSR. Tekhnicheskaya Kibernetika, no. 2, pp. 47–53 (1973). (in Russian)
3. Bocharov, P.P., D'Apice, C., Pechinkin, A.V., Salerno, S.: Queueing Theory. VSP, Utrecht, Boston (2004)
4. Chen, X., Stidwell, A.G., Harris, M.B.: Radio telecommunications apparatus and method for communications internet data packets containing different types of data. US Patent No. 7,558,240 B2 (2009). https://patents.google.com/patent/US7558240B2/en
5. Kim, H.-K.: System and method for processing multimedia packets for a network. US Patent No. 7,236,481 B2 (2002). https://patents.google.com/patent/US7236481B2/en
6. Morozov, E., Nekrasova, R., Potakhina, L., Tikhonenko, O.: Asymptotic analysis of queueing systems with finite buffer space. In: Kwiecień, A., Gaj, P., Stera, P. (eds.) CN 2014. CCIS, vol. 431, pp. 223–232. Springer, Cham (2014). https://doi.org/10.1007/978-3-319-07941-7_23
7. Schwartz, M.: Computer-communication Network Design and Analysis. Prentice-Hall, Englewood Cliffs, New York (1977)
8. Schwartz, M.: Telecommunication Networks: Protocols, Modeling and Analysis. Addison-Wesley Publishing Company, New York (1987)
9. Sengupta, B.: The spatial requirement of an M/G/1 queue, or: how to design for buffer space. In: Baccelli, F., Fayolle, G. (eds.) Modelling and Performance Evaluation Methodology. LNCIS, vol. 60, pp. 547–562. Springer, Heidelberg (1984). https://doi.org/10.1007/BFb0005191
10. Tikhonenko, O.: Computer Systems Probability Analysis. Akademicka Oficyna Wydawnicza EXIT, Warsaw (2006). (in Polish)
11. Tikhonenko, O.M.: Destricted capacity queueing systems: determination of their characteristics. Autom. Remote Control 58(6), 969–972 (1997)
12. Tikhonenko, O.M.: Generalized Erlang problem for service systems with finite total capacity. Prob. Inf. Transm. 41(3), 243–253 (2005). https://doi.org/10.1007/s11122-005-0029-z
13. Tikhonenko, O.M., Klimovich, K.G.: Queuing systems for random-length arrivals with limited cumulative volume. Prob. Inf. Transm. 37(1), 70–79 (2001). https://doi.org/10.1023/A:1010451827648
14. Tikhonenko, O.: Queueing systems with common buffer: a theoretical treatment. In: Kwiecień, A., Gaj, P., Stera, P. (eds.) CN 2011. CCIS, vol. 160, pp. 61–69. Springer, Heidelberg (2011). https://doi.org/10.1007/978-3-642-21771-5_8
15. Tikhonenko, O.M.: Queuing systems with processor sharing and limited resources. Autom. Remote Control 71(5), 803–815 (2010). https://doi.org/10.1134/S0005117910050073
16. Tikhonenko, O., Ziółkowski, M.: Single-server queueing system with external and internal customers. Bull. Polish Acad. Sci. Tech. Sci. 66(4), 539–551 (2018). https://doi.org/10.24425/124270
17. Ziółkowski, M.: M/G/n/0 Erlang queueing system with heterogeneous servers and non-homogeneous customers. Bull. Polish Acad. Sci. Tech. Sci. 66(1), 59–66 (2018). https://doi.org/10.24425/119059

18. Ziółkowski, M.: Some practical applications of generating functions and LSTS. Sci. Issues. Mathematics **17**, 97–110 (2012). Jan Długosz University in Częstochowa

19. Ziółkowski, M., Tikhonenko, O.: Multiserver queueing system with non-homogeneous customers and sectorized memory space. In: Gaj, P., Sawicki, M., Suchacka, G., Kwiecień, A. (eds.) CN 2018. CCIS, vol. 860, pp. 272–285. Springer, Cham (2018). https://doi.org/10.1007/978-3-319-92459-5_22

Transient Solution of a Heterogeneous Queuing System with Balking and Retention of Reneging Customers

Rakesh Kumar[1(✉)], Sapana Sharma[1], and Vladimir Rykov[2]

[1] School of Mathematics, Shri Mata Vaishno Devi University,
Katra, Jammu and Kashmir, India
`rakesh.kumar@smvdu.ac.in`, `sapanasharma736@gmail.com`
[2] Department of Applied Mathematics and Computer Modeling,
Gubkin Russian State University of Oil and Gas, Moscow, Russia
`vladmir_rykov@mail.ru`

Abstract. We study a multiple heterogeneous servers queuing model with customers' impatience and retention of impatient customers where the service is provided by multiple heterogeneous servers. The transient solution of the model is derived by using probability generating function technique. Some important measures of performance are studied. A numerical example is provided to study the stability analysis of the system. Further, a comparative analysis is performed between the heterogeneous servers queuing model and the homogeneous servers queuing model and it is found that the heterogeneous servers queuing model performs better. The stationary probabilities of the number of customers in the system are also derived. At the end, some particular cases of the model are discussed.

Keywords: Heterogeneous servers · Time-dependent solution ·
Balking · Retention of reneging customers ·
Probability generating function

1 Introduction

Queuing theory has been playing a vital role in the design and analysis of many real life queuing systems. In particular, the systems like telecommunication systems with impatient phone calls, call centers with impatient calling customers, computer networks with time-out mechanisms, and inventories with perishable items can be modeled by queuing systems with impatient customers. Since in most of service and manufacturing systems the servers work with different speeds, so the queuing systems with identical servers are not effective anymore. The queuing systems with servers possessing identical service rate are appropriated only to the automatic systems. The heterogeneity in service was first considered by Morse [22]. Saaty [26] obtained the stationary distribution

© Springer Nature Switzerland AG 2019
P. Gaj et al. (Eds.): CN 2019, CCIS 1039, pp. 330–346, 2019.
https://doi.org/10.1007/978-3-030-21952-9_25

of a two-heterogeneous servers' queuing model. For more literature on heterogeneous servers' queuing systems one may refer Krishnamoorthy [13], Singh [28], Sharma and Dass [27], Dharmaraja [6], Kumar and Arivudainambi [14], Ammar [2], and Dharmaraja and Kumar [7].

Queuing models with impatient customers have attracted the attention of queuing modelers in the recent past. The research work in queuing theory on customers's impatience commenced with the work of Haight [9,10]. Ancker and Gafarian [3] studied a limited capacity queuing model with reneging and balking. Ancker and Gafarian [4] extended their own work and studied a pure balking system. In [5] Baccelli et al. considered the queuing system with impatient customers where the customer leaves the system whenever his patience time is greater than a random threshold. Montazer-Haghighi et al. [21] derived the stationary distribution of a multi-server Markovian queuing model with reneging and balking. In [24], Rykov studied a heterogeneous multi-server controllable queuing systems. A retrial queuing system which are used in the analysis of aircraft landing process is studied by Koba and Kovalenko [12]. The long run behavior of a queuing system with multiple heterogeneous servers' is studied by Efrosinin and Rykov [8]. The time-dependent behavior of a multiple server queuing model with impatient customers are obtained by Al-Seedy et al. in [1]. In [11] Kapodistria considered a queuing system with two abandonment scenarios. Rykov [25] considered a slow server queuing problem and observed that the optimal policy possesses a monotone property. Vijaya Laxmi and Jyothsna [30] studied a renewal input multiple vacations heterogeneous servers queuing model with impatient customers and obtained the stationary probabilities of this model.

Customers' impatience is a serious problem for any firm. Kumar and Sharma [15] took this idea into account and study a limited capacity single server Markovian queuing model with customers' impatience and retention. They analysed the steady-state behavior of the model. Kumar [16] obtained the time-dependent probabilities of a multi-server queuing model using matrix method. Madheshwari et al. [20] discussed a queuing system wit retrials and derived the steady-state probabilities of the system using the probability generating function technique. Tiwari et al. [29] extended the concept of retention of reneging customer to $M/D/1$ feedback queuing model and derived its steady-state solution. Lee [19] introduced the concept of retention of reneging customer in a discrete-time queue. Vijayalaxmi and Kassahun [31] obtained the stationary probabilities of a multi-server infinite capacity Markovian feedback queue with reneging, balking and retention of reneged customers iteratively. The time-dependent analysis of heterogeneous servers' queuing system with retention of impatient customers is studied by Kumar and Sharma [17].

The queuing model discussed in this paper has an application in dynamic routing problem in computer communication networks. Messages arrive at the buffer have to be routed over one of the several communication lines. The communication lines are heterogeneous servers which transmit the messages with different speeds. If the time taken to transmit the message is too long the sender may give up and leave the system. This situation is analogous to that of a customers' impatience in queuing system. The strategies which minimize the overall delay of the messages can be thought of customer retention strategies.

Recently, Kumar and Sharma [18] performed the time-dependent analysis of a multi-server queuing model with customers' impatience and retention is obtained in an explicit way. But they considered the case of homogeneous servers. Usually in most of the queuing systems the servers work with different rates. Thus the heterogeneity in service must be taken into account for better results. Therefore, we study a multiple heterogeneous servers' system with customers' impatience and retention, and obtain its time-dependent solution.

Rest of the paper is arranged as follows: the queuing model is discussed in Sect. 2. The time-dependent analysis is performed in Sect. 3. The transient numerical results are presented in Sect. 4. In Sect. 5, we discuss the stationary probabilities of the system size. Some particular cases of the queuing system are provided in Sect. 6. Finally, the conclusion is given in Sect. 7.

2   System Model

The arrivals to the queuing system occur in a Poisson stream with mean rate λ. The inter-arrival times are independently, identically and exponentially distributed with parameter λ. If an arriving customer finds all the servers busy, then he joins the system with probability ν, and otherwise balks with probability $1-\nu$. The system has a single queue and multiple servers who serve the customers with different rates μ_i $(i = 1, 2, \ldots, c)$ for each of the c servers, and the service times at each server are exponentially distributed. The capacity of the system (buffer) is infinite and the queue discipline is first-come, first-served (FCFS). The servers are ordered in decreasing service speed. It means that the customers are served by the fastest servers i.e., when $k < c$ customers are present, servers $1, 2, \ldots, k$ are used. That is, during his service, a customer may switch to a fastest server (without any delay) when such a server becomes available and there are no other customers waiting. Upon joining the queuing system and waiting for some time a customer may become impatient and abandon the queue with probability p or gets retained with complementary probability. The probability distribution for reneging times is assumed to be negative exponential with parameter ξ. Initial condition: $P_0(0) = 1$.

The differential equations of the model are:

$$\frac{dP_0(t)}{dt} = -\lambda P_0(t) + \mu_1 P_1(t), \tag{1}$$

$$\frac{dP_n(t)}{dt} = -\left(\lambda + \sum_{i=1}^{n} \mu_i\right) P_n(t) + \lambda P_{n-1}(t) + \sum_{i=1}^{n+1} \mu_i P_{n+1}(t), 1 \le n < c, \tag{2}$$

$$\frac{dP_c(t)}{dt} = -\left(\nu\lambda + \sum_{i=1}^{c} \mu_i\right) P_c(t) + \lambda P_{c-1}(t) + \left(\sum_{i=1}^{c} \mu_i + \xi p\right) P_{c+1}(t), \tag{3}$$

$$\frac{dP_n(t)}{dt} = - \left(\nu\lambda + \sum_{i=1}^{c}\mu_i + (n-c)\xi p \right) P_n(t) + \nu\lambda P_{n-1}(t)$$

$$+ \left(\sum_{i=1}^{c}\mu_i + (n-c+1)\xi p \right) P_{n+1}(t), n > c. \tag{4}$$

3 Time-Dependent Behavior of the Model

In order to analyse the time-dependent behavior of the queuing system we employ probability generating function technique.

Let us denote the probability generating function $P(z,t)$ by

$$P(z,t) = \sum_{n=0}^{c-1} P_n(t) + \sum_{n=0}^{\infty} P_{n+c}(t)z^{n+1}; \ P(z,0) = 1 \tag{5}$$

with

$$\sum_{n=0}^{c-1} P_n(t) = M_{c-1}(t). \tag{6}$$

Adding (1) and (2), we have

$$\frac{d(M_{c-1}(t))}{dt} = -\lambda P_{c-1}(t) + \sum_{i=1}^{c}\mu_i P_c(t). \tag{7}$$

The system of Eqs. (1)–(4) yield

$$\frac{\partial P(z,t)}{\partial t} - \xi p(1-z)\frac{\partial P(z,t)}{\partial z} = \left[\nu\lambda(z-1) + \left(\sum_{i=1}^{c}\mu_i - \xi p \right)(z^{-1}-1) \right]$$
$$\times [P(z,t) - M_{c-1}(t)] + \lambda(z-1)P_{c-1}(t). \tag{8}$$

On solving (8), we obtain

$$P(z,t) = \exp\left\{ \left[\left(\sum_{i=1}^{c}\mu_i - \xi p \right)(\frac{1}{z}-1) + \nu\lambda(z-1) \right]t \right\} + \int_0^t \left[\lambda(z-1) \right.$$
$$\times P_{c-1}(u) - \left(\left(\sum_{i=1}^{c}\mu_i - \xi p \right)(\frac{1}{z}-1) + \nu\lambda(z-1) \right) M_{c-1}(u) \right]$$
$$\times \exp\left\{ \left[\left(\sum_{i=1}^{c}\mu_i - \xi p \right)(\frac{1}{z}-1) + \nu\lambda(z-1) \right](t-u) \right\} du. \tag{9}$$

By using the modified Bessel function of first kind $I_n(.)$, assume

$$x = 2\sqrt{\nu\lambda\left(\sum_{i=1}^{c}\mu_i - \xi p \right)} \text{ and } y = \sqrt{\nu\lambda/\left(\sum_{i=1}^{c}\mu_i - \xi p \right)}, \text{ we obtain}$$

$$\exp\left\{ \left(\nu\lambda z + \frac{\sum_{i=1}^{c}\mu_i - \xi p}{z} \right)t \right\} = \sum_{n=-\infty}^{\infty} (yz)^n I_n(xt). \tag{10}$$

Using (10) in (9), we have

$$
P(z,t) = \exp\left\{ -\left(\nu\lambda + \sum_{i=1}^{c}\mu_i - \xi p\right)t\right\}\sum_{n=-\infty}^{\infty}(yz)^n I_n(xt)
$$

$$
+ \lambda\int_0^t P_{c-1}(u)\exp\left\{ -\left(\nu\lambda + \sum_{i=1}^{c}\mu_i - \xi p\right)(t-u)\right\}
$$

$$
\times \sum_{n=-\infty}^{\infty}(yz)^n [y^{-1}I_{n-1}(x(t-u)) - I_n(x(t-u))]du
$$

$$
+ \int_0^t M_{c-1}(u)\exp\left\{ -\left(\nu\lambda + \sum_{i=1}^{c}\mu_i - \xi p\right)(t-u)\right\}
$$

$$
\times \sum_{n=-\infty}^{\infty}(yz)^n\left[-\nu\lambda y^{-1}I_{n-1}(x(t-u)) + \left(\nu\lambda + \sum_{i=1}^{c}\mu_i - \xi p\right)\right.
$$

$$
\left.\times I_n(x(t-u)) - \left(\sum_{i=1}^{c}\mu_i - \xi p\right)yI_{n+1}(x(t-u))\right]du. \tag{11}
$$

Comparing the coefficients of z^n on both sides of Eq. (11), we have

$$
P_{n+c-1}(t) = \exp\left\{ -\left(\nu\lambda + \sum_{i=1}^{c}\mu_i - \xi p\right)t\right\}y^n I_n(xt)
$$

$$
+ \lambda\int_0^t \exp\left\{ -\left(\nu\lambda + \sum_{i=1}^{c}\mu_i - \xi p\right)(t-u)\right\}\left[I_{n-1}(x(t-u))\right.
$$

$$
\times y^{n-1} - I_n(x(t-u))y^n\right]P_{c-1}(u)du - \int_0^t\exp\left\{ -\left(\nu\lambda + \sum_{i=1}^{c}\mu_i - \xi p\right)\right.
$$

$$
\left.\times (t-u)\right\}M_{c-1}(u)\left[\nu\lambda I_{n-1}(x(t-u))y^{n-1} - \left(\nu\lambda + \sum_{i=1}^{c}\mu_i - \xi p\right)\right.
$$

$$
\left.\times I_n(x(t-u))y^n + \left(\sum_{i=1}^{c}\mu_i - \xi p\right)I_{n+1}(x(t-u)y^{n+1})\right]du;\, n = 1,2,\ldots \tag{12}
$$

for $n = 0$,

$$
M_{c-1}(t) = \exp\left\{ -\left(\nu\lambda + \sum_{i=1}^{c}\mu_i - \xi p\right)t\right\}I_0(xt)
$$

$$
+ \lambda\int_0^t\exp\left\{ -\left(\nu\lambda + \sum_{i=1}^{c}\mu_i - \xi p\right)(t-u)\right\}\times P_{c-1}(u)\left[I_1(x(t-u))y^{-1}\right.
$$

$$
- I_0(x(t-u))]\,du - \int_0^t\exp\left\{ -\left(\nu\lambda + \sum_{i=1}^{c}\mu_i - \xi p\right)(t-u)\right\}
$$

$$
\times M_{c-1}(u)\left[xI_1(x(t-u)) - (\nu\lambda + \sum_{i=1}^{c}\mu_i - \xi p)I_0(x(t-u))\right]du. \tag{13}
$$

Since the negative powers of z are not included in the probability generating function $P(z,t)$ so the right hand side of (12) with n replaced by $-n$ must be zero. Also by using $I_{-n}(.) = I_n(.)$, we obtain

$$
\int_0^t \exp\left\{-\left(\nu\lambda + \sum_{i=1}^c \mu_i - \xi p\right)(t-u)\right\} M_{c-1}(u)\left[\nu\lambda I_{n+1}(x(t-u))y^{n-1}\right.
$$

$$
-\left(\nu\lambda + \sum_{i=1}^c \mu_i - \xi p\right) I_n(x(t-u))y^n + \left(\sum_{i=1}^c \mu_i - \xi p\right) I_{n-1}(x(t-u))y^{n+1}\bigg] du
$$

$$
= \exp\left\{-\left(\nu\lambda + \sum_{i=1}^c \mu_i - \xi p\right)t\right\} I_n(xt)y^n + \lambda \int_0^t \exp\left\{-\left(\nu\lambda + \sum_{i=1}^c \mu_i - \xi p\right)\right.
$$

$$
\times (t-u)\bigg\} P_{c-1}(u)\left[I_{n+1}(x(t-u))y^{n-1} - I_n(x(t-u))y^n\right] du. \tag{14}
$$

The usage of (14) in (12) considerably simplifies the work and results in a simple expression for $P_n(t)$. This yields, for $n = 1, 2, \ldots,$

$$
P_{n+c-1}(t) = ny^n \int_0^t \exp\{-(\nu\lambda + \sum_{i=1}^c \mu_i - \xi p)(t-u)\} \frac{I_n(x(t-u))}{(t-u)} P_{c-1}(u)du. \tag{15}
$$

The Eq. (15) generates the probabilities $P_c(t), P_{c+1}(t), P_{c+2}(t), \ldots,$ where c represents the number of servers.

Now, we write (1) and (2) in matrix form as:

$$
\frac{d\mathbf{P}(t)}{dt} = Q\mathbf{P}(t) + \sum_{i=1}^{c-1} \mu_i P_{c-1}(t)\mathbf{e_1}, \tag{16}
$$

where the matrix $Q = (b_{k,j})_{c-1 \times c-1} =$

$$
\begin{bmatrix}
-(\lambda) & \mu & \cdots & 0 \\
\lambda & -(\lambda + \mu) & \cdots & 0 \\
\vdots & \vdots & \vdots & \vdots \\
0 & 0 & \cdots & -(\lambda + (c-2)\mu)
\end{bmatrix},
$$

$\mathbf{P}(t) = (P_0(t)\ \ P_1(t)\ \ \cdots\ \ P_{c-2}(t))^T_{c-1 \times 1}$, $\mathbf{e_1} = (0\ \ 0\ \ \cdots\ \ 1)^T_{c-1 \times 1}$.

Let $\mathbf{P}^*(s) = (P_0^*(s)\ \ P_1^*(s)\ \ \cdots\ \ P_{c-2}^*(s))^T$ represents the Laplace transform of $\mathbf{P}(t)$, then applying Laplace transform on (16) we get

$$
\mathbf{P}^*(\mathbf{s}) = \frac{1}{(sI - Q)}\left\{P_{c-1}^*(s)\mathbf{e_1} + \mathbf{P}(0)\right\} \tag{17}
$$

with $\mathbf{P}(0) = (1\ \ 0\ \ \cdots\ \ 0)^T$. Now, if $\mathbf{e_2} = (1\ \ 1\ \ \cdots\ \ 1)^T_{c-1 \times 1}$, then

$$
M_{c-1}^*(s) = \mathbf{e_2}^T\mathbf{P}^*(\mathbf{s}) + P_{c-1}^*(s). \tag{18}
$$

Define

$$\Theta(s) = \left[\left(s + \nu\lambda + \left(\sum_{i=1}^{c}\mu_i - \xi p\right)\right) - \sqrt{\left(s + \nu\lambda + \left(\sum_{i=1}^{c}\mu_i - \xi p\right)\right)^2 - x^2}\right].$$

The Laplace transform of (13) gives

$$sM_{c-1}^*(s) = 1 + P_{c-1}^*(s)\left[\frac{1}{2}\{\Theta(s)\} - \lambda\right]. \tag{19}$$

Substituting (19) in (18), we get

$$P_{c-1}^* = \frac{1 - se_2^T(sI - Q)^{-1}\mathbf{P}(0)}{\{(s + \nu\lambda) - \frac{1}{2}[\Theta(s)] + \sum_{i=1}^{c-1}\mu_i se_2^T(sI - Q)^{-1}\mathbf{e_1}\}}. \tag{20}$$

In Eqs. (17) and (20), $(sI - Q)^{-1}$ has to be found. Assume

$$(sI - Q)^{-1} = (b_{kj}^*(s))_{c-1 \times c-1}.$$

We observe that $(sI - Q)^{-1}$ is almost lower triangular. From Raju and Bhat [23] we obtain, for $k = 0, 1, \ldots, c - 2$,

$$b_{kj}^*(s) = \begin{cases} \frac{1}{\sum_{i=1}^{j+1}\mu_i}\frac{u_{c-1,j+1}(s)u_{k,0}(s) - u_{k,j+1}(s)u_{c-1,0}(s)}{u_{c-1,0}(s)}, & j = 0, 1, \ldots, c-3, \\ \frac{u_{k,0}(s)}{u_{c-1,0}(s)}, & j = c - 2. \end{cases} \tag{21}$$

where $u_{k,j}(s)$ are recursively given as

$$u_{k,k}(s) = 1, \qquad\qquad\qquad k = 0, 1, \ldots, c - 2,$$
$$u_{k+1,k}(s) = \frac{s + \lambda + \sum_{i=1}^{k}\mu_i}{\sum_{i=1}^{k+1}\mu_i}, \qquad k = 0, 1, \ldots, c - 3,$$

$$u_{k+1,k-j}(s) = \frac{\left(s + \lambda + \sum_{i=1}^{k}\mu_i\right)u_{k,k-j} - \lambda u_{k-1,k-j}}{\sum_{i=1}^{k+1}\mu_i}, j \le k, k = 1, 2, 3, \ldots, c - 3,$$

$$u_{c-1,j}(s) = \begin{cases} [s + \lambda + \sum_{i=1}^{c-2}\mu_i]u_{c-2,j} - \lambda u_{c-3,j}, & j = 0, 1, \ldots, c - 3, \\ s + \lambda + \sum_{i=1}^{c-2}\mu_i, & j = c - 2. \end{cases} \tag{22}$$

and
$$u_{k,j}(s) = 0, \text{ for other } k \text{ and } j. \text{ Using these in (20), we get}$$

$$P_{c-1}^*(s) = \frac{\{1 - s\sum_{k=0}^{c-2}b_{k,0}^*(s)\}}{\{(s + \nu\lambda) - \frac{1}{2}[\Theta(s)] + \sum_{i=1}^{c-1}\mu_i s\sum_{k=0}^{c-2}b_{k,c-2}^*(s)\}}, \tag{23}$$

and from Eq. (17), for $k = 0, 1, \ldots, c - 2$ we get

$$P_k^*(s) = b_{k,0}^*(s) + \sum_{i=1}^{c-1} \mu_i b_{k,c-2}^*(s) P_{c-1}^*(s). \tag{24}$$

$b_{k,j}^*(s)$ are rational algebraic functions in s. The cofactor of the $(k, j)^{th}$ element of $(sI - Q)$ is a polynomial of degree $c - 2 - \mid k - j \mid$. The inverse Laplace transform $b_{k,j}(t)$ of $b_{k,j}^*(s)$ are obtained by partial fractions. Let the matrix Q has characteristic roots $s_k, k = 0, 1, \ldots, c - 2$. Then after simplifying (23) gives

$$P_{c-1}^*(s) = U^*(s) 2[p + \sqrt{p^2 - \alpha^2}]^{-1}$$

$$\times \left[1 - \frac{2\left\{ (\sum_{i=1}^{c} \mu_i - \xi p) \left(1 - \frac{\sum_{i=1}^{c} \mu_i}{\sum_{i=1}^{c} \mu_i - \xi p} V^*(s) \right) \right\}}{\{p + \sqrt{p^2 - \alpha^2}\}} \right]^{-1}, \tag{25}$$

where

$$U^*(s) = \sum_{i=0}^{c-2} \frac{U_i}{s - s_i}, \tag{26}$$

$$V^*(s) = \sum_{i=0}^{c-2} \frac{V_i}{s - s_i} \tag{27}$$

with constants U_i and V_i given by

$$U_i = \lim_{s \to s_i} (s - s_i) \left[1 - \sum_{l=0}^{c-2} s b_{l,0}^*(s) \right], \tag{28}$$

$$V_i = \lim_{s \to s_i} (s - s_i) \left[\sum_{l=0}^{c-2} s b_{l,c-2}^*(s) \right]. \tag{29}$$

Hence, (25) becomes

$$P_{c-1}^*(s) = \sum_{n=0}^{\infty} \sum_{m=0}^{n} \frac{(-1)^m}{\Omega} \left(\frac{2\Omega}{x} \right)^{n+1} (n + 1) \binom{n}{m}$$

$$\times \left(\frac{\sum_{i=1}^{c-1} \mu_i}{\Omega} \right)^m U^*(s)(V^*(s))^m$$

$$\times \frac{\left[(s + \nu\lambda + \sum_{i=1}^{c} \mu_i - \xi p) - \sqrt{(s + \nu\lambda + \sum_{i=1}^{c} \mu_i - \xi p)^2 - x^2} \right]^{n+1}}{(n + 1)x^{n+1}}. \tag{30}$$

The Laplace inversion of (30) gives

$$
P_{c-1}(t) = \sum_{n=0}^{\infty} \sum_{m=0}^{n} \frac{(-1)^m}{\Omega} \left(\frac{2\Omega}{x}\right)^{n+1} \left(\frac{\sum_{i=1}^{c-1} \mu_i}{\Omega}\right)^m (n+1)\binom{n}{m}\left[\int_0^t U(t-u)\right.
$$

$$
\times \int_0^u V^{C(m)}(u-v)\exp\{-(\nu\lambda + \sum_{i=1}^{c} \mu_i - \xi p)v\}\frac{I_{n+1}(xv)}{v}dudv\Bigg],
$$

$$(31)$$

where $V^{C(m)}(t)$ is $m-fold$ convolution of $V(t)$ with itself with $V^{C(0)} = \delta(t)$, the Dirac delta function, $\Omega = \sum_{i=1}^{c} \mu_i - \xi p$, $x = 2\sqrt{\nu\lambda(\sum_{i=1}^{c} \mu_i - \xi p)}$, $y = \sqrt{\frac{\nu\lambda}{\sum_{i=1}^{c} \mu_i - \xi p}}$, and $U(t-u)$ is a function of u at a particular value of t which is obtained from the convolution of two functions $U(t)$ and $V(t)$, where $U(t) = L^{-1}(U^*(s))$ and $V(t) = L^{-1}(V^*(s))$.

The Eq. (31) gives us the probability of having $c-1$ customers in the system at time t.

By taking the Laplace inverse of (24) we obtain

$$
P_k(t) = b_{k,0}(t) + \sum_{i=1}^{c-1} \mu_i \int_0^t b_{k,c-2}(u)P_{c-1}(t-u)du, \quad k = 0, 1, \ldots, c-2 \quad (32)
$$

where $b_{k,j}(t)$ represents the laplace inverse of $b_{k,j}^*(s)$ and $b_{k,j}^*(s)$ are the elements of matrix $(sI - Q)^{-1}$. $P_{c-1}(u)$ can be computed with the help of Eq. (31).

Therefore all the time-dependent probabilities are explicitly obtained in (15), (31) and (32).

4   Numerical Example

The Quality of Service characteristics of the model are as follows:

1. Average number of customers in the system $(L_s(t))$

$$
L_s(t) = \sum_{n=0}^{\infty} nP_n(t)
$$

2. Average Reneging Rate $(R_r(t))$

$$
R_r(t) = \sum_{n=c}^{\infty} (n-c)\xi pP_n(t)
$$

3. Average Retention Rate $(R_R(t))$

$$
R_R(t) = \sum_{n=c}^{\infty} (n-c)\xi qP_n(t)
$$

The probabilities $P_n(t)$ are given in (15), (31) and (32).

4.1   Stability Analysis

In this sub-section, the stability analysis is performed. Time-dependent variations in Quality of Service characteristics with respect to system's parameters
are shown in Figs. 1, 2, 3, 4 and 5. In Figs. 1, 2 and 3, the variations in QoS
characteristics with respect to the mean arrival rate (λ) are shown. An increase
in λ increases the average number of customers in the system. The average rate
of reneging and the average rate of retention also show an increasing trend with
the increase in λ and they reach stable values with the passage of time. In Figs. 4

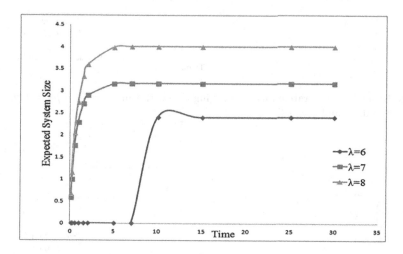

Fig. 1. Effect of arrival rate on mean system size with time ($\nu = 0.99, \mu_1 = 4, \mu_2 = 3$,
$\mu_3 = 2, \xi = 0.05$, and $q = 0.4$)

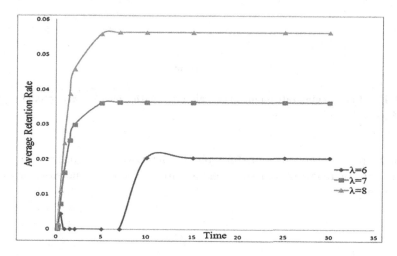

Fig. 2. Effect of arrival rate on mean retention rate with time ($\nu = 0.99, \mu_1 = 4, \mu_2 = 3$,
$\mu_3 = 2, \xi = 0.05$, and $q = 0.4$)

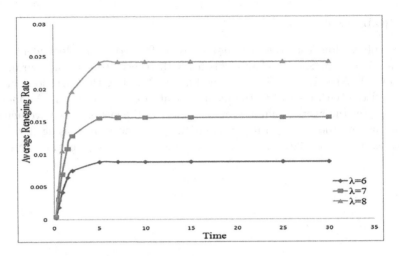

Fig. 3. Effect of arrival rate on mean reneging rate with time ($\nu = 0.99, \mu_1 = 4, \mu_2 = 3$, $\mu_3 = 2, \xi = 0.05$, and $q = 0.4$)

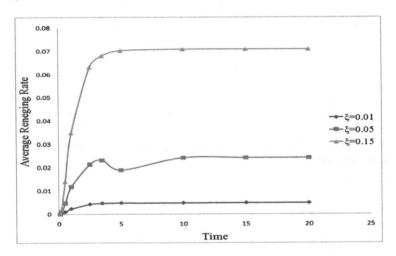

Fig. 4. Effect of reneging rate on mean reneging rate with time ($\lambda = 8, \nu = 0.99, \mu_1 = 4$, $\mu_2 = 3, \mu_3 = 2, \xi = 0.05$)

and 5, we present the variations in QoS characteristics with respect to ξ. The average rate of reneging and the average rate of retention increases with the increase in ξ.

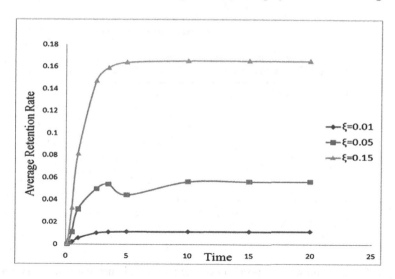

Fig. 5. Effect of reneging rate on mean retention rate with time ($\lambda = 8, \nu = 0.99$, $\mu_1 = 4, \mu_2 = 3, \mu_3 = 2, \xi = 0.05$)

4.2 Comparative Analysis of Two Queuing Models with and Without Heterogeneous Servers

In this sub-section, we compare an $M/M/c$ (homogeneous servers) queuing model with retention of impatient customers to that of an $M/M/c$ (heterogeneous servers) queuing model with retention of impatient customers. We study the variations in expected system size and the average reneging rate of these queuing systems with respect to time as shown in Figs. 6 and 7. The parameters' values for these models are as follows:

$M/M/c$ queuing system with homogeneous servers: $\lambda = 8, \nu = 0.99, \mu_1 = \mu_2 = \mu_3 = 3, \xi = 0.05, p = 0.3$
$M/M/c$ queuing system with heterogeneous servers: $\lambda = 8, \nu = 0.99, \mu_1 = 4, \mu_2 = 3, \mu_3 = 2, \xi = 0.05, p = 0.3$.

From Fig. 6, one can see that the average number of customers in the system is lesser in case of heterogeneous servers' queuing model than in case of homogeneous servers' model. Further, Fig. 7 shows that the average reneging rate is always lower in case of heterogeneous servers' queuing model than that of its homogeneous counterpart. Thus, the servers with heterogeneous rates reduce the average reneging rate.

From the analysis of Figs. 6 and 7 we can analyze that the queuing model with heterogeneous servers' performs better as compared to multiple homogeneous servers' model.

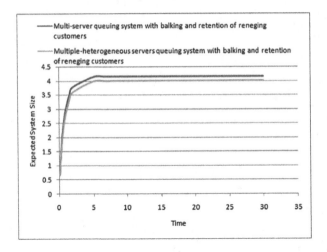

Fig. 6. Comparison with respect to average number of customers in the system

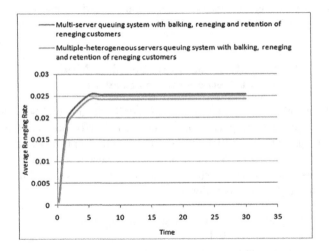

Fig. 7. Comparison with respect to average reneging rate

5   Stationary Probabilities

Let P_n be the steady-state probability that there are n customers in the system. In steady-state $\lim_{t \to \infty} P_n(t) = P_n$, and $\lim_{t \to \infty} \dfrac{dP_n(t)}{dt} = 0$. Therefore, the Eqs. (1)–(4) in steady-state becomes:

$$0 = -\lambda P_0 + \mu_1 P_1, \tag{33}$$

$$0 = -\left(\lambda + \sum_{i=1}^{n} \mu_i\right) P_n + \lambda P_{n-1} + \sum_{i=1}^{n+1} \mu_i P_{n+1}, 1 \le n < c, \tag{34}$$

$$0 = -\left(\nu\lambda + \sum_{i=1}^{c}\mu_i\right)P_c + \lambda P_{c-1} + \left(\sum_{i=1}^{c}\mu_i + \xi p\right)P_{c+1}, \tag{35}$$

$$0 = -\left(\nu\lambda + \sum_{i=1}^{c}\mu_i + (n-c)\xi p\right)P_n + \nu\lambda P_{n-1}$$

$$+ \left(\sum_{i=1}^{c}\mu_i + (n-c+1)\xi p\right)P_{n+1}, n > c. \tag{36}$$

From Eq. (33), we have

$$P_1 = \frac{\lambda}{\mu_1}P_0.$$

Re-write Eq. (34) as

$$\left(\sum_{i=1}^{n+1}\mu_i\right)P_{n+1} = \left(\lambda + \sum_{i=1}^{n}\mu_i\right)P_n - \lambda P_{n-1}. \tag{37}$$

Now substitute $n = 1, 2, 3, ...c - 1$ in (37), we get

$$P_n = \frac{\lambda^n}{\prod_{j=1}^{n}\left(\sum_{i=1}^{j}\mu_i\right)}P_0, \ 1 \le n \le c. \tag{38}$$

Also, Eq. (36) can be written as

$$\left(\sum_{i=1}^{c}\mu_i + (n-c+1)\xi p\right)P_{n+1} = -\left(\lambda + \sum_{i=1}^{c}\mu_i + (n-c)\xi p\right)P_n + \lambda P_{n-1}. \tag{39}$$

Now, on substituting $n = c, c + 1, ...$ in Eq. (39), we get

$$P_n = \frac{\lambda^n}{\prod_{j=1}^{c}\left(\sum_{i=1}^{j}\mu_i\right)\prod_{k=1}^{n-c}\left(\sum_{i=1}^{c}\mu_i + k\xi p\right)}P_0, \ n \ge c + 1. \tag{40}$$

Using normalization condition, $\sum_{n=0}^{\infty}P_n = 1$, the probability P_0 can be obtained as

$$P_0 = \left[1 + \sum_{n=1}^{c}\left(\frac{\lambda^n}{\prod_{j=1}^{n}\left(\sum_{i=1}^{j}\mu_i\right)}\right) + \sum_{n=c+1}^{\infty}\left(\frac{\lambda^n}{\prod_{j=1}^{c}\left(\sum_{i=1}^{j}\mu_i\right)\prod_{k=1}^{n-c}\left(\sum_{i=1}^{c}\mu_i + k\xi p\right)}\right)\right]^{-1} \tag{41}$$

5.1 Stationary Characteristics

Following are the stationary characteristics of the model:

1. Average number of customers in the system (L_s)

$$L_s = \sum_{n=0}^{\infty} n P_n.$$

2. Average Reneging Rate (R_r)

$$R_r = \sum_{n=c}^{\infty} (n - c)\xi p P_n.$$

3. Average Retention Rate (R_R)

$$R_R = \sum_{n=c}^{\infty} (n - c)\xi q P_n,$$

where P_n are given in (36), (40) and (41).

6 Particular Cases

Case 1 When all the servers are homogeneous, then the results of this paper are identical with the one obtained by Kumar and Sharma (2018).

Case 2 Our model reduces to the one studied by Kumar and Arivudainambi (2001), when there is no impatience.

7 Conclusions

The time-dependent behavior of a queuing model with multiple-heterogeneous servers and retention of reneging customers is studied. The stability analysis and comparative analysis of the model are also performed. Finally, the stationary probabilities of the system size are also derived.

Acknowledgments. Dr. Rakesh Kumar acknowledges the financial assistance provided by UGC, New Delhi, India under the Major Research Project vide letter no. F.-43-434/2014(SR) for preparation of this paper, and Prof. Rykov thanks Russian Fund of Fundamental Investigations for the support in framework of Grants No. 17-01-00633 and No. 17-07-00142.

References

1. Al-Seedy, R.O., El-Sherbiny, A.A., El-Shehawy, S.A., Ammar, S.I.: Transient solution of the M/M/c queue with balking and reneging. Comput. Math. Appl. **57**, 1280–1285 (2009)
2. Ammar, S.I.: Transient analysis of a two-heterogeneous servers queue with impatient behaviour. J. Egypt. Math. Soc. **22**, 90–95 (2014)
3. Ancker, C.J., Gafarian, A.V.: Some queueing problems with balking and reneging I. Oper. Res. **11**, 88–100 (1963a)
4. Ancker, C.J., Gafarian, A.V.: Some queueing problems with balking and reneging II. Oper. Res. **11**, 928–937 (1963b)
5. Baccelli, F., Boyer, I.P., Hebuterne, G.: Single-server queues with impatient customers. Adv. Appl. Prob. **16**, 887–905 (1984)
6. Dharmaraja, S.: Transient solution of a two-processor heterogeneous system. Math. Comput. Modell. **32**, 1117–1123 (2000)
7. Dharmaraja, S., Kumar, R.: Transient solution of a Markovian queuing model with heterogeneous servers and catastrophes. Opsearch **52**, 810–826 (2015)
8. Efrosinin, D., Rykov, V.: On performance characteristics for queueing systems with heterogeneous servers. Autom. Remote Control **69**, 61–75 (2008)
9. Haight, F.A.: Queueing with balking. Biometrika **44**, 360–369 (1957)
10. Haight, F.A.: Queueing with reneging. Metrika **2**, 186–197 (1959)
11. Kapodistria, S.: The M/M/1 queue with synchronized abandonments. Queueing Syst. **68**, 79–109 (2011)
12. Koba, E.V., Kovalenko, I.N.: Three retrial queueing systems representing some special features of aircraft landing. J. Autom. Inf. Sci. 4 (2002). https://doi.org/10.1615/JAutomatInfScien.v34.i4.10
13. Krishanamoorthy, B.: On Poisson queues with heterogeneous servers. Oper. Res. **11**, 321–330 (1963)
14. Kumar, B.K., Arivudainambi, D.: Transient solution of an M/M/c queue with heterogeneous servers and balking. Inf. Manage. Sci. **12**, 15–27 (2001)
15. Kumar, R., Sharma, S.K.: M/M/1/N queuing system with retention of reneged customers. Pak. J. Stat. Oper. Res. **8**, 859–866 (2012)
16. Kumar, R.: Economic analysis of an M/M/c/N queuing model with balking, reneging and retention of reneged customers. Opsearch **50**, 383–403 (2013)
17. Kumar, R., Sharma, S.: Transient solution of a two-heterogeneous servers' queuing system with retention of reneging customers. Bull. Malays. Math. Sci. Soc. (2017). https://doi.org/10.1007/s40840-017-0482-z
18. Kumar, R., Sharma, S.: Transient analysis of M/M/c queuing system with balking and retention of reneging customers. Commun. Stat. Theor. Methods **47**, 1318–1327 (2018)
19. Lee, Y.: Discrete-time queue with batch geometric arrivals and state dependent retention of reneging customers. Int. J. New Technol. Res. **3**, 32–33 (2017)
20. Madheswari, S.P., Suganthi, P., Josephine, S.A.: Retrial queueing system with retention of reneging customers. Int. J. Pure Appl. Math. **106**, 11–20 (2016)
21. Montazer-Hagighi, A., Medhi, J., Mohanty, S.G.: On a multi server Markovian queueing system with balking and reneging. Comput. Oper. Res. **13**, 421–425 (1986)
22. Morse, P.M.: Queues, Inventories and Maintenance. Wiley, New York (1958)
23. Raju, S.N., Bhat, U.N.: A computationally oriented analysis of the G/M/1 queue. Opsearch **19**, 67–83 (1982)

24. Rykov, V.: Monotone control of queueing systems with heterogeneous servers. Queueing Syst. **37**, 391–403 (2001)
25. Rykov, V.: On a slow server problem. In: Li, H., Li, X. (eds.) Stochastic Orders in Reliability and Risk. LNS, vol. 208. Springer, New York (2013). https://doi.org/10.1007/978-1-4614-6892-9_18
26. Saaty, T.L.: Elements of Queuing Theory with Applications. McGraw Hill, New York (1961)
27. Sharma, O.P., Dass, J.: Initial busy period analysis for a multichannel Markovian queue. Optimization **20**, 317–323 (1989)
28. Singh, V.P.: Two-server Markovian queues with balking: heterogeneous vs. homogeneous servers. Oper. Res. **18**, 145–159 (1970)
29. Tiwari, S.K., Gupta, V.K., Joshi, T.N.: M/D/1 feedback queueing models with retention of reneged customers. Int. J. Sci. Res. **5**, 405–408 (2016)
30. Vijaya Laxmi, P., Jyothsna, K.: Balking and reneging multiple working vacations queue with heterogeneous servers. J. Math. Modell. Algorithms **14**, 267–285 (2015)
31. Vijaya Laxmi, P., Kassahun, T.W.: A Multi-server infinite capacity Markovian feedback queue with balking, reneging and retention of reneged customers. Int. J. Math. Arch. **8**, 53–59 (2017)

Cluster-Based Web System Models
for Different Classes of Clients in QPN

Tomasz Rak[(✉)]

Department of Computer and Control Engineering,
Rzeszow University of Technology, Al. Powstancow Warszawy 12,
35-959 Rzeszow, Poland
trak@prz.edu.pl
https://trak.kia.prz.edu.pl

Abstract. Simulation studies for Web systems have been carried out by many academic researchers and practitioners. Models are often less time-consuming to develop and run production system. Performance Engineering is done to determine the system performance. In the paper various performance models of Cluster-based Web Systems are discussed, as well as their influence on response time. The Queueing Petri Nets simulations are based on different loads, but also on changing environmental parameters and system structures. A novelty in this approach is the use of two client-classes related to customer behavior and routes in the system. In all cases Web system architectures include clusters are taken into consideration. Simulation results obtained from this models are compared with data from a real system and show good accuracy.

Keywords: Cluster-based Web Systems · Response Time Analysis ·
Queueing Petri Nets · Performance Engineering

1 Introduction

The popularity of the Internet as the medium of information still grows. The increase of amount of Internet clients brought problems with service of large number of HTTP requests. It is result of rapid development of Web system technology. Providing high quality of Web systems is now very important for most of services. The most common way to build highly efficient Internet system is to use a cluster of servers, which gives the opportunity to scale the system to the nowadays needs [1]. Clustering is a methodology of grouping independent servers for providing better flexibility, scalability, availability and performance. Understanding of Cluster-based Web Systems (CWS) is not simple. This approach easily makes a possibility of construction CWS. Cluster structures are used to achieve greater performance. We use the cluster structures to build Web system layers[1].

---

[1] We have the same number corresponding to each other logical layers and physical tiers.

© Springer Nature Switzerland AG 2019
P. Gaj et al. (Eds.): CN 2019, CCIS 1039, pp. 347–365, 2019.
https://doi.org/10.1007/978-3-030-21952-9_26

The main aim of this research is to develop a method for predicting the performance of CWS by using Queueing Petri Net (QPN) modeling and simulation environment [2]. As a formal model, QPN inherits many merits of Petri Nets and Queueing Nets [Appendix A]. Additionally, they have more powerful modeling capabilities, which make QPN widely used in many fields. What's more, QPN can also be used to analyze performance based on simulation [2–5]. The method of performance analysis using QPN was systematically proposed in [6,7], for defining ways of how to model, simulate and analyze.

This method is a little bit different from the traditional methods on systems analysis. It is suitable for performance analysis of many systems with different structures and various workloads. All of the above works indicate that the method of performance analysis (part of Performance Engineering (PE)) based on QPN is feasible and has the great practical value. In comparison with other languages, QPN is easier to program and offers a variety of other analysis techniques. Based on QPN modeling and analyzing performance, we can provide quantitative and qualitative performance metrics and predictions, which are helpful to guide planning capacity and system optimization.

The motivation why we explore the QPN method for PE of Web systems, is that we haven't found any other similar method yet. The second motivation is that our QPN method helps to improve the previous work based on Time Colored Petri Nets [8,9].

In this paper, we study parameters of systems widespread in CWS and propose the QPN method for analyzing their performance, which can be easily implemented in Queueing Petri net Modeling Environment (QPME) software [10] by potential system architects. We also propose a hierarchical modeling technique for reduce the model size. As a model-based performance analysis method, this method is low cost and highly flexible. Earlier experimental results show that this method is feasible and can be applied to more complex and large-scale systems (e.g. CWS).

This paper makes two main contributions:

- Firstly, we propose the QPN method for modeling CWS and analyzing their performance.
- Secondly, we implement QPN models with many parameters using QPME tool [10] and conduct simulations to verify system performance.

In this paper, we propose one PE technique that can be analyzed with a simulation.

The rest of this paper is structured as follows. Section 2 introduces some related works. Section 3 presents an overview of CWS architecture. Section 4 describes proposed modeling approach and elaborates proposed PE method for CWS. Section 5 presents results of performance analysis. Section 6 presents simulation results. Section 7 provides conclusions and future implications for this work.[2] Appendix A presents the QPN model for CWS.

---

[2] We assumed that the reader is familiar with PE and QPN formalism.

2   Related Works

There are two mathematical formalisms currently used for solving problems of performance evaluation and analysis: Queueing Nets (QN) and Petri Nets (PN). We review some related works in the area of PE for Web systems. Several approaches have been proposed for performance analysis. Some formal languages [11] are analyzed for systems consisting of concurrently operating units. Some of works use variants of the PN, but they are applied in a generic context for stream processing [12] or distributed systems [6]. In [13] PN models for cooperation modes of emergency actions using resources are proposed. The work [14] was devoted to the Apache Storm performance evaluation, combining a genuine UML profile and Generalized Stochastic PN. Analyzed technologies like Apache Storm [15] can enormously help in context of Information Systems performance, that today faces the era of Big Data and complex systems. Paper [16] proposes arrival stream modeling and analysis. According to a distributed system modeling method introduced in [9] the modeled system (Platform as a Service designed to host Web applications) has a structure similar to an open queueing system. Analysis of distribution of data packages within the system makes it possible to find out potential bottlenecks. In paper [17], they propose new Colored PN method for modeling job scheduling systems in High Performance Clusters and analyzing their performance. In [18] Mironescu et al. proposed the development of model and simulation technique for prediction of the performance of concurrent application running on cluster architecture. Models for the hardware and software were developed using the PN formalism and then coupled with the simulation. Mathematical models for predicting the performance of Web system are introduced in [8,9].

Closed queueing model comprising of client, application server cluster, database server and production line stations is described in [4]. Sometimes elements of the control theory are used to manage the movement of packages in Web servers [9]. In [19] for multi-tiered applications, author uses QN and Mean Value Analysis models. The article compares performance of a multi-server MVA model with actual performance testing measurements.

Recent publications [6,7] use QPN for modeling and performance evaluation of Web structure.

Experiments are the second way [1] for PE. Applying experiment and simulation models greatly influences validity of the systems developed.

Our approach described in this paper may be treated as extension of selected solutions summed up in [6,7], where we propose the QPN models for Internet system. In these papers we proposed QPN models for one kind of distributed Web system with one class of requests [20].

3   Cluster-Based Web System Architecture

A multi-tier architecture is a type of architecture which is composed of tiers or layers of logical computing. These architectures provide many benefits for production environments by modularizing the user interface, business logic, and data storage layers, e.g.: scalability, performance, and availability. By separating different layers we can scale each of them independently according to the need at any given time. By having separate layers we can also increase performance. Typically, distributed Web systems are composed of layers where each layer consists of a set of servers – cluster. Layers are dedicated for proper tasks and exchange requests between each other. Cluster nodes (servers) are used to receive requests submitted by clients. In some cases a master node, designed as a global scheduler allocates the jobs to every node. There are also systems with uniform task distribution, without designating a master node.

In our approach presented architecture has been simplified into two layers [7]:

– Front-End (FE) layer,
– Back-End (BE) layer.

Clustering mechanism is used in the FE layer. Architecture composed of these layers is used for e-business systems.

The Closed QN model with services as shown in Fig. 1a presents typical CWS architecture. Each of application and database servers have individual queues.

Additionaly modeled system follows the paradigms of component-based system development using transitions, places and exchange of tokens.

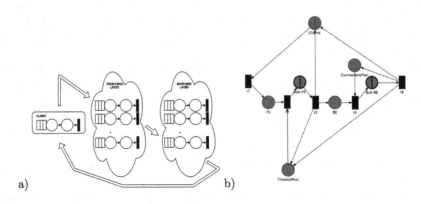

a)   b)

Fig. 1. CWS model: (a) QPN hardware and software, (b) QPN main page

4   Formal Model

QPN models for software and hardware elements of CWS were developed. Models make possible the description of hardware/software mapping and can imitate behavior of the system in a realistic manner. Some concepts involved in these models are described in details as follow.

4.1   Combination of Queueing Nets and Petri Nets

QN (quantitative analysis, e.g. evaluation of logical system correctness) has a queue. If the times between arrivals of requests are generally short and the service times are long it is clear that the queue will build up. QN usually consists of set of connected queueing systems. Various queue systems represent computer or system components. To analyze any queue system it is necessary to determine:

- Arrival process e.g. Poisson[3], Erlang, Hyper-exponential, General.
- Service process is time which each request spends at the station e.g. Logarithmic, Chi-square, Hyper-exponential, Exponential[4]. Service times are Independent and Identically Distributed (IID).
- Scheduling strategies (service/queueing disciplines) e.g.: First In First Out (FIFO), Last In First Out (LIFO), Last In First Out with Preempt and Resume (LIFO-PR), Round Robin (RR) with a fixed quantum, Small Quantum $\Rightarrow$ Processor Sharing (PS), Infinite Server (IS) = fixed delay, Shortest Processing Time first (SPT)[5].
- Number of servers[6].
- Number of buffers (waiting room size[7]). Queue length includes requests currently receiving service as well as those waiting in the queue.

PN (qualitative analysis, e.g. assessment of system performance, network efficiency) has tokens representing requests. PN describes dynamics based on rules of tokens flow. PN is referred as a connection between the engineering description and theoretical approach. To analyze any PN it is necessary to determine:

- Set of places.
- Set of transactions.
- Initial marking (number of tokens).
- Incidence function (routing probability) assigns natural numbers to arcs (weights of arcs).
- Token color function (for Colored PN).

QPN adds queue and time aspects to model.

4.2   Theoretical Introduction

This approach to design (part of PE[8]) join Load Testing (LT) – experiments – and Performance Modeling (PM) – simulations. PE provides some recommendations to realize the required performance level. LT is the process of putting

---

[3] We analysed queueing systems with the Poisson clients arrival process.

[4] We analysed queueing systems with the exponential clients service process.

[5] We used IS for clients station, PS for FE servers and FIFO for BE server.

[6] This paper considers single server queue.

[7] Size of the queue is infinite.

[8] PE analyzes the expected performance characteristics of system during the different phases of its life cycle.

demand on existing system and measuring its response. PM is used to predict performance of the system under the construction.

Performance estimation of most implemented systems is typically performed using LT techniques, but we also use PM at the state of system design [21]. In order to perform formal analysis of CWS, we specify the notations in Tables 1, 2 and 3[9].

Table 1. General notations for mathematical model

Symbol	Notation
X_o	Throughput of the system
N	Total number of clients[a] in the system
NC	Number of clients in the client layer
NS	Number of clients in the system layers
R	Response time of the system (sum of residence times)
TT	Think time in *Clients* queueing place

[a] The client class job is identified with the request.

Fundamental laws applicable to our PM have been proposed using the operational metrics from this section.

4.3   Mathematical Client Model

The characteristic feature of Web systems is a large number of clients using the Internet services at the same time. The process of requests arrival is modeled with client think time TT in *Clients* queueing place.

In our analysis, we consider closed system with two client-classes at *Clients* station. The system is a collection of resources. There are no outside arrivals or departures. Total number of requests in the system is constant. Each resource i has exponentially distributed service rate μ_i. A customer departing resource i choose next resource i' with probability $RO_k(i, i')$.

The workload is composed of requests. In some researches a recorded data stream is used as an input to the system model under performance evaluation. However usually the arrival stream generator (*Clients*) produces an input data stream in a random order according to predefined arrival distribution [8,16, 20,22]. It has been assumed that every client can be attributed to a certain client class. We have obtained marking a set of client-classes (types) as N_k and assuming that the client types number is k (1).

$$N_k = \{(C_1)_k, (C_2)_k, \ldots, (C_j)_k\} \tag{1}$$

Particular client has been marked as C_j. Model of the particular client type depends on numerous parameters, such as sort of service, time limits imposed on

---

[9] Classes are distinguished by different routing probabilities of requests in the model.

Table 2. Notations for mathematical model at resource i

Symbol	Notation
X_i	Throughput
S_i	Service time (residence time)
U_i	Utilization
V_i	Number of clients visit[a]
SD_i	Service demand[b]
WT_i	Waiting time (queueing time)[c]
λ_i	Arrival rate
μ_i	Processing rate
DD_i	Departure discipline
SS_i	Scheduling strategy
NS_i	Number of servers
CS_i	Client service

[a] Queueing station or server.
[b] Service demands don't include any queueing time and it is just service.
[c] Waiting times depend on the load (arrival rate of requests) and on the service demand.

Table 3. Extended notations for mathematical model (at resource $i = \{FE, BE, FE\_CPU_n, BE\_I/O\}$ where $n = \{1, 2, 3\}$) per client class k

Symbol	Notation
$(X_i)_k$	Throughput
$(S_i)_k$	Service time
$(U_i)_k$	Utilization
$(V_i)_k$	Number of client visits
TT_k	Think time
$(SD_i)_k$	Service demand
$(\lambda_i)_k$	Arrival rate
$(\mu_i)_k$	Processing rate
N_k	Number of clients
F_k	Function describing arrivals
$(C_j)_k$	Client j
$RO_k(i, i')$	Rounting function

collective time of service execution and intensity of clients arrival distribution. Therefore model of client class k can be defined as following (2).

$$(C_j)_k = \{CS_i, (\lambda_i, \mu_i)_k, RO_k(i)\} \tag{2}$$

Depending on client class k we can determine different kinds of servicing based on client service on resource i (3).

$$CS_i = \{DD_i, SS_i, NS_i\} \tag{3}$$

where:

- $DD_i = \{NORMAL, FIFO\}^{10}$,
- $SS_i = \{PS, FIFO, IS\}$,
- $NS_i = \{1, 2, ...\}$.

4.4   Mathematical System Model

Utilization Law (UL): Utilization (4) is the fraction of time that resource is busy, measured in Requests Per Second [RPS].

$$U_i = X_i \times S_i \tag{4}$$

Service demand parameter is required by most performance models. QPN requires service demands SD_i as an input, which can be derived using the mean utilization U_i and the throughput X_o (7).

Service Demand Law (SDL): Total average service time required by clients at resource i, denoted SD_i [ms] (5). We make use of the SDL to designate service demands required in simulation analysis.

$$SD_i = V_i \times S_i \tag{5}$$

Typically average system throughput X_o [RPS] defined as the number of requests that complete per unit of time and utilization U_i of resource i defined as the fraction of time that the resource is busy are easier to obtain than S_i and V_i.

Forced Flow Law (FFL): The average number of visits V_i, each completing request has to pass V_i times, on the average, by resource i. If X_o requests, complete per unit of time (6)

$$X_i = V_i X_o \tag{6}$$

will visit resource i and then SDL (5) is equal (7).

$$SD_i = V_i \times S_i = X_i/X_o \times U_i/X_i = U_i/X_o \tag{7}$$

---

[10] Departure disciplines are an extension to the QPN modeling formalism: $NORMAL$ implies that tokens become available for output transitions immediately upon arrival, $FIFO$ implies that tokens become available for output transitions in the order of their arrival, i.e., a token can leave the place/depository only after all tokens that have arrived before it have left. The departure discipline of an ordinary place or depository determines the order in which arriving tokens become available for output transitions.

We assume that utilization is going to one (100% utilization) (8).

$$(S_i)_k = (U_i)_k/(X_i)_k = 1/(X_i)_k \tag{8}$$

Little's Law (LL): This law relates the number of clients in a system with throughput and response times. If there are simultaneously present N clients in the system (9), each with think time TT and the software application processes at the throughput rate X_o producing a response time R. Think time TT_k is the same for two classes.

$$N = X_o(R + TT) \tag{9}$$

Simplification for LL $N = X_o R = \lambda R$.

Interactive Response Time Law (IRTL): System throughput is equal total number of requests X_o divide into total time (10).

$$\begin{aligned}
N &= NC + NS \\
NC &= X_o \cdot TT \\
NS &= X_o \cdot R, \\
\Rightarrow N &= X_o(R + TT) \\
\Rightarrow R &= (N/X_o) - TT
\end{aligned} \tag{10}$$

The response time was chosen to analyze from many PE parameters. The response time (11) is equal to a sum of service times in individual resources (residence time (12)). The service time in the resource (residence time) is the sum of time spent in the queue (queueing time (13)) and average service time (service demand (14)). Total response time is a sum of all individual response times of queues and depositories in the simulation model without the client queue response time (client think time). Service times in all queueing places are modeled by service demand SD_i.

Response time:

$$R = \sum_{i=1}^{m} S_i \tag{11}$$

Service time (Residence time):

$$S_i = WT_i + SD_i \tag{12}$$

Queueing time:

$$WT_i = \sum_{a=1}^{b} WT_a \tag{13}$$

Service demand:

$$SD_i = \sum_{a=1}^{b} SD_a \tag{14}$$

QPN model fo two client-classes (17) in [Appendix A].

4.5   Mathematical Requests Model

For every queueing system we have to specify both the arrival and the service processes. In many uses it is necessary for clients to be served in allotted period of time. Checking if the time limits are fulfilled by Web clients can be carried out in certain cases with the application of performance analysis.

Inter-arrival times have a Poisson distribution[11] and service times have a exponential distribution[12]. Poisson processes assume the occurrence of random, discrete and independent events within a time frame where the average amount of occurrences is known. This implies a memoryless process. Thus, the Poisson distribution describes the probability of a certain number of events for a defined interval length. The exponential distribution assumes a Poisson process and describes the probability density of the time between the occurrence of two events.

Average service intensity $\mu_i = \mu_{Clients} = 1/TT$ for clients and $\mu_i = 1/SD_i$ for system resources ($\mu_{FE\_CPU} = 1/SD_{FE\_CPU}$ and $\mu_{BE\_I/O} = 1/SD_{BE\_I/O}$).

5   Performance Enginnering with QPME Tool

In this section, we will propose a performance analysis approach from PE based on simulations. Simulation was the main mechanism used to do analysis of the constructed models.

QPN net (Fig. 1b) is used to predict the system response time.

5.1   Model Construction

As response time is very important in Web systems and many other distributed systems, therefor many works focus performance analysis on it. Here, QPN models are used to predict the response time for CWS. To explain approach to CWS modeling a typical structure will be modeled and simulated. QPN consists of incorporating queues and scheduling strategies into places forming queueing places on the main page (Fig. 1b). Each queueing place on subpage $Sub-FE$ and $Sub-BE$ is described by arrival process, waiting room, service process and additionally depository. We used several queueing systems most frequently useing to represent properties of Web system components. Resources such as queueing places in the system under simulation may be hardware elements (CPU and disk I/O) and resources such as places may be software elements (threads and connection pools).

Requests are sent to the system and then can be processed in both layers. The first layer (FE) is responsible for presentation and processing of client requests. Client requests C_j arrive to FE layer (Fig. 1b). FE place is used to stop incoming requests when they await application server threads tp. Application server

---

[11] Inter-arrival time (arrival process) $1/\lambda_i$ with mean arrival rate per unit time parameter λ_i.

[12] Service time $1/\mu_i$ with service rate per unit time parameter μ_i.

threads (application server) are modeled by *ThreadsPool* place. Requests are placed in FE queueing places to get service and after service they can be forwarded to next layer. FE servers are modeled by *FE_CPU* queueing places. The second layer (BE) is responsible for data layer jobs and implements system data handling. Next client requests arrive to BE layer (Fig. 1b). *BE* place is used to stop incoming requests when they await database server connections *cp*. Database server processes (database server) are modeled by *ConnectionsPool* place. Requests are placed in BE queues to get service and after service they can be forwarded to the client. BE server is modeled by *BE_I/O* queueing place.

We have many types[13] of tokens:

- client-classes ($N_1 = (C_j)_1$, $N_2 = (C_{j'})_2$),
- application server threads *tp*,
- database server connections *cp*.

We use queueing systems with one server to model:

- Clients node is modeled by queueing place with Infinite Server scheduling strategy (-/M/1/PS/∞ queueing system).
- Nodes of FE layer are modeled by queueing places with Processor Sharing scheduling strategy (-/M/1/PS/∞ queueing systems).
- Node of BE layer is modeled by queueing place with First In First Out scheduling strategy (-/M/1/FIFO/∞ queueing system).

Consequently, an executable (in a simulation sense) QPN model is obtained. Tokens are transferred in sequence by models FE layer and by BE layer. To effectively model two client-classes, a separate token colors (types) have been used. Routing of clients class N_1 contains all system resources in both layers. Routing of clients class N_2 contains only resources in FE layer. Probability $RO_k(i, i')$ on routing path for one client class is always equal 1, only for resources in FE layer $RO_k(FE, FE\_CPU_n) = 1/3$ because there are three resources $n = 3$ in the FE cluster. The queue mean service time, the service time probability distribution function and the number of servicing units defined for each queueing system in the model are the main parameters of the modeled system. The amounts of tokens destroyed and created are specified by the arc weights. The weights in the model are set to 1.

Experimental results presented in [6], [20] have shown that X_i for FE layer $X_{FE} = 1400[RPS]$ and respectively, the mean measured X_i for BE layer $X_{BE} = 7500[RPS]$. We use experimental results in our simulations. Service demand was determined based on (7): $SD_{FE\_CPU} = 1/X_{FE} = 0.714[ms]$, $SD_{BE\_I/O} = 1/X_{BE} = 0.133[ms]$.

In the demonstrated model it has been assumed that queues belonging to a FE layer have identical parameters (homogeneous cluster).

5.2   Input and Workload Simulation Parameters

Explored branches conducted by author of this paper are different configuration parameters of client and system (Table 4).

---

[13] A color specifies a type of tokens that can be resided in the place.

Table 4. Input parameters of simulations (client and system)

Parameter	One class (N_1)	Two classes (N_1 and N_2)
Clients queueing place	500	$250 + 250^{(a)}$
$X_{Clients}$ [RPS]	15; 30;	7.5 and 7.5; 15 and 15;
	45; 60	22.5 and 22.5; 30 and 30
ThreadsPool place	30; 60; $90^{(b)}$	30; 60; 90
ConnectionsPool place	$40^{(c)}$	40
Simulation time [s]	300	300

[a] Each request emulates a specific type of client session with multiple round-trips over the system. Transaction are selected by the driver based on the mix (*Browse* 50%, *Purchase* 25% and *Manage* 25%). *Browse* requests are placed in FE layer and *Purchase/Manage* requests are placed in FE and BE layer, that why we use 50%/50% division for client-classes (N_1 and N_2).
[b] Threads for FE nodes respectively – Initial marking per node (1, 2, 3).
[c] Connections for BE node – Initial marking per node (1).

Typically during the simulation of a stream of data (tokens) is produced in the part of the model called stream generator (*Clients* queueing place), as an input for the latter part of the model. We have several cluster nodes (one, two and three) in FE layer.

During the simulation, a number of requests is used, consisting of requests from individual classes, the number of clients in the class and the number of RPS (Tables 5 and 6).

Table 5. Think time for two client-classes (client)

$(X_{Clients})_k [RPS]$	$(X_{Clients})_k [RPMS]$ [b]	$TT_k [ms]$ [a]
7.5; 7.5	0.0075; 0.0075	133.33; 133.33
15; 15	0.015; 0.015	66.67; 66.67
22.5; 22.5	0.0225; 0.0225	44.44; 44.44
30; 30	0.03; 0.03	33.33; 33.33

[a] Formula (8).
[b] RPMS – Requests Per MilliSecond.

Client model is a source of requests (different types of requests arrival). Different client-classes are different clients behavior. Individual client class uses some hardware and software elements (network routing). Other important parameters are presented in QPN model [Appendix A].

Table 6. Service demand for two client-classes (system)

Resource i	$(X_i)_1[RPS]$	$(X_i)_2[RPS]$	$(X_i)_1[RPMS]$	$(X_i)_2[RPMS]$	$(SD_i)_k[ms]$
FE_CPU_n	1,400	1,400	1.4	1.4	0.714
BE_I/O	7,500	7,500	7.5	7.5	0.133

5.3 Number of Requests in Two Classes

We analyse how the number of requests influences the response time of a particular class. This part of simulation is a comparison (Fig. 2) one class (case0: $N_1 = 500$) and different number of requests for two classes (case1: $N_1 = 100, N_2 = 400$, case2: $N_1 = 250, N_2 = 250$, case3: $N_1 = 400, N_2 = 100$).

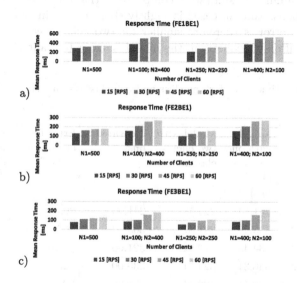

Fig. 2. Response time for one and two client-classes (500 clients, different workload [RPS]): (a) FE1, (b) FE2, (c) FE3

You can see that the results in case1 and case3 generates almost 50% error compared to case0. It doesn't mean that the results in case2 are ideal, therefore some correction will be doubtless needed, but we used this case in simulations.

6 Simulations

QPN model was used to predict the performance of the system for the scenarios mentioned above and it was developed using QPME 2.0.

The SimQPN simulator as a part of QPME tool can be used for calculating e.g. [6]: utilization, throughput, token residence time, token population.

The scenarios involve appointment of the response time of CWS. The initial parameters used in the model are: types of tokens, the way of releasing tokens, initial marking, scheduling strategies, number of servers and service times [7]. Previous works have been based on simulation models with: several nodes, one-way flow, feedback loop and one class of requests.

6.1 Validation of Simulation Results

Table 7 presents results of the analysis. We present here results for one such variation in which FE cluster consists of one node (first part), two nodes (second part) and three nodes (third part) (Table 7). It is possible to find the difference between the response time for case0 with one class of clients (second column), case2 with two client-classes (third column) and results of the real system experiment (fourth column). Results were compared in the last two columns. As you can see the response time in case2 is slightly shorter than in case0. The calculated simulation error is therefore smaller for two client-classes, which means that the distribution of requests in the model is closer to reality.

Table 7. Response time error for one class and two classes - one, two and three nodes in FE layer

Client think time [ms]	Model with one client class [ms]	Model with two client-classes [ms]	Measured [ms]	Error for one client class [%]	Error for two client-classes [%]
133.33 and 133.33	291.14	224.58	241.00	20.63	6.94
66.67 and 66.67	323.46	291.87	303.00	6.55	3.85
44.44 and 44.44	334.45	312.62	321.00	4.16	2.64
33.33 and 33.33	340.31	324.71	330.00	2.98	1.74
133.33 and 133.33	128.58	102.11	106.66	20.55	4.27
66.67 and 66.67	162.36	128.44	135.01	20.26	4.87
44.44 and 44.44	173.49	150.63	159.42	8.83	5.51
33.33 and 33.33	178.48	162.33	175.43	1.74	7.47
133.33 and 133.33	76.46	56.23	65.12	17.41	13.65
66.67 and 66.67	110.32	76.78	85.28	29.36	9.96
44.44 and 44.44	121.23	99.38	110.83	9.38	10.34
33.33 and 33.33	126.59	109.76	120.94	4.68	9.25

This is also visually confirmed by charts (Fig. 3) presenting the data from tables for four different system loads (15 [RPS], 30 [RPS], 45 [RPS], 60 [RPS]), results for one class of requests ($N_1 = 500$), two request classes ($N_1 = 250$ and

$N_2 = 250$) and experiments. The system with many client-classes was used for experiments, but as evidenced in [7], only two of them are significant, therefore, results of experiments were used for comparison.

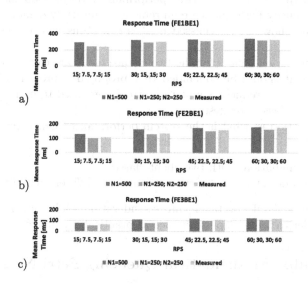

Fig. 3. Response time error (500 clients, different RPS workload) for: (a) one node in FE layer, (b), two nodes in FE layer, (c) three nodes in FE layer

The convergence of simulation results with the real systems results confirms a correctness of modeling methods. Validation results show that the model with two client-classes is better and will be able to predict the performance.

7 Summary and Future Work

It is still an open issue how to obtain an appropriate Web system. Our earlier works propose PE frameworks [7,8] to evaluate performance during different phases of their life cycle.

The demonstrated research results are an attempt to apply QPN formalism to the development of a software tool that can support CWS design. The main aim of my work was to develop models of CWS. In this paper, the analysis of the influence of different client-classes (two classes of requests) on the response time of CWS system is presented. The method presented in this article is intended to be used in the context of PE. We demonstrate the effective of the formal performance model. We choose the QPN variant and its associated environment (QPME tool) because it presents some advantages in performance analysis, mainly concerning flexibility in use and data generation and extraction capabilities. In the presented approach an modified implementation of QPN net has been proposed. Rules of modeling and analysis of CWS were introduced.

Comparison to experimental results [6] shows that our method is feasible and can be used to make prediction about key performance metric. At present, we have demonstrated potential in modeling CWS using QPN, understood as monolithic modeling technique used to assess the performance. We develop the framework that helps to identify performance requirements. The study demonstrates modeling power and shows how discussed models can be used to represent system behavior.

Because simulation results for the two client-classes are closer to real experimental results, it should be concluded that proposed approach is better. Use of request classes has reduced simulation errors compared to earlier simulation models with one class of requests.

Still the question is: what is the system performance? Our future works will concern:

- Large scale simulations with two client-classes.
- Experimental designation of service demand for different request classes.

Described results should be taken into consideration while modeling real Web systems in order to improve performance as practical use of the analysis.

A   Appendix: Mathematical Queueing Petri Nets Model

QPN is an tuple (16), where CPN is Colored Petri Net (15) [7].

$$CPN = (PL, TR, CO, IN, MA) \tag{15}$$

where:

- $PL = \{p_1, p_2, ..., p_i\}$ is a finite and non-empty set of places,
- $TR = \{t_1, t_2, ..., t_j\}$ is a finite and non-empty set of transitions,
- $PL \cap TR = \varnothing$,
- CO is a color function defined from $PL \cup TR$ into finite and non-empty sets (specify the types of tokens that can reside in the place and allow transitions to fire in different modes),
- $IN(p, t)$ are the backward and forward incidence functions defined on $PL \times TR$, such that $IN(p, t) \in [CO(t) \longrightarrow CO(p)_{MS}], \forall (p, t) \in PL \times TR^{14}$ (specify the interconnections between places and transitions),
- $MA(p)$ is an initial marking defined on PL such that $MA(p) \in CO(p), \forall p \in PL$ (specify how many tokens are contained in each place).

$$QPN = (CPN, QU, WE) \tag{16}$$

where:

- $QU = (QU_1, QU_2, (q_1, ..., q_{|PL|}))$, where:

---

[14] The subscript MS denotes multisets. $CO(p)_{MS}$ denotes the set of all finite multisets of $CO(p)$.

- $QU_1 \subseteq PL$ is a set of timed queueing places,
- $QU_2 \subseteq PL$ is a set of immediate queueing places,
- $QU_1 \cap QU_2 = \varnothing$,
- $(q_1, ..., q_{|PL|})$ is an array with description of places (if p_i is a queueing place q_i denotes the description of a queue with all colors of $CO(p_i)$ into consideration or if p_i is the ordinary place (p_i) equals $null$).

- $WE = (WE_1, WE_2, (w_1, ..., w_{|TR|}))$, where:
  - $WE_1 \subseteq TR$ is a set of timed transitions,
  - $WE_2 \subseteq TR$ is a set of immediate transitions,
  - $WE_1 \cap WE_2 = \varnothing$, $WE_1 \cup WE_2 = TR$,
  - $(w_1, ..., w_{|TR|})$ is an array (entry $w_j \in [CO(t_j) \longmapsto \mathbb{R}^+]$ such that $\forall c \in CO(t_j) : w_j(c) \in \mathbb{R}^+$) of:
    * rate of a negative exponential distribution specifying the firing delay due to color, if $t_j \in WE_1$,
    * firing weight specifying the relative firing frequency due to color, if $t_j \in WE_2$.

Based on definition (16) we define following QPN model (17) of CWS.

$$QPN = (PL, TR, CO, IN, MA, QU, WE) \tag{17}$$

where:

- $PL = \{FE, BE, ThreadsPool, ConnectionsPool\}$,
- $TR = \{t_1, t_2, t_3, t_4, t_5\}$,
- $CO(p_i)$ for $c = \{K_1, K_2, tp, cp\}$, where:
  - $K_1 = 250$ and $K_2 = 250$ - client-classes,
  - $th = \{30, 60, 90\}$ - threads,
  - $cp = 40$ - connections,
- $IN(p, t)$,
- $MA(p) = \{Clients(250, 250), ThreadsPool(30, 60, 90), ConnectionsPool (40)\}$,
- $QU = (QU_1, QU_2, (-/M/\infty/IS_{Clients}, null,$
  $-/M/1/PS_{Sub-FE}, null,$
  $-/M/1/FIFO_{Sub-BE}, null, null))$, where:
  - $QU_1 = \{Clients, FE\_CPU_n, BE\_I/O\}$, where $n = \{1, 2, 3\}$,
  - $QU_2 = \varnothing$,
- $WE = (WE_1, WE_2)$, where:
  - $WE_1 = \varnothing$,
  - $WE_2 = TR$,
  - $\forall c \in CO(t_j) : w_j(c) := 1$ (all transition firings are equally likely).

References

1. Zatwarnicki, K., Płatek, M., Zatwarnicka, A.: A cluster-based quality aware web system. In: Grzech, A., Borzemski, L., Świątek, J., Wilimowska, Z. (eds.) Part II. AISC, vol. 430, pp. 15–24. Springer, Cham (2016). https://doi.org/10.1007/978-3-319-28561-0_2
2. Kounev, S., Buchmann, A.: Performance modelling of distributed e-business applications using queuing Petri nets. In: 2003 IEEE International Symposium on Performance Analysis of Systems and Software, ISPASS 2003, pp. 143–155, March 2003
3. Coulden, D., Osman, R., Knottenbelt, W.J.: Performance modelling of database contention using queueing Petri nets. In: Proceedings of the 4th ACM/SPEC International Conference on Performance Engineering, ICPE 2013, pp. 331–334. ACM, New York (2013)
4. Kounev, S., Buchmann, A.: On the use of queueing petri nets for modeling and performance analysis of distributed systems. In: Kordic, V. (ed.) Petri Net. IntechOpen, Rijeka (2008)
5. Muller, C., Rygielski, P., Spinner, S., Kounev, S.: Enabling fluid analysis for queueing Petri nets via model transformation. Electron. Notes Theor. Comput. Sci. **327**, 71–91 (2016). The 8th International Workshop on Practical Application of Stochastic Modeling, PASM 2016
6. Rak, T.: Performance modeling using queueing Petri nets. In: Gaj, P., Kwiecień, A., Sawicki, M. (eds.) CN 2017. CCIS, vol. 718, pp. 321–335. Springer, Cham (2017). https://doi.org/10.1007/978-3-319-59767-6_26
7. Rak, T.: Response time analysis of distributed web systems using QPNs. Math. Probl. Eng. **2015**, 10 (2015)
8. Rak, T., Werewka, J.: Performance analysis of interactive internet systems for a class of systems with dynamically changing offers. In: Szmuc, T., Szpyrka, M., Zendulka, J. (eds.) CEE-SET 2009. LNCS, vol. 7054, pp. 109–123. Springer, Heidelberg (2012). https://doi.org/10.1007/978-3-642-28038-2_9
9. Rak, T., Samolej, S.: Distributed internet systems modeling using TCPNs, vol. 3, pp. 559–566 (2008). cited By 7
10. Buchmann, A., Dutz, C., Kounev, S., Buchmann, A., Dutz, C., Kounev, S.: QPME - queueing Petri net modeling environment. In: Third International Conference on the Quantitative Evaluation of Systems - (QEST 2006), pp. 115–116, September 2006
11. Szpyrka, M., Podolski, Ł., Wypych, M.: Modelling and verification of real-time systems with Alvis. In: Kosiuczenko, P., Madeyski, L. (eds.) Towards a Synergistic Combination of Research and Practice in Software Engineering. SCI, vol. 733, pp. 165–178. Springer, Cham (2018). https://doi.org/10.1007/978-3-319-65208-5_12
12. Nalepa, F., Batko, M., Zezula, P.: Performance analysis of distributed stream processing applications through colored Petri nets. In: Kofroň, J., Vojnar, T. (eds.) MEMICS 2015. LNCS, vol. 9548, pp. 93–106. Springer, Cham (2016). https://doi.org/10.1007/978-3-319-29817-7_9
13. Zhou, J., Reniers, G.: Petri-net based modeling and queuing analysis for resource-oriented cooperation of emergency response actions. Process Saf. Environ. Prot. **102**, 567–576 (2016)
14. Requeno, J., Merseguer, J., Bernardi, S.: Performance analysis of apache storm applications using stochastic Petri nets. In: 2017 IEEE International Conference on Information Reuse and Integration (IRI), pp. 411–418, August 2017

15. Requeno, J.I., Merseguer, J., Bernardi, S., Perez-Palacin, D., Giotis, G., Papaniko-laou, V.: Quantitative analysis of apache storm applications: the newsasset case study. Inf. Syst. Front. **21**, 67–85 (2019)
16. Rzonca, D., Rząsa, W., Samolej, S.: Consequences of the form of restrictions in coloured Petri net models for behaviour of arrival stream generator used in per-formance evaluation. In: Gaj, P., Sawicki, M., Suchacka, G., Kwiecień, A. (eds.) CN 2018. CCIS, vol. 860, pp. 300–310. Springer, Cham (2018). https://doi.org/10.1007/978-3-319-92459-5_24
17. Li, Z., Jiao, L., Hu, X.: Performance analysis for job scheduling in hierarchical HPC systems: a coloured Petri nets method. In: Wang, G., Zomaya, A., Perez, G.M., Li, K. (eds.) ICA3PP 2015. LNCS, vol. 9531, pp. 259–280. Springer, Cham (2015). https://doi.org/10.1007/978-3-319-27140-8_19
18. Mironescu, I.D., Vintan, L.: Performance prediction for parallel applications run-ning on HPC architectures through Petri net modelling and simulation. In: 2013 IEEE 9th International Conference on Intelligent Computer Communication and Processing (ICCP), pp. 267–270 (2013)
19. Kattepur, A., Nambiar, M.: Service demand modeling and performance prediction with single-user tests. Perform. Eval. **110**, 1–21 (2017)
20. Rak, T.: Performance analysis of distributed internet system models using QPN simulation. Institute of Electrical and Electronics Engineers Inc., pp. 769–774 (2014)
21. Menascé, D.: Load testing of web sites. IEEE Internet Comput. **6**(4), 70–74 (2002)
22. Rak, T.: Performance analysis of cluster-based web system using the QPN models. In: Czachórski, T., Gelenbe, E., Lent, R. (eds.) Information Sciences and Sys-tems 2014, pp. 239–247. Springer, Cham (2014). https://doi.org/10.1007/978-3-319-09465-6_25

A Queueing Model and Performance Analysis of UPnP/HTTP Client Server Interactions in Networked Control Systems

Marek Fiuk[1] and Tadeusz Czachórski[2]($\boxtimes$)

[1] Gliwice, Poland
[2] Institute of Theoretical and Applied Informatics, Polish Academy of Sciences, Baltycka 5, 44–100 Gliwice, Poland
marekjfiuk@gmail.com

Abstract. HTTP based communication protocols are ubiquitous in modern distributed control/supervision systems, therefore it becomes important to understand the performance of the HTTP protocol working with this type of applications and to develop techniques to investigate the impact of the protocol on the overall performance of these systems. A distributed control/supervision system based on HTTP usually consists of multiple nodes sending control requests or measurements to one or more recipient nodes; these data are used to perform control functions, and are stored or presented to supervising personnel. Typically multiple, concurrent flows of data are carried over HTTP to a given recipient. The stochastic characteristics of these flows are often different from the bursty nature of the ordinary Web Browser – Web Server interactions. Furthermore, in order to efficiently carry the randomly arriving data, the number of overlapping (concurrent) HTTP request/response sessions utilized by a sending node is higher than in typical Web Browser – Web Server communications.

We investigate how the characteristics of nodes participating in the distributed control system communications (number of nodes, service times, number of service stations, queue lengths) influence performance of the employed HTTP protocol. The main focus is on end-to-end delays in transmissions, the most important factor in control. The presented approach can be used to select configuration parameters for the nodes of a realistic control/supervision system having in mind the best communication related performance of such system.

In this work we propose a queuing model of HTTP client/server interactions which we subsequently simplify to be represented by a numerically solvable Markov chain. We compare the results obtained with the simplified model with the output of a discrete event simulation based on the original full model.

The initial motivation for this research came from a study of UPnP based remote PTZ camera control mechanism used by a leading manufacturer of large scale video monitoring systems.

M. Fiuk—Private.

P. Gaj et al. (Eds.): CN 2019, CCIS 1039, pp. 366–386, 2019.
https://doi.org/10.1007/978-3-030-21952-9_27

Keywords: HTTP performance · UPnP performance ·
Queueing models

1 Introduction

The rapid growth of the Internet created new opportunities for network-based
Automation, Control and Monitoring systems. Attracted by the broad range
of open standards and protocols (TCP/IP, HTTP, XML. SOAP), numerous
teams from both research and industrial communities have made attempts to
use Internet technologies in Networked Control Systems.

HTTP is clearly the dominant application protocol of today's Internet. When
used in a distributed control systems, the HTTP operates in an environment
that differs in several aspects from that of a typical Web browser – Web server
scenario. Firstly, the characteristics of the load are different. Control commands,
sensor measurements or statuses sent over the HTTP from a given node will
likely be more evenly distributed in time – as opposed to heavily bunched (bulk)
traffic characteristic for the typical Web page interactions. Additionally, both
the HTTP requests and responses will typically be very short, whereas the Web
page fetch responses tend to be significant in size, often carrying images or other
types of media. It can also be assumed that stream of HTTP request arriving
at the HTTP server in a given time interval will consist of requests that not
only are distributed in time but also are distributed in space, e.g. will consist of
(interleaved) requests originating from different nodes.

Implementations of higher level communication protocols used in distributed
control (UPnP being a good example) typically also accommodate the situation
when a node needs to send multiple requests in a rapid succession by issuing
multiple HTTP requests.

A system of nodes communicating using the HTTP protocol is often modeled
as a network of queues with multiple, parallel service stations that process the
queued requests (service stations represent worker threads). Due to the limited
communication resources in the system (limited number of worker threads, finite
queues), a blocking mechanism is often employed in the queuing model, which
technically corresponds to the HTTP client's thread waiting to establish a TCP
connection to a server that has temporarily suspended accepting incoming connections due to the lack of space in the incoming request queue. In addition to
representing (multiple instances of) the HTTP protocol stack, the model typically also includes representation of communication channels between servers
and clients. Such model (queueing network) can predict the latency, throughput
and utilization of the system and its components.

A number of queueing models of Web server performance exists in the literature. Slothouber [16] models a Web server as an open queueing network, ignoring
lower level details of HTTP and TCP/IP protocols. Dilley [4] presents layered
queueing model of an HTTP server (dealing with the HTTP communications
at the request/response level) together with a framework to collect and analyze empirical data. Van der Mei and Reeser [17] propose a detailed queueing

model of multiple protocol stack and software levels of a Web server. The client is represented only as a simple source of request, without any mechanism to queue-up blocked ones. The model reflects some of the relevant communication mechanisms at both the TCP and the HTTP levels. Authors successfully used the model (which is a multi-stage tandem queueing network) to predict Web server performance metrics – this was validated through measurements and simulations. Reeser and Hariharan [13] propose a similar model, which however more faithfully represents the blocking and queueing at various stages of HTTP request processing.

Heidemann [5] presents analytic models for the interaction of the HTTP protocol with several different transport layers such as TCP and UDP, including the impact of slow start algorithms. Cao and Nyberg [3] investigate overload control for Web servers, they show that performance models predict improvements when using a certain overload control strategy. Liu [8] uses a multi-station queuing center to model each of the servers in the 3-tiered Web services architecture, e.g. the Web, application and database servers. Menascé and Almeida [9] provide an analysis of the Web application and a characterization of workload as both impact the overall performance. They also propose queueing models of various Web server hardware components (CPU, hard disk etc.). Tien Van Do et al. present a mathematically tractable analytic queueing model based on Markovian assumptions and decomposition approach to the client population, the Web server and the TCP transport system, predicting the performance of an Apache server with its load-dependent dynamic pool, [15]. The arrival process of TCP connection requests demanding the support of a HTTP service process follows there a Markov-Modulated Compound Poisson Process (MMCPP).

Although the papers mentioned here give significant insights into the performance of Web servers, none of them provides a model that captures all the relevant aspects of the end-to-end HTTP communications between Web clients and Web servers. That is, while the above discussed models represent processing of HTTP requests at the server reasonably well, they largely fail to model detail operations of the HTTP stack at the client. In particular, they do not take into account that sending a request to the server may get blocked until a response is received for the previously sent request.

2   The UPnP Performance

Universal Plug and Play (UPnP) is an architecture for peer-to-peer network connectivity of intelligent appliances, wireless devices, and PCs. It leverages TCP/IP and the Web technologies (IP, TCP, UDP, HTTP and XML) to enable seamless proximity networking in addition to control and data transfer among networked devices in the home, office, and public spaces. UPnP is formally specified in the document Universal Plug and Play Device Architecture, available from the UPnP Forum at http://www.upnp.org/resources/documents.asp. The model presented in this work is based on the analysis of a concrete implementation of the UPnP protocol - the "Intel Software Development KIT (SDK) DK for UPnP Devices Version 1.2.1", which is a C language implementation of the UPnP stack. Since it is available as a source code, it allows for detailed analysis of its functioning and consequently for creation of a model of its operations. It comes bundled with a sample application which implements UPnP based remote TV control. This application and the SDK together represent a real instance of an HTTP based control system.

The UPnP document "Universal Plug and Play Device Architecture" was used as a guide to the source code provided in the SDK. The detailed analysis of the source code guided by the full description of the protocol stack provided in "Universal Plug and Play Device Architecture" resulted in the following observations:

- as the request flows through the stages of the protocol stack, it is only at a very few of them that the processing of the request takes any significant time (under the typical conditions). Thus for the analysis of the time the request spends in the system, only these stages need to be taken into account, the other can be ignored,
- at several stages new requests or responses are generated when processing the control command request,
- some processing stages are blocked (after finishing processing the request) until that request reaches some further stage in the flow.

These observations were used to generalize to operations of the UPnP/HTTP protocol stack and to express them in terms of typical queuing model constructs – queues, service stations, requests and jobs.

2.1   The Control Point – Control Requests Generation

An application implemented at the UPnP Control Point is assumed to generate Control Requests (UPnP actions) at the rate λ_r (requests per second), each destined to a single UPnP Device. This is modeled with the source G_r that generates requests tagged for a specific Device. These requests may all be destined for the same Device or may be distributed among several different ones. Control Requests generated by the source G_r are queued before being handled by the concurrent worker threads (transmission tasks) $T_{cr1} - T_{crK}$. Each of the tasks fetches the next request from the queue, sends it to the appropriate Device and then waits for the corresponding Control Response to arrive from that Device – which indicates that the request was received and fully processed there. Only after that it is ready to fetch the next request from the queue.

The Control Point is capable of concurrently handling (or keeping track of) K Control Requests. The transmission tasks are modeled with K parallel service stations, each requiring on average time t_{cr} to service (process) a job. Additionally, the service stations implement a unique blocking mechanism, where the service station, after it finishes servicing the job, is blocked until the corresponding Control Response is received.

2.2   The Device – Control Requests Processing

Control Requests received from Control Points are queued before being processed by the concurrent working threads (service tasks) $T_{sr1} - T_{srM}$. Each of the tasks fetches the next request from the queue and then processes it accordingly to the command carried in the request. It is assumed that this processing requires time t_{sr} to complete. The Device is capable of processing M request simultaneously. Once finished with processing the request, the service task sends the Control Response to the Control Point that generated that Control Request. The service tasks are modeled with M parallel service stations, each requiring on average time t_{sr} to service (process) a job. Additionally, after job servicing is finished, a Control Response is sent to the corresponding Control Point.

Some of the requests may be used to change value of UPnP State Variables at the Device, which in turn may result in sending the State Variable Change Event Notification Request to all Control Points that registered for that particular event (e.g. changing the value of a particular State Variable). It is assumed that there exists a correlation factor (coefficient) Ms between the number of received Control Requests and the number of resulting State Variable Change Event Notification Requests. The Notification Request generation is modeled by generating (with probability Ms) the request for each job processed by the service stations.

2.3   The Device – Event Notification Requests Generation

In addition to the State Variable Change Event Notification Requests, the Device can also generate (in response to the operations of the Plant) the Plant State Change Event Notification Requests. These are generated (at the assumed rate λ_e) as the Plant's state changes – either as a result of processing the Control Requests or as a result of the external environment affecting the Plant. The State Plant Change notifications are sent to the Control Points that registered for them. This is modeled with the source G_e that generates notifications requests, each tagged for a specific Control Point. These requests may all be destined for the same Control Point or may be distributed among several different ones.

All the Event Notification Requests generated at the Device are queued before being handled by the concurrent working threads (transmission tasks) $T_{se1} - T_{seN}$. Each of those tasks fetches the next request from the queue, sends it to the appropriate Control Point and then waits for the Event Notification Response to arrive from that Control Point – which indicates that the request was received and fully processed there. The transmission tasks are modeled with N parallel service stations, each requiring on average time t_{se} to service (process) a job. Additionally, the service stations implement a unique blocking mechanism, where the service station, after it finishes servicing the job, is blocked until the corresponding Event Notification Response is received.

2.4   The Control Point – Event Notification Requests Processing

Event Notification Requests received from Devices are queued before being processed by the concurrent working threads (service tasks) $T_{ce1} - T_{ceT}$. Each of those tasks fetches the next request from the queue and then processes it. It is assumed that this processing requires time t_{ce} to complete. The Control Point is capable of processing T request simultaneously. Once finished with processing the request, the service task sends the Event Notification Response to the Control Point that generated the Event Notification Request. The service tasks are modeled with T parallel service stations, each requiring on average time t_{ce} to service (process) a job. Additionally, after job servicing is finished, an Event Notification Response is sent to the corresponding Device.

A single UPnP request/response pair (sequence) results in a corresponding HTTP request/response pair. In the UPnP protocol stack, the Control Phase HTTP traffic is carried by the standard TCP protocol, Persistent Connection mode is not utilized. As it is well known, under such assumptions, an HTTP request/response sequence results in three network round trips, of which the first two contribute to the minimum time necessary to complete the operation at the HTTP level, Fig. 1. Each trip over the network could be modeled as a single service station (possibly with exponential service time distribution) but this would lead to a chain of service stations (or a single one with Erlang-n distribution) and therefore would significantly increase complexity of the model. In order to avoid that, the average network transmission times are simply added to service times of service stations present in the model. Specifically three one-way trip times are added to t_{cr} and t_{se} and one trip time is added to t_{sr} and t_{ce}.

Fig. 1. HTTP request/response pair – resulting network traffic

3   UPnP/HTTP Queueing Model

The above description directly defines the queuing model of UPnP/HTTP protocol stack operations, Fig. 2. It captures the generation, sending and processing of the UPnP requests/responses (messages) and thus the inner workings of the UPnP protocol during the Control Phase.

The model consists of Control Point elements (numbered $1 \div P$), Device elements (numbered $1 \div Q$) and a single Network element. Control Points and Devices are all connected to the Network with connections symbolizing either the UPnP Request flow (solid line) or the UPnP Response flow (dashed line). The element labeled "Network" represents all aspects of transmitting the UPnP messages over the network.

It is clear that the parameters of both the HTTP protocol stack model (characterized by the number of clients and servers, number of service stations, queue lengths) and the communication channel model (representing the connection speed and message delivery time) contribute to the overall latency. This work concentrates on the analysis of the former, assuming very simple treatment of the actual network delivery aspects. In other words, the attention of this effort is particularly focused on investigating how does the latency in an HTTP based control system depends on the architectural parameters of the HTTP stack implementation (number of client and server nodes, number of worker threads, sizes of queues) and the utilization.

The proposed queuing model of the UPnP network is characterized by the following parameters

- P – number of Control Points,
- Q – number of Devices,
- K – number of Control Point's service stations sending out the Control requests,
- T – number of Control Point's service stations processing incoming Event Notification requests,
- M – number of Device's service stations processing incoming Control requests,
- N – number of Device's service stations sending out the Event Notification requests,
- characteristics of sources G_r and G_e (given as distributions of inter-event times)
- distribution of the recipients of requests generated at a given node
- value of the coefficient Ms,
- processing times t_{cr}, t_{ce}, t_{sr} and t_{se}.

This model was programmed with the use of discrete event system OMNET++ [10] end used to play with parameters of the investigated system. Some other options were to use ns-3 or a system using Petri nets as an interface to define the simulated network, e.g. [12,14].

4   Simplifications of the Model

However, we are also interested in an analytical model based on Markov chain approach. The complexity of the model makes its formal analysis and consequently finding a numerical solution difficult. To address that and to provide a way to compute values of the essential performance metrics, an approach consisting of the following steps is proposed and outlined in the paragraphs below:

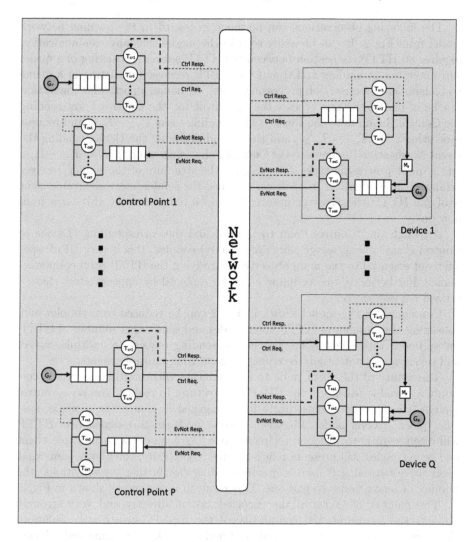

Fig. 2. Queuing model of the UPnP/HTTP network

- simplify the model by identifying and eliminating aspects non-essential for the analysis of the HTTP protocol performance
- further reduce the model of the network with multiple client nodes to a presumably near equivalent tandem queue model consisting of only one client node
- construct an algorithm to efficiently compute numerical solutions for this model
- find ways to compute (estimate) solutions for the model with multiple client nodes from the numerical results obtained for the tandem queue model

The following observations can be made in regard to the queuing network model from Fig. 2. It can be easily seen in the model that any communication requires an HTTP connection between the HTTP sender (consisting of a queue and a number of sending tasks) and the HTTP receiver (consisting of a queue and a number of request/job processing tasks). Although a particular connection can be of one of the two types - the first involving the Control Point sending the Control Requests to the Device (with sending tasks $T_{cr1} - T_{crK}$ and request processing tasks $T_{sr1} - T_{srM}$) and the second involving the Device sending the Event Notification Requests to the Control Point (with sending tasks $T_{se1} - T_{seN}$ and request processing tasks $T_{ce1} - T_{ceT}$), the structure of the sender/receiver arrangement is identical in both cases. Since the sender/receiver arrangement is of the HTTP client/server nature, it will be referred to by this term from now on.

Although the "Control Point to Device" and the corresponding "Device to Control Point" client/server pairs are (weakly) coupled, this is very UPnP specific, not essential to the main objective of studying the HTTP protocol performance. Furthermore, this coupling could be replaced by appropriately characterized request source.

Consequently, the model shown in Fig. 2 can be reduced to a simpler one, consisting of a number of HTTP client nodes and some other number of HTTP server nodes with client nodes potentially sending request to multiple servers and server nodes potentially receiving requests from multiple clients.

The ability of the client node to communicate with multiple servers is certainly a valuable feature of the UPnP architecture. In reality however a control system (client) node is most likely to be engaged in controlling a single, specific "plant" (server node). Likewise, a sensor node sending signals over HTTP will often send it to just one recipient node. As a model with multiple client and server nodes still presents relatively high complexity, the above, somewhat speculative reasoning is used to justify limiting (for further investigations) the number of server nodes to just one. This result in a new model shown in Fig. 3.

The number of nodes in the modeled system directly and very strongly impacts the computational complexity of the model's steady state solution, therefore it severely limits the number of nodes for which to numerical solution can be obtained. In order to limit the number of client nodes (and consequently constrain the computational complexity growth) we propose the following two-step approach.

First, construct a system with one sending node that is near-equivalent (in terms of capacity and throughput) to the system with multiple sending nodes. To solve this system (which is a tandem queue with a unique blocking mechanism) we have developed an optimized, Gaussian elimination based algorithm.

Second, investigate a number of approaches that allow to obtain results for the system with multiple sending nodes utilizing the solution for the near-equivalent, one sending node system. The concept of anear-equivalency for a queue system with one server node and multiple client nodes assumes that replacing a number of client nodes (characterized by cumulative values of capacity, number of service stations and throughput) with different number of nodes (in particular – with one) that possesses the same cumulative capacity, number of service stations and throughput will result in a system that is not very dissimilar from the original one – at least in terms of essential performance metrics – number of jobs in the system, sojourn time, etc. For example, a system with 4 nodes each with $m_{c\_t}/4$ service stations, queue length of $L_{c\_t}/4$ and intensity of incoming requests $\lambda_t/4$ is near-equivalent to the system with 2 nodes each with $m_{c\_t}/2$ service stations, queue length of $L_{c\_t}/2$ and intensity of incoming requests $\lambda_t/2$, and is also near-equivalent to the system with 1 node with m_{c_t} service stations, queue length of L_{c_t} and intensity of incoming requests λ_t. This is illustrated in Figs. 3 and 4.

Furthermore, it is assumed that a method can be devised to obtain a reasonable estimation of the vital metrics for the system with multiple client nodes utilizing the results obtained for the near-equivalent system.

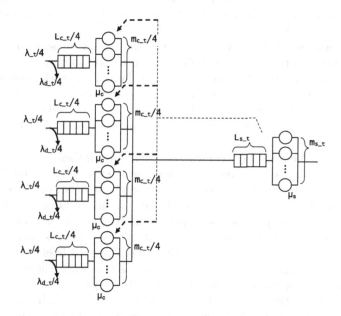

Fig. 3. One server node UPnP network

The resulting near-equivalent, one client node model in Fig. 5 is a tandem-queue and is simple enough to be solved numerically. An efficient method to obtain such numerical solution is presented in [6].

The model is characterized by following parameters:

- S_c – number of jobs currently serviced at the client, $0 \leq S_c + S_b \leq m_c$
- S_s – number of jobs currently serviced at the server

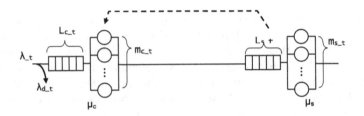

Fig. 4. Nearly equivalent system with one control node

- K_c – total number of jobs at the client, $K_c = S_c + Q_c$
- Ks – total number of jobs at the server, $K_s = S_s + Q_s$
- S_b – number of client service stations blocked by jobs, $S_b = K_s$ waiting/serviced at the server and jobs in the network

The blocking mechanism present in the system makes the ratio between the number of service stations at the client (mc) and the total capacity of the server ($L_s + m_s$) significant. We assume that the system is balanced in respect to that ratio, e.g. $m_c = L_s + m_s$.

The rational here is that in a system with $m_c < L_s + m_s$ some of the server resources (queue, service stations) will simply never get fully utilized, therefore such system will effectively get reduced to a balanced one with smaller Ls and (possibly) ms. It is important to emphasize the consequence of this condition – despite the queue on the server side being easily expandable to practically infinite size (as it is with the UPnP implementation) it is in effect truncated to the length equal $m_c - m_s$.

On the other hand, in a system with $m_c > L_s + m_s$, if there are $L_s + m_s$ client service stations "in use" - either processing the request – which mostly stands for sending it over the network (as it was already explained) or blocked due to the request waiting or being served on the server, any additional client service station attempting to process the request (e.g. send it over the network) will not find space on the server and therefore will not be able to establish the TCP session (the TCP/IP protocol stack on the server will most likely not accept the TCP CONNECT signal). The end effect of this is that while there are $L_s + m_s$ client service stations "in use", the additional service stations will never get used (or more precisely, their work will never produce any practically useful results – in this case will not transmit the request over the network), thus effectively reducing the number of service stations on the client side to $L_s + m_s$.

5 Solution to the Simplified Model, Transitions Graph and Balance Equations

The behavior of the above defined tandem queue system can be fully represented by a set of states and transitions between them. Although the state of such system can be defined in many different ways, probably the simplest and the most intuitive one is to express it as a combination of the number of jobs at the client K_c and the number of jobs at the server K_s.

The transitions between states are directly determined by the topology and the parameters of the tandem queue system (L_c, m_c, L_s, m_s, λ, μ_c and μ_s) as well as it rules of operation (FIFO queuing discipline, blocking mechanism, balanced system requirement).

The set of states and transitions representing a queuing system are commonly captured in the state transition diagram, where circles represent the individual states and arrows represent the transitions between them. The labels within the circles represent the state (as defined above), the labels associated with arrows represent the intensity of the transition. Figure 6 shows the state transition diagram for a system with parameters L_c, m_c, L_s, m_s.

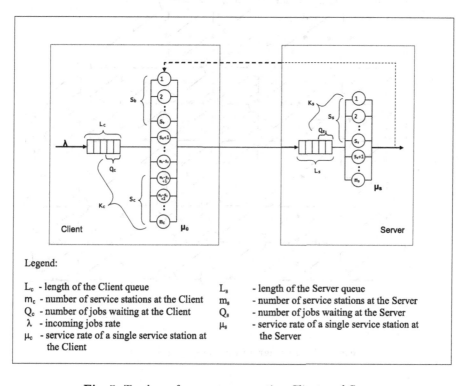

Legend:

L_c - length of the Client queue	L_s - length of the Server queue
m_c - number of service stations at the Client	m_s - number of service stations at the Server
Q_c - number of jobs waiting at the Client	Q_s - number of jobs waiting at the Server
λ - incoming jobs rate	μ_s - service rate of a single service station at
μ_c - service rate of a single service station at the Client	the Server

Fig. 5. Tandem of queues representing Client and Server

The number N of states is [6]: $N = (m_c + 1)(m_c/2 + L_c + 1)$. The set of global balance (Chapman-Kolmogorov) equations for a system represented by such state transition diagram can be written in a matrix form as:

$$\Pi Q = 0$$

where Π is a horizontal vector of steady state probabilities $\pi_{i,j}$ of system being in particular state (i.e. having a particular combination of $K_c = i$ and $K_s = j$)

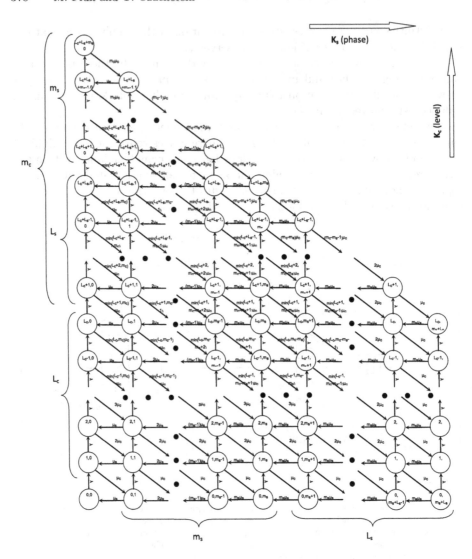

Fig. 6. State diagram for the system in Fig. 5

and $\mathbf{Q}$ is state transition rate matrix corresponding to the state flows in the system. Solving this set of global balance equations produces a set of (steady) state probabilities $\pi_{ij} = \mathrm{Prob}[K_c = i, K_s = j]$

The average system occupancy $E[K]$ given by

$$E[K] = \sum_i \sum_j (i + j)\pi_{ij}$$

and the average occupancies $E[K_c]$ and $E[K_s]$ are given by

$$E[K_c] = \sum_i \sum_j i\pi_{ij}, \quad E[K_s] = \sum_i \sum_j j\pi_{ij}$$

An exemplary 3D plot $[\text{Prob}(K_c, K_s), K_c, K_s]$ and 2D contour plot for $(L_c = 256, m_c = 32, m_s = 16, L_s = 16, uc\_tn = 1.55, us\_tn = 1.55)$ is presented in Fig. 7. We may compare it with simulation results for the same set of parameters of the more detailed model, Fig. 8.

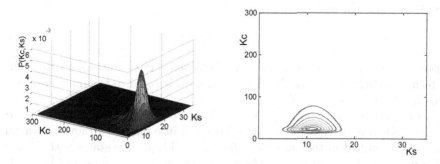

Fig. 7. State probability distribution for $n_c = 1, n_s = 1, L_c = 256, m_c = 32, m_s = 16, L_s = 16, uc\_tn = 1.55, us\_tn = 1.55$, numerical results

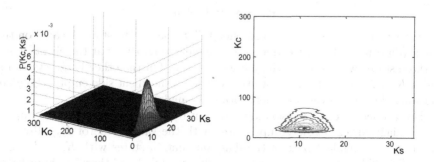

Fig. 8. State probability distribution for $n_c = 1, n_s = 1, L_c = 256, m_c = 32, m_s = 16, L_s = 16, uc\_tn = 1.55, us\_tn = 1.55$), simulation results

5.1   Comments on the Numerical Solution

The tandem queue model presented in Fig. 5 incorporates a unique blocking mechanism where a service station on the client side is blocked while the last job it has serviced is present on the server side (either being serviced or waiting in the queue). Thus, the duration of blocking of the service station on the clients side is a function of the sojourn time of the job on the server side. This makes this blocking mechanism different from the blocking mechanisms usually considered in the literature – e.g. Blocking Before Service (also called Type 1 Blocking) or Blocking After Service (also called Type 2 Blocking), where blocking is a result of unavailability of space on the destination node, [2,11]. Such blocking has an effect of modulating the total service rate at the client, it would be wrong, however to model this effect with client service stations that have server side occupancy dependent service rate $\mu_c(K_s)$, as a service station with service rate equal $n\mu_c$ is not equivalent to n service stations with service rate μ_c. Due to this blocking mechanism, the client and server side occupancies (K_c and K_s) must satisfy a simple constrain:

$$K_c + K_s \leq m_c + L_c, \tag{1}$$

where K_s stands here for the number of blocked service stations on the client (which by definition of this blocking mechanism is equal to the number of jobs present on the server, e.g. K_s).

In respect to the client and server side occupancy (K_c and K_s), the states of the tandem queue system can be subdivided into distinct groups, which can be easily identified in the state flow diagram in Fig. 6. The states with $K_s = 0$ have no associated intensity of jobs exiting the system (represented by a left-pointing arrow in the diagram), these are the states located on the left edge of the diagram. The states with $K_c = 0$ have no associated intensity of jobs moving from client to server (represented by a down-and-right-pointing arrow in the diagram), these are the states located on the bottom edge of the diagram. The states with

$$K_c + K_s = L_c + m_c \tag{2}$$

have the waiting queue on the client side full, these are the states located on the upper (diagonal) edge of the diagram. Any new job arriving at the client in any of those states will result in the job getting lost. On the other hand, states for which $K_c + K_s \leq m_c$ have the waiting queue on the client side empty, these are the states located in the lower-left corner (triangle) of the diagram.

Few properties can be observed for the entire "rows" of states (as they appear in the diagram, each row groups states with the same value of K_c). Rows in the upper (triangular) part of the diagram contain states with $K_c > L_c$ (e.g. some client side service stations are servicing the jobs) and therefore in order to satisfy Eq. 1, K_s may only assume values progressively (with each row) smaller than $L_s + m_s$. Additionally it can be observed that if $L_c > m_c$ then rows in the diagram satisfying the constrain

$$L_c > K_c \geq m_c \tag{3}$$

are identical to each other.

The two-node tandem queue systems are known to be described by Quasi-Birth-Death (QBD) process. The level partitioned matrix $\mathbf{Q}$ may be rewritten in the form

$$
\begin{bmatrix}
\mathbf{L}_0 & \mathbf{F}_0 \\
\mathbf{B}_1 & \mathbf{L}_1 & \mathbf{F}_0 \\
& & \vdots \\
\cdots & \mathbf{B}_{L_c-1} & \mathbf{L}_{L_c-1} & \mathbf{F}_{L_c-1} \\
& & \mathbf{B}_{L_c} & \mathbf{L}_{L_c} & \mathbf{F}_{L_c} \\
& & & & \vdots \\
& & & & \mathbf{B}_{L_c+m_c-1} & \mathbf{L}_{L_c+m_c-1} & \mathbf{F}_{L_c+m_c-1} \\
& & & & & \mathbf{L}_{L_c+m_c} & \mathbf{F}_{L_c+m_c}
\end{bmatrix}
$$

giving an evidence that the tandem queue system is described by the QBD process as well. Furthermore, since the tuples $\mathbf{F}_r, \mathbf{L}_r, \mathbf{B}_r$ generally change from level to level, the describing Markov process can be more precisely characterized as the truncated Level-Dependent-Quasi-Birth–Death (LDQBD) process. The relatively complex nature of the state transition rate matrix $\mathbf{Q}$ can be simplified by ordering the states in the matrix in a different way. Specifically, grouping states into differently defined levels – such that a level with index r contains all states satisfying the condition:

$$K_c + K_s = r = \text{total number of jobs in system}$$

produces simpler rules for constructing matrices $\mathbf{F}_r, \mathbf{L}_r, \mathbf{B}_r$. The states within a level are ordered by K_s.

Solving an equation set with the Gaussian elimination method usually consists of two steps – matrix triangularization with arithmetic complexity of $N^3/3$ and the backward substitution with arithmetic complexity of $N^2/2$, where N is the size of the set. Thus, it can be said that the direct Gaussian Elimination method of solving the global balance equations set for the two-node tandem queue system has computational complexity of $O(N^3)$. As the two-node tandem queue system is used to approximate the multi node UPnP system, the required values for L_c and m_c are rather high, say 256 and 32 for small to medium UPnP networks. This makes reducing the required computational complexity a very desirable goal. To that end, two computational complexity reduction techniques have been employed – one to reduce the number of elements that have to be eliminated in the matrix triangularization process, second to avoid fetching and processing zero valued elements of matrix $\mathbf{Q}$. Additionally, in order to conserve the required storage space, only a small part of the matrix $\mathbf{Q}$ is kept in memory at any time. Furthermore, the need to generate matrix $\mathbf{Q}$ on disk and then read it in is eliminated altogether – the required pieces of the matrix are generated on the fly. To simplify that process the Alternate State Ordering is used for the computations as it simplifies the structure of the matrix $\mathbf{Q}$.

Matrix $\mathbf{Q}$ is a form of a band-diagonal (or nearly block-tridiagonal) matrix. The actual amount of computations needed for the triangularization of $\mathbf{Q}^T$ with the Gauss algorithm is proportional to the width (measured along rows) of the non-zero elements band. This corresponds to the height (measured along columns) of the non-zero elements band in the matrix $\mathbf{Q}$. One way to reduce the width of the non-zero elements band in $\mathbf{Q}^T$ is to eliminate the leftmost non-zero element in each row of matrix $\mathbf{Q}^T$ thus producing new (transformed) matrix $\mathbf{Q}^{T*}$. This is accomplished by (conceptually) adding to each row the row that precedes it, starting with the second row and advancing to the last one, and then, for each $t_{phase\_r} \leq m_c$, restoring the last rows in that t_{phase} to its original form.

The second technique of reducing the matrix $\mathbf{Q}^T$ triangularization complexity is to avoid fetching and processing zero valued elements in the matrix $\mathbf{Q}^{T*}$. Owing to the structural regularity of the phase/level process' state flow diagram, the structure of the matrix $\mathbf{Q}^{T*}$ exhibits strong patterns. The pattern of interest here is the number of elements preceding and following the diagonal one in a particular row, expressed as a function of the t_{level} this row belongs to and of the position of the row within that t_{level}. Combining the generalization of matrix $\mathbf{Q}$ (ordered per ASO) with the transformation of matrix $\mathbf{Q}$ into matrix $\mathbf{Q}^T$, a generalized version of matrix $\mathbf{Q}^T$ was constructed and rules for the number of elements preceding and following the diagonal one were obtained.

6   System Occupancy as a Function of Essential System Parameters

Occupancy in queuing systems is commonly presented as a function of vital system parameters – for example utilization. Given the parameters of the generalized UPnP Queuing Network Model defined in 0, it seems to be useful to present the computed occupancy values $E[K] = E[K_c] + E[K_s]$ as a function of the following system parameters:

- number of service stations at the client node m_c,
- number of service stations at the server node: m_s
- combined (for all service stations) client "side" utilization $\varrho = \lambda/\mu_{c\_t}$ which can also be expressed using the normalized (for $\lambda = 1$) combined client side service rate $1/\mu_{c\_tn}$: $\varrho = 1/\mu\_c_{tn}$
- "server side" to "client side" utilization ratio: $(\lambda/\mu s_t)/(\lambda/\mu c_t) = \mu c_t n/\mu s_t n$

Using results from the numerical computations, multiple line chart figures were generated, each containing a family of plots $E[K] = f(\varrho) = f(1/\mu_{c\_tn})$ for several different values of $\mu c_{tn}/\mu s_{tn}$. A few results are presented in Figs. 9 and 10. They concern occupancy K but the transmission delays are immediately obtained with the use of Little's formula [7]: $E[W] = E[K]/\lambda$ where E[W] is the average waiting (sojourn) time, which in the context of this work represents the transmission delay.

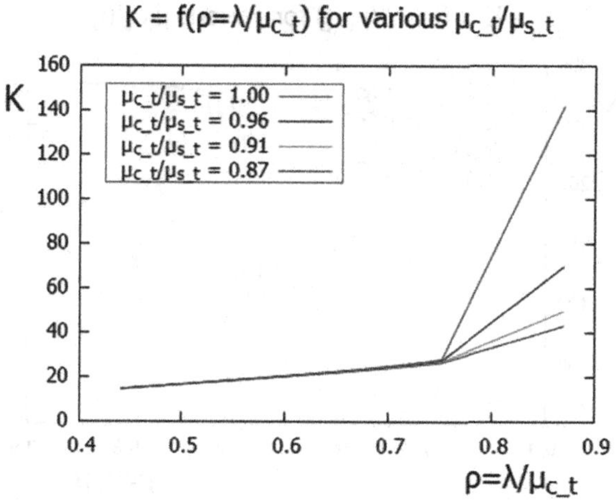

Fig. 9. System occupancy for $m_c = 32$ and $m_s = 1$

Occupancy can be viewed as a function of certain system parameters (like ϱ or the $\mu c_t/\mu s_t$ ratio), it exhibits some interesting properties:

- system occupancy – when plotted as function of ϱ, has a shape of a slope connecting two flat regions – Fig. 11. Thus it divides the plot into three regions (the two flat ones and the slope) which have natural interpretation related to the utilization factor ϱ:
  - saturation region, where occupancy is approaching the system capacity (e.g. the system is "full"). This region corresponds to the highest values of the utilization factor ϱ.
  - equilibrium region, where the occupancy is rapidly changing from high to low. This region corresponds to the moderate values of ϱ.
  - under saturation region, where occupancy is very low (e.g. the system is almost "empty"). This region corresponds to the lowest values of the utilization factor ϱ.

  Long term occupancy "oscillations" were also observed in some simulation runs.
- for certain combinations of system parameters, the saturation region and possibly a portion of the equilibrium region are not present (e.g. they lay outside of the chosen range of values of ϱ.
- occupancy plots appear to get horizontally "shifted" as the ratio $\mu c_t/\mu s_t$ changes (this can be particularly clearly observed in systems with higher number of server side service stations -m_s) - Fig. 12.

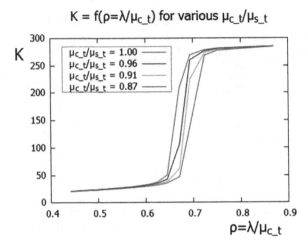

Fig. 10. System occupancy for $m_c = 32$ and $m_s = 16$

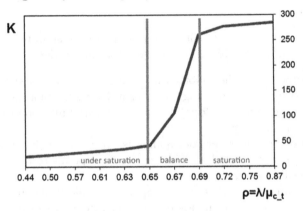

Fig. 11. System occupancy – the three distinct regions

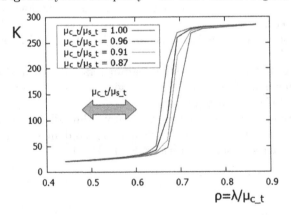

Fig. 12. Horizontal shifting of occupancy plots

7   Conclusions

This works makes several conceptual contributions, including:

- a unique model of end-to-end HTTP communications in the distributed control system, reflecting the actual implementation of the HTTP stack in systems based on the UPnP protocol. The model employs blocking mechanism of that was not considered in previous queueing models.
- methods of reducing (collapsing) a 'multi-client node, single server node' HTTP communications model into 'single client node, single server node' model.
- an efficient numerical algorithm for computing distribution of occupancy probabilities in the collapsed model.

Additionally, this work makes an empirical contribution by performing simulations of the full UPnP system model in the multidimensional model parameters space.

References

1. Andersson, M., Cao, J., Kihl, M., Nyberg, C.: Performance modeling of an apache web server with bursty arrival traffic. In: Proceedings of the International Conference on Internet Computing, Las Vegas, USA (2003)
2. Balsamo, S., de Nitto Person'e, V., Onvural, R.: Analysis of Queueing Network with Blocking. Springer, Boston (2001). distributed by Kluwer Academic Publishers
3. Cao, J., Andersson, M., Nyberg, C., Kihl, M.: Web server performance modeling using an M/G/1/K*PS Queue. In: 10th International Conference on Telecommunications, pp. 1501–1506 (2003)
4. Dilley, J., Friechich, R., Jin, T., Rolia, J.: Web server performance measurement and modeling techniques. Perform. Eval. **33**, 27–44 (1998)
5. Heidemann, J., Obraczka, K.: Modeling the performance of HTTP over several transport protocols. IEEE/ACM Trans. Networking **5**(5), 616–630 (1997)
6. Performance Analysis of the HTTP Protocol in Networked Control Systems, Ph.D. thesis, submitted at Silesian University of Technology, Gliwice (2019)
7. Kleinrock, L.: Queueing Systems. Volume I: Theory. Wiley Interscience, New York (1975)
8. Liu, X., Heo, J., Sha, L.: Modelling 3-tiered web applications. In: Proceedings of the 13th IEEE International Symposium on Modeling, Analysis, and Simulation of Computer and Telecommunication Systems, pp. 307–310 (2005)
9. Menascé, D., Almeida, V.: Capacity Planning for Web Services: Metrics, Models, and Methods. Prentice Hall, Upper Saddle River (2001). ISBN 10: 0130659037
10. OMNET++ Discrete Event Simulator. https://omnetpp.org/
11. Perros, H.: Queueing Networks with Blocking. Oxford University Press Inc., Oxford (1994)
12. Rak, T.: Performance modeling using queueing petri nets. In: Gaj, P., Kwiecień, A., Sawicki, M. (eds.) CN 2017. CCIS, vol. 718, pp. 321–335. Springer, Cham (2017). https://doi.org/10.1007/978-3-319-59767-6_26

13. Reeser, P., Hariharan, R.: Analytic model of web servers in distributed environments. In: Proceedings of the 2nd International Workshop on Software and Performance, pp. 158–167. ACM, New York (2000)

14. Rzonca, D., Rząsa, W., Samolej, S.: Consequences of the form of restrictions in coloured petri net models for behaviour of arrival stream generator used in performance evaluation. In: Gaj, P., Sawicki, M., Suchacka, G., Kwiecień, A. (eds.) CN 2018. CCIS, vol. 860, pp. 300–310. Springer, Cham (2018). https://doi.org/10.1007/978-3-319-92459-5_24

15. Van Do, T., et al.: Performance modeling of an apache web server with a dynamic pool of service processes. Telecommun. Syst. **39**(2), 117–129 (2008)

16. Slothouber, L.: A model of web server performance. In: 5th International World Wide Web Conference, Paris, France, (1996)

17. van der Mei, R., Hariharan, R., Reeser, P.: Web server performance modeling. Telecommun. Syst. **16**(3–4), 361–378 (2001)

18. Tari, Z., Phan, A.K.A., Jayasinghe, M., Abhaya, V.G.: On the Performance of Web Services. Springer, Boston (2011). 2011 edition

A System with Warm Standby

Galina Zverkina[1,2(✉)]

[1] Russian University of Transport (RUT - MIIT), 9b9 Obrazcova Street,
Moscow 127994, Russia
[2] V. A. Trapeznikov Institute of Control Sciences of Russian Academy of Sciences,
65 Profsoyuznaya Street, Moscow 117997, Russia
zverkina@gmail.com

Abstract. The mathematical model of the restorable system with warm reserve considered, in the case when all working and repair times are bounded by exponential random variable (upper and lower), and working and repair times can be dependent. The exponential upper bounds for the convergence rate of the distribution of this system. The bounds for the convergence rate of the availability factor are estimated.

Keywords: Mathematical reliability theory ·
Mathematical model of restorable element with a warm reserve ·
Convergence rate of distribution · Availability factor ·
Markov processes · Coupling method

1 Introduction

One of the most important problems of the reliability theory is to determine the availability factor, or to find the probability that the restorable element will be operable at a given time t. In particular, it is important to know the rate of convergence of a availability factor to its stationary value for the case when the exact solution is hard to calculate. The results of asymptotic analysis of queueing systems can be applied for reliability theory, because reliability and queueing theory use the same methods from probability theory – see [2,4]. It is well-known that the convergence rate of many characteristics of the reliability system with the light-tailed distributed working and repair times is exponential. We are interested in finding the parameters of this exponent.

The availability factor of the single element with an exponential distribution of the working and recovery times, with parameters λ and μ, accordingly, was studied (see, e.g., [3]) during the initial development of mathematical methods in the reliability theory.

Supported by RFBR (project No 17-01-00633 A).

P. Gaj et al. (Eds.): CN 2019, CCIS 1039, pp. 387–399, 2019.
https://doi.org/10.1007/978-3-030-21952-9_28

1.1   Classic Model of Restorable Element

For one restorable element with cumulative distribution function (c.d.f.) of working time $F(s) = 1 - e^{-\lambda s}$, and c.d.f. of recovery time $G(s) = 1 - e^{-\mu s}$, it is well known that in the case of $A_1(0) = 1$ (see, e.g., [3, §2.3]) $A_1(t) \overset{\text{def}}{=}$ $P\{$at the time t the element works$\} = \frac{\mu + \lambda e^{-(\lambda+\mu)t}}{\mu + \lambda}$.

This result is obtained from the solution of the differential equation

$$\dot{A}_1(t) = A_1(t)(1 - \lambda) + (1 - A_1(t))\mu \tag{1}$$

with an initial value $A_1(0) = 1$. In this case,xλ and μ are *constant intensities* of the failure and the repair correspondingly.

Accordingly, it is easy to find the convergence rate of availability factor $A_1(t)$ to its stationary value $A_1(\infty) \overset{\text{def}}{=} \frac{\mu}{\mu+\lambda}$, and this rate is exponential: $A_1(t) - A_1(\infty) = \frac{\lambda e^{-(\lambda+\mu)t}}{\mu+\lambda}$.

It is very important to study the failure of the element described above for technical applications. This element is reserved by additional one.

In this paper, the situation when main element has redundant element which is in "warm" mode will be studied.

1.2   Markov Mathematical Model of Restorable Element with One Warm Standby Element

This is a mathematical model of the restorable element with a warm reserve. Its behaviour is described by the *intensities* of the failures and the repairs. Firstly, let us describe the simplest case of the reliability system with "warm" reserve.

Denote the state of the reliability system as the pair (i, j) of the numbers $\{0, 1\}$:

$$i \overset{\text{def}}{=} \begin{cases} 0, \text{ if the main element is in working state;} \\ 1, \text{ if it is repaired;} \end{cases}$$

$$j \overset{\text{def}}{=} \begin{cases} 0, \text{ if the reserve element is in working state;} \\ 1, \text{ if it is repaired,} \end{cases}$$

Obviously, the work and repair times of the reserve element depends on the state of the main element.

Suppose that the *intensities* of the failure and the repair of both elements depend on the state (i, j) of the system. And in all (different) states of the reliability system, the intensities are constant. Thus, the probability of the change of the elements of the pair (i, j) in the little time period $\Delta > 0$ are follows:

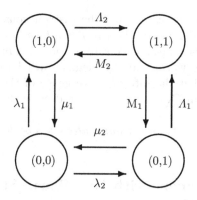

Fig. 1. If the main element is in working state, then the reserve element has the intensities of the failure and of the repair λ_2 and μ_2 accordingly. If the main element is in the repair state, then these intensities are Λ_2 and M_2 accordingly. Analogously, if the reserve element is working, then the intensities of the failure and repair of the main element are λ_1 and μ_1, and Λ_1 and M_1 otherwise.

$\mathbf{P}\{$at the time $t + \Delta, i = 1$ given $i = 0$ and $j = 0$ at the time $t\} = \lambda_1 + o(\Delta)$;
$\mathbf{P}\{$at the time $t + \Delta, i = 0$ given $i = 1$ and $j = 0$ at the time $t\} = \mu_1 + o(\Delta)$;
$\mathbf{P}\{$at the time $t + \Delta, j = 1$ given $i = 0$ and $j = 0$ at the time $t\} = \lambda_2 + o(\Delta)$;
$\mathbf{P}\{$at the time $t + \Delta, j = 0$ given $i = 0$ and $j = 1$ at the time $t\} = \mu_2 + o(\Delta)$;
$\mathbf{P}\{$at the time $t + \Delta, i = 0$ given $i = 1$ and $j = 1$ at the time $t\} = M_1 + o(\Delta)$;
$\mathbf{P}\{$at the time $t + \Delta, i = 1$ given $i = 0$ and $j = 1$ at the time $t\} = \Lambda_1 + o(\Delta)$;
$\mathbf{P}\{$at the time $t + \Delta, j = 0$ given $i = 1$ and $j = 1$ at the time$t\} = M_2 + o(\Delta)$;
$\mathbf{P}\{$at the time $t + \Delta, j = 1$ given $i = 1$ and $j = 0$ at the time$t\} = \Lambda_2 + o(\Delta)$;

In other words, in this case we have the continuous Markov chain with transition intensities λ_k, μ_k, Λ_k, $\mathbf{M}_k$, $k = 1, 2$ (see Fig. 1).

Denote $p_{i,j}(t) \stackrel{\text{def}}{=} \mathbf{P}\{$At the time t, the state of the reliability system is $(i, j)\}$ $(i, j \in \{1, 2\})$. Note, the availability factor of this system is

$$A_2(t) = p_{0,0} + p_{0,1} + p_{1,0} = 1 - p_{1,1}$$

. It is easy to write the system of Kolomogorov differential equation for $p_{i,j}$:

$$\begin{cases} \dot{p}_{0,0}(t) = (1 - \lambda_1 - \lambda_2)p_{0,0} + \mu_2 p_{0,1} + \mu_1 p_{1,0}; \\ \dot{p}_{0,1}(t) = (1 - \mu_2 - \Lambda_1)p_{0,1} + M_1 p_{1,1} + \lambda_2 p_{0,0}; \\ \dot{p}_{1,0}(t) = (1 - \Lambda_2 - \mu_1)p_{1,0} + M_2 p_{1,1} + \lambda_1 p_{0,0}; \\ \dot{p}_{1,1}(t) = (1 - M_1 - M_2)p_{1,1} + \Lambda_1 p_{0,1} + \Lambda_2 p_{1,0}. \end{cases}$$

One of eigenvalues of this system is $\mu_1 + \lambda_1$ as $p_{0,0} + p_{0,1} = A_1(t)$ is the solution of (1). The other eigenvalues can be found by solution of the cubic equation by Cardano's formulae.

However, this is a very rare case, when all distributions in the reliability system are exponential.

So, we will consider the case when the distributions have an exponential moments, and they are defined by the *variable intensities*.

And in this paper the convergence rate of the availability factor of the reliability system with one warm standby element will be estimated. Note that the calculations show that the rate of convergence of the availability factor in this case can be less than $C \exp\left(-t(\lambda_1 + \mu_1)\right)$, despite the presence of a reserve element.

In this paper, the case of non-constant intensities of the failure and the repair is considered.

2   Restorable Element with One Warm Standby Element: Semi-Markov Case

Obviously, the classic assumption about exponential type of the distributions of the failure and repair times is wrong in real practice. All elements of the complicated systems interact with each other in reality. Therefore, the characteristics of system parts change over time and depend on the state of other elements.

One such system is considered in this paper.

2.1   Generalised Mathematical Model of Restorable Element with One Warm Standby Element

This system consists of the main element and redundant element with reliability characteristics that depend on each other and on the time.

So, the intensities of failure and repair of both elements are the functions of their states, or the *full system state*. This *full state of the system* is described by pairs: $X_t \stackrel{\text{def}}{=} ((i_t, x_t), (j_t, y_t))$, where $i_t = 0$ or $i_t = 1$ if the first (main) element is working or not at the time t correspondingly. And the value x_t is equal to the time elapsed from the last change in the state of i_t of the first (main) element to the time t. The pair (j_t, y_t) describes the state of the reserve element at the time t in the same way.

Suppose that the intensities of failure and repair are the function $\lambda_k(X_t)$ and $\mu_k(X_t)$ (k is the number of element). I.e.

$$\begin{cases} \mathbf{P}\{i_{t+\Delta} = 1, x_{t+\Delta} \in [0, \Delta), j_{t+\Delta} = j_t, y_{t+\Delta} = y_t + \Delta | i_t = 0, X_t\} \\ \qquad\qquad\qquad\qquad = \lambda_1(X_t)\Delta + o(\Delta); \\ \mathbf{P}\{i_{t+\Delta} = 0, x_{t+\Delta} \in [0, \Delta), j_{t+\Delta} = j_t, y_{t+\Delta} = y_t + \Delta | i_t = 1, X_t\} \\ \qquad\qquad\qquad\qquad = \mu_1(X_t)\Delta + o(\Delta); \\ \mathbf{P}\{j_{t+\Delta} = 0, y_{t+\Delta} \in [0, \Delta), i_{t+\Delta} = i_t, x_{t+\Delta} = x_t + \Delta | j_t = 0, X_t\} \\ \qquad\qquad\qquad\qquad = \lambda_2(X_t)\Delta + o(\Delta); \\ \mathbf{P}\{j_{t+\Delta} = 1, y_{t+\Delta} \in [0, \Delta), i_{t+\Delta} = i_t, x_{t+\Delta} = x_t + \Delta | j_t = 1, X_t\} \\ \qquad\qquad\qquad\qquad = \mu_2(X_t)\Delta + o(\Delta). \end{cases} \qquad (2)$$

The availability factor $A_2(t) \stackrel{\text{def}}{=} \mathbf{P}\{i_t + j_t < 2\}$. In general case, it is impossible to calculate $A_2(t)$ and its convergence rate to the stationary value. So, as $A_2(t) =$

$\mathbf{P}\{i_t + j_t < 2\}$ is probability of some set, it is enough to estimate the convergence rate of the *distribution* of the process X_t to the stationary distribution.

Therefore, we can use the theory of stochastic processes, to find the bounds of the convergence rate of the distribution of X_t. Note that the process X_t is defined on the state space $\mathcal{X} \stackrel{\text{def}}{=} \{0,1\} \times R_+ \times \{0,1\} \times R_+$ with the standard Borel σ-algebra. Moreover, (2) implies that X_t is a Markov process on the state space $\mathcal{X}$. So, we can use the theory of Markov processes. But this process is not regenerative, and it is impossible to use regenerative process theory as for different queueing system (see, e.g., [1,8,9,14]).

The reliability system "close" to the system described on the Fig. 1 is considered here. Thus, that the distributions of the work and repair periods of both elements have an *light tails*, i.e. they have finite exponential moments with some parameter $\alpha > 0$. For this aim, suppose, that there exists some $\lambda_k^- > 0$, $\mu_k^- > 0$, λ_k^+, μ_k^+ such that for all t,

$$\lambda_k^- \le \lambda_k(X_t) \le \lambda_k^+ \text{ and } \mu_k^- \le \lambda_k(X_t) \le \mu_k^+. \tag{3}$$

Denote $\lambda_0 \stackrel{\text{def}}{=} \min\{\lambda_1^-, \lambda_2^-\}$, $\Lambda \stackrel{\text{def}}{=} \max\{\lambda_1^-, \lambda_2^-\}$, $\mu_0 \stackrel{\text{def}}{=} \min\{\mu_1^-, \mu_2^-\}$, $M \stackrel{\text{def}}{=} \max\{\mu_1^-, \mu_2^-\}$.

The distribution functions of the work and repair times of the studied system can be reconstructed by their intensities $\lambda_k(X_t)$ and $\mu_k(X_t)$. If i-th element is in work state at the time t_0, then residual period of its work is $\xi_k(X_{t_0})$. If i-th element is in repair state at the time t_0, then residual period of its repair is $\zeta_k(X_{t_0})$, and from (2),

$$F_{t_0}^k(s) = \mathbf{P}\{\xi_k(X_{t_0}) \le s\} = 1 - \exp\left(-\int_0^s \lambda_k(X_{t_0+u}) \, du\right);$$

$$(F_{t_0}^k)'(s) = f_{t_0}^k(s) = \lambda_k(X_{t_0+u})(s) \exp\left(-\int_0^s \lambda_k(X_{t_0+u})(s) \, du\right);$$

$$G_{t_0}^k(s) = \mathbf{P}\{\zeta_k(X_{t_0}) \le s\} = 1 - \exp\left(-\int_0^s \mu_k(X_{t_0+u}) \, du\right);$$

$$(G_{t_0}^k)'(s) = g_{t_0}^k(s) = \mu_k(X_{t_0+u})(s) \exp\left(-\int_0^s \mu_k(X_{t_0+u})(s) \, du\right).$$

Remark 1. From (3), for all $t_0 \ge 0$,

$$\lambda_k^- e^{-\lambda_k^+ s} \le f_{t_0}^k \le \lambda_k^+ e^{-\lambda_k^- s}; \mu_k^- e^{-\mu_k^+ s} \le g_{t_0}^k \le \mu_k^+ e^{-\mu_k^- s};$$

and $\mathbf{E} \, e^{\alpha \xi_k(t_0)} = \int_0^\infty e^{\alpha t} f_{t_0}(s) \, ds \le \int_0^\infty e^{\alpha t} \lambda_k^+ e^{-\lambda_k^- s} \, ds \le \dfrac{\lambda_k^+}{\lambda_k^- - \alpha} < \infty$, if $\alpha < \lambda_k^-$.

Analogously, $\mathbf{E} \, e^{\alpha \zeta_k(t_0)} \le \dfrac{\mu_k^+}{\mu_k^- - \alpha} < \infty$, if $\alpha < \mu_k^-$. Thus, the distributions of work and repair times, and its residual times have *light tales*.

Also, $\mathbf{E} \, \xi_k(t_0) = \int_0^\infty \exp\left(-\int_0^s \lambda_k(X_{t_0+u})(s) \, du\right) ds \le \int_0^\infty e^{-\lambda_k^- s} ds = \dfrac{1}{\lambda_k^-}$.

Analogously, $\mathbf{E} \, \zeta_k(t_0) \le \dfrac{1}{\mu_k^-}$.

Lemma 1. *In conditions (3), the process X_t is ergodic, i.e. there exists the invariant probabilistic distribution $\mathcal{P}$ such that for any initial states of the process X_t, the distribution $\mathcal{P}_t$ of the process X_t at the time t weak converges to the distribution $\mathcal{P}$. I.e. for all set $\mathcal{S} \in \sigma(\mathcal{X})$, $\mathcal{P}_t(\mathcal{S}) \to \mathcal{P}$ as $t \to \infty$.*
In particular, $A_2(t) = \mathbf{P}\{i_i + j_t < 2\}$ has a limit.

Here we skip the proof of the Lemma 1. It is based on the Doeblin-Doob condition ([7, Ch. 6, §2]) and Doob Theorem ([7, Theorem 2.1]).

Now, we will to find the bounds of convergence $\mathcal{P}_t \to \mathcal{P}$.

2.2 Coupling Method and Its Modification

The bounds of the convergence of the distributions needs the definition of the metric on the space of the distributions. We will use the total *variation distance*, i.e. for two measures $\mathcal{D}_1$ and $\mathcal{D}_2$ on the measurable space $(\Omega, \sigma(\Omega))$,

$$\|\mathcal{D}_1 - \mathcal{D}_2\|_{TV} \stackrel{\text{def}}{=} \sup_{\mathcal{A} \in \sigma(\Omega)} |\mathcal{D}_1(\mathcal{A}) - \mathcal{D}_2(\mathcal{A})|.$$

If $\|\mathcal{P}_t - \mathcal{P}\| \leq \phi(t)$ for studied model, then $|A_2(t) - A_2| \leq \phi(t)$, where $A_2(t) = \mathbf{P}\{i_t + j_t < 2\}$ – availability factor at the time t, and A_2 is its limit value.

For estimate the convergence rate of stochastic processes, we can use the *coupling method* invented in [6] – see, e.g. [6,11]. See also [13,14] for application of the coupling method in Queuing theory.

Idea of the coupling method. Consider two *independent Markov processes* with different initial states X_0 and $\widehat{X}_0$ and with the same transition function. Denote these processes by X_t and $\widehat{X}_t$ correspondingly. Let we can find the time τ where they are coincided. The time τ is called *coupling epoch* and it depends on X_0 and $\widehat{X}_0$. After the time $\tau(X_0, \widehat{X}_0)$, the distributions of the processes X_0 and $\widehat{X}_0$ are coincided – by Markov property. Thus, for all $t \geq \tau(X_0, \widehat{X}_0)$, and for all set $\mathcal{A} \in \sigma(\mathcal{X})$, $\mathbf{P}\{X_t \in \mathcal{A}\} = \mathbf{P}\{\widehat{X}_t \in \mathcal{A}\}$. It implies the basic coupling inequality:

$$|\mathbf{P}\{X_t \in \mathcal{A}\} - \mathbf{P}\{\widehat{X}_t \in \mathcal{A}\}| = |\mathbf{P}\{X_t \in \mathcal{A} \,\&\, \tau > t\} - \mathbf{P}\{\widehat{X}_t \in \mathcal{A} \,\&\, \tau > t\}$$
$$+ |\mathbf{P}\{X_t \in \mathcal{A} \,\&\, \tau \leq t\} - \mathbf{P}\{\widehat{X}_t \in \mathcal{A} \,\&\, \tau \leq t\}|$$
$$= |\mathbf{P}\{X_t \in \mathcal{A} \,\&\, \tau > t\} - \mathbf{P}\{\widehat{X}_t \in \mathcal{A} \,\&\, \tau > t\}| \leq \mathbf{P}\{\tau > t\}.$$

Then, if it possible to find the increasing positive function $\varphi(\tau)$ such that $\mathbf{E}\,\varphi(\tau(X_0, \widehat{X}_0)) < \infty$, then by Markov inequality, $\mathbf{P}\{\tau(X_0, \widehat{X}_0) \geq t\} = \mathbf{P}\{\varphi(\tau(X_0, \widehat{X}_0)) \geq \varphi(t)\} \leq \frac{\varphi(\tau(X_0, \widehat{X}_0))}{\varphi(t)}$. From the last inequality the bounds for convergence of the distribution $\mathcal{P}_t$ can be obtained.

This schema can be used for discrete Markov chain and for Markov chain in continuous time. But for studied process X_t (see Sect. 2.1) the "direct" coupling method is impossible, because for different values $X_0 \neq \widehat{X}_0$, $\mathbf{P}\{\tau(X_0, \widehat{X}_0) < \infty\} = 0$. Thus, the modification of coupling method, or *successful coupling* will be used.

Successful coupling (see [5]). Let X_t and $\widehat{X}_t$ be two independent Markov processes with the same transition function (2), but with different initial states at time $t = 0$.

Suppose that (*dependent*) processes $Y_t = ((i_t, x_t), (j_t, y_t))$ and $\widehat{Y}_t = ((\widehat{i}_t, \widehat{x}_t), (\widehat{j}_t, \widehat{y}_t))$ are constructed on some probability space, in such a way that:

1. $Y_t \overset{D}{=} X_t$ and $\widehat{Y}_t \overset{D}{=} \widehat{X}_t$ for all *non-random* t;
2. $\mathbf{P}\{\tau(X_0, \widehat{X}_0) < \infty\} = 1$, where $\tau(X_0, \widehat{X}_0) = \tau(Y_0, \widehat{Y}_0) = \inf\{t > 0 : Y_t = \widehat{Y}_t\}$.

This pair of processes $Y_t = ((i_t, x_t), (j_t, y_t))$ and $\widehat{Y}_t = ((\widehat{i}_t, \widehat{x}_t), (\widehat{j}_t, \widehat{y}_t))$ is called *successful coupling* for the processes X_t and $\widehat{Y}_t$, and $\tau(X_0, \widehat{X}_0)$ is called *coupling epoch*.

For successful coupling, the basic coupling inequality can be applied as:

$$
\begin{aligned}
|\mathbf{P}\{X_t \in \mathcal{A}\} - \mathbf{P}\{\widehat{X}_t \in \mathcal{A}\}| &= |\mathbf{P}\{Y_t \in \mathcal{A}\} - \mathbf{P}\{\widehat{Y}_t \in \mathcal{A}\}| \\
&= |\mathbf{P}\{Y_t \in \mathcal{A} \ \& \ \tau > t\} - \mathbf{P}\{\widehat{Y}_t \in \mathcal{A} \ \& \ \tau > t\} \\
&+ |\mathbf{P}\{Y_t \in \mathcal{A} \ \& \ \tau \le t\} - \mathbf{P}\{\widehat{Y}_t \in \mathcal{A} \ \& \ \tau \le t\}| \\
&= |\mathbf{P}\{Y_t \in \mathcal{A} \ \& \ \tau > t\} - \mathbf{P}\{\widehat{Y}_t \in \mathcal{A} \ \& \ \tau > t\}| \le \mathbf{P}\{\tau > t\}
\end{aligned} \tag{4}
$$

for any set $\mathcal{A} \in \sigma(\mathcal{X})$. Here, identical distribution of pairs $Y_t \overset{D}{=} X_t$ and $\widehat{Y}_t \overset{D}{=} \widehat{X}_t$ means only a coincidence of distributions in any time, but not the coincidence of finite-dimensional distributions of these processes.

Now, our goal is a construction of the successful coupling and an estimation of exponential moments of a random variable $\tau(X_0, \widehat{X}_0)$. For this construction the Basic Coupling Lemma is needed.

3. Basic Coupling Lemma (see, e.g., [10,12]). Here the simplest formulation of the Basic Coupling Lemma is given (see, e.g., [15]).

Lemma 2. *If the random variable ϑ_1 and ϑ_2 have c.d.f. $\Phi_1(s)$ and $\Phi_2(s)$ correspondingly, and their common part $\kappa \overset{\text{def}}{=} \int_{\mathbf{R}} \min\{\Phi_1'(s), \Phi_2'(s)\}\,\mathrm{d}s > 0$, then it can construct (on some probability space) the random variables $\widehat{\vartheta}_1$ and $\widehat{\vartheta}_2$ such, that*

1. $\widehat{\vartheta}_1 \overset{D}{=} \vartheta_1, \ \widehat{\vartheta}_2 \overset{D}{=} \vartheta_2;$ 2. $\mathbf{P}\{\widehat{\vartheta}_1 = \widehat{\vartheta}_2\} = \kappa.$ ▷

The statement of Lemma 2 is naturally transferred to any finite number of random variables.

Lemma 3. *Let $\vartheta_1, \vartheta_2, \dots, \vartheta_n$ be the random variable with probability densities $\varphi_1(s), \varphi_2(s), \dots, \varphi_n(s)$ correspondingly, and $\kappa \overset{\text{def}}{=} \int_{\mathbf{R}} \min_{i=1,\dots,n}\{\varphi_i(s)\}\,\mathrm{d}s > 0$. Then on some probabilistic space it is possible to construct the random variables $\widehat{\vartheta}_1(s), \widehat{\vartheta}_2(s), \dots, \widehat{\vartheta}_n(s)$ such that*

1. $\widehat{\vartheta}_i \overset{D}{=} \vartheta_i, \ i = 1, 2, \dots n;$ 2. $\mathbf{P}\{\widehat{\vartheta}_1 = \widehat{\vartheta}_2 = \dots = \widehat{\vartheta}_n\} = \kappa.$

Proof. Let $\kappa < 1$ (the proof in the case $\kappa = 1$ is very simple). Consider a probability space $(\Omega, \sigma(\Omega), \mathbf{P})$, where $\Omega = [0;1)^{n+1} = [0;1)_1 \times [0;1)_2 \times \cdots \times [0;1)_{n+1}$, $\sigma(\Omega)$ is its Borel σ-algebra, and $\mathbf{P}$ is Lebesgue measure on Ω. Let $\mathcal{U}_i$ be the random variable with continuous uniform distribution on $[0;1)_i$, $i = 1, \ldots, (n+1)$. Let $\varphi(s) \stackrel{\text{def}}{=} \min_{i=1,\ldots,n} \varphi_i(s)$, and $\Psi(s) \stackrel{\text{def}}{=} \frac{1}{\kappa} \int_{-\infty}^{s} \varphi(u) \, du$, $\Psi_i(s) \stackrel{\text{def}}{=} \frac{1}{1-\kappa} \int_{-\infty}^{s} (\varphi_i(u) - \varphi(u)) \, du$. Ψ and Ψ_i are the distribution functions.

Put $\widehat{\vartheta}_i \stackrel{\text{def}}{=} \Psi_i^{-1}(\mathcal{U}_i) \times \mathbf{1}(\mathcal{U}_{n+1} > \kappa) + \Psi^{-1}(\mathcal{U}_i) \times \mathbf{1}(\mathcal{U}_{n+1} \leq \kappa)$, $i = 1, \ldots, n$. It is easy to see that the random variables $\widehat{\vartheta}_i$ satisfy the conditions 1 and 2 of Lemma 3. $\triangleright$

Remark 2. Let us construct successful coupling, i.e. the processes $Y_t = ((i_t, x_t), (j_t, y_t)) \stackrel{\mathcal{D}}{=} X_t$ and $\widehat{Y}_t = ((\widehat{i}_t, \widehat{x}_t), (\widehat{j}_t, \widehat{y}_t)) \stackrel{\mathcal{D}}{=} \widehat{X}_t$. Let the processes Y_t and $\widehat{Y}_t$ be constructed on an interval $[0; t_0]$. And let the values of i_{t_0} and $\widehat{i}_{t_0}$ coincide, and the values of j_{t_0} and $\widehat{j}_{t_0}$ coincide the time t_0. Therefore, at this time t_0 it can construct (on some probability space) the residual times of the stay of both processes in this states using Lemma 3 by such a way, that all these random values are equal with positive probability. I.e. the event $Y_t = ((i', 0), (j', 0)) = Y_t = ((\widehat{i}', 0), (\widehat{j}', 0))$ will happen with positive probability at the next change of at least one of the parameters i, j, $\widehat{i}$, $\widehat{j}$ (here $r' = 1$ if $r = 0$, and $r' = 0$ if $r = 1$). Really, the positive value of the probability is a consequence of Lemma 3 and Remark 1:

1. The probability of the transition from the pair $(Y_t, \widehat{Y}_t)$ from $((0, x_{t_0}), (1, y_{t_0})) \times ((0, \widehat{x}_{t_0}), (1, \widehat{y}_{t_0}))$ to $((1, 0), (0, 0)) \times ((0, 0), (0, 0))$ is

$$\kappa_{0,1} \stackrel{\text{def}}{=} \int_{\theta_0}^{\infty} \min\{f_{t_0}^1(s), f_{t_0}^2(s), g_{t_0}^1(s), g_{t_0}^2(s)\} \, ds$$

$$\geq \int_{0}^{\infty} \min(\lambda_0, \mu_0) \exp\left(-s \max\{\Lambda, \mathrm{M}\}\right) ds = \frac{\min(\lambda_0, \mu_0)}{\max\{\Lambda, \mathrm{M}\}}.$$

2. Analogously, the probability of transition from $((1, x_{t_0}), (0, y_{t_0})) \times ((1, \widehat{x}_{t_0}), (0, \widehat{y}_{t_0}))$ to $((0, 0), (0, 0)) \times ((0, 0), (0, 0))$ is $\kappa_{0,1} \geq \frac{\min(\lambda_0, \mu_0)}{\max\{\Lambda, \mathrm{M}\}}$.

3. And the probability of transition from $((0, x_{t_0}), (0, y_{t_0})) \times ((0, \widehat{x}_{t_0}), (0, \widehat{y}_{t_0}))$ to $((1, 0), (1, 0)) \times ((1, 0), (1, 0))$ is $\kappa_{0,0} \geq \frac{\lambda_0}{\Lambda}$.

4. Also, the probability of transition from $((1, x_{t_0}), (1, y_{t_0})) \times ((1, \widehat{x}_{t_0}), (1, \widehat{y}_{t_0}))$ to $((0, 0), (0, 0)) \times ((0, 0), (0, 0))$ is $\kappa_{1,1} \geq \frac{\mu_0}{\mathrm{M}}$.

Some Denotations and Auxiliary Considerations.

Definition 1. *The random variable η does not exceed the random variable θ by* **distribution** *if the inequality $F_\eta(s) = \mathbf{P}\{\eta \leq s\} \geq \mathbf{P}\{\theta \leq s\} = F_\theta(s)$ is true for all $s \in \mathbf{R}$.*

In other words, c.d.f. of the random variable θ does not exceed c.d.f. of the random variable η. It is the order in the set of the random variables. Denote it by $\eta \prec \theta$ – see [16]. $\triangleright$

Remark 3. If the intensity $\phi(s)$ of the end of the random period $\mathcal{T}$ satisfies the inequality $0 < c < \phi(s) < C < \infty$, then

$$\mathcal{T}_- \prec \mathcal{T} \prec \mathcal{T}_+, \tag{5}$$

where $\mathbf{P}\{\mathcal{T}_- \leq s\} = 1 - e^{-Cs}$, and $\mathbf{P}\{\mathcal{T}_- \leq s\} = 1 - e^{-cs}$. Correspondingly, $\mathbf{E}\,\mathcal{T} \leq \frac{1}{c}$. Indeed, in accordance with Remark 1, $\mathbf{P}\{\mathcal{T}_- \leq s\} = 1 - \exp(-Cs)$, and $\mathbf{P}\{\mathcal{T}_- \leq s\} = 1 - \exp(-cs)$, whence it follows (5).   ▷

Let ξ_m^+, ξ_m^-, ζ_m^+, ζ_m^- be an additional random variables with c.d.f.:

$$\mathbf{P}\{\xi_m^+ \leq s\} = 1 - \exp(\int\limits_0^s -\lambda_m^+(u)\,du); \quad \mathbf{P}\{\xi_m^- \leq s\} = 1 - \exp(\int\limits_0^s -\lambda_m^-(u)\,du);$$

$$\mathbf{P}\{\zeta_m^+ \leq s\} = 1 - \exp(\int\limits_0^s -\mu_m^+(u)\,du); \quad \mathbf{P}\{\zeta_m^- \leq s\} = 1 - \exp(\int\limits_0^s -\mu_m^-(u)\,du).$$

In the future, the sets of such random variables $\xi_m^+(k)$, $\xi_m^-(k)$, $\zeta_m^+(k)$, $\zeta_m^-(k)$ ($k \in \mathbf{N}$) will be used, assuming that these random variables are mutually independent.

Remark 4. Denote: $\xi_m(k)$ is k-th working time of m-th element ($m = 1, 2$), and $\zeta_m(k)$ is k-th recovery time of m-th element.

According to Remark 3, $\zeta_m(k) \prec \zeta_m^-$, $\xi_m(k) \prec \xi_m^-$. Also $\mathbf{E}\,e^{\alpha\zeta_m(k)} \leq \mathbf{E}\,e^{\alpha\zeta_m^-}$ and $\mathbf{E}\,e^{\alpha\xi_m(k)} \leq \mathbf{E}\,e^{\alpha\xi_m^-}$ for $\alpha > 0$.

2.3   Construction of Successful Coupling, I.e. the Processes Y_t and $\widehat{Y}_t$.

In ordered to construct the processes $Y_t = ((i_t, x_t), (j_t, y_t))$ and $\widehat{Y}_t = ((\widehat{i}_t, \widehat{x}_t), (\widehat{j}_t, \widehat{y}_t))$ let us consider a probability space $\prod\limits_{k=0}^{\infty} (\Omega_k, \mathbf{P}_k, \mathcal{B}_k)$, where $(\Omega_k, \mathbf{P}_k, \mathcal{B}_k)$ are some probability spaces. For simplicity, let Ω_k be $[0; 1)^5$, $\mathbf{P}_k$ be the Lebesgue measure, $\mathcal{B}_k$ be the statdard Borel σ-algebra on Ω_k.

Remind that all interesting for us random variables (residual times of work or repair) are absolutely continuous and have positive derivative on the positive arguments. The construction from the Lemma 3 is possible at any time.

So, we will proceed "step by step" subsequent stopping times t_1, t_2, ..., where one of the components $i, j, \widehat{i}, \widehat{j}$ of the pair $(Y_t, \widehat{Y}_t)$ changes. The times t_i are the times when at least one of elements of studied system changes its first component.

Let the state of both processes be known at the moment t_k. Then, we know the intensities of the change of the components $i_t, j_t, \widehat{i}_t, \widehat{j}_t$ at the time t_k, and the values $i_{t_k}, j_{t_k}, \widehat{i}_{t_k}, \widehat{j}_{t_k}$ at the moment t_k. So, joint probability distributions of the residual times of the stay of the elements at their states $i, j, \widehat{i}, \widehat{j}$ are known.

At this step, one of the probability spaces $(\Omega_k, \mathbf{P}_k, \mathcal{B}_k)$ is used. On this probability space, we construct our four random variables in such a way that their joint probability distribution coincides with the described above distribution (Lemma 3). And in any steps we use the construction from the proof of the

Lemma 3. Now, let us define the minimal value of these four random variables. Denote it by ς_k.

At the moment $t_{k+1} \overset{\text{def}}{=\!=} t_k + \varsigma_k$, the next change of at least one of the components $i, j, \widehat{i}, \widehat{j}$ occurs. And, by construction of these random variables, at the time t_{k+1} all values $i, j, \widehat{i}, \widehat{j}$ can be changed with positive probability.

If at the time t_k all the first component of both processes was equal, i.e. $i_{t_k} = \widehat{i}_{t_k}$ and $j_{t_k} = \widehat{j}_{t_k}$, – then in the case of changing all four components $i, j, \widehat{i}, \widehat{j}$ at the same time, $Y_{t_{k+1}} = ((i_{t_{k+1}}, 0), (j_{t_{k+1}}, 0)) = \widehat{Y}_{t_{k+1}}$.

Thus, for construction of successful coupling, it need to find the times, when $i_t = \widehat{i}_t$ and $j_t = \widehat{j}_t$.

Here, we can not study all possibilities of the coincidence $i_t = \widehat{i}_t$ and $j_t = \widehat{j}_t$. Consider the times θ_k:

$$\theta_1 \overset{\text{def}}{=\!=} \inf\{t > 0 : i_t = 0\}; \qquad \theta_1' \overset{\text{def}}{=\!=} \inf\{t > \theta_1 : i_t = 1\};$$
$$\theta_2 \overset{\text{def}}{=\!=} \inf\{t > \theta_1' : i_t = 0\}; \qquad \theta_2' \overset{\text{def}}{=\!=} \inf\{t > \theta_2 : i_t = 1\}; \qquad \ldots ;$$
$$\theta_i \overset{\text{def}}{=\!=} \inf\{t > \theta_{i-1}' : i_t = 0\}; \quad \theta_i' \overset{\text{def}}{=\!=} \inf\{t > \theta_i : i_t = 1\}; \text{ etc.}$$

Let us fix some $\varepsilon > 0$, $\delta > 0$. The choice of values ε and δ affects further calculations. The main result (see belove) can be optimized by the choice of ε and δ.

At the time θ_k, the main element is a work state ($i_{\theta_k} = 0$), and two situations can be: $j_{\theta_k} = 0$ or $j_{\theta_k} = 1$.

1. If $j_{\theta_k} = 1$, i.e. the second element is not working, then the residual time of its stay in not-working state *by distribution* is less than the random variable ζ_2^-, So, the second element will hit to the working state for a time less than ε with the probability greater then $\varpi_1 \overset{\text{def}}{=\!=} 1 - e^{-\varepsilon\mu_2^-}$. Thus, at the time $\widehat{\theta}_k$, $j_{\widehat{\theta}_k} = 1$, where $\theta_k < \widehat{\theta}_k < \theta_k + \varepsilon$. And work period of the main element is greater *by distribution* then the random variable ξ_m^+. Thus this work period greater than ε with probability greater then $e^{-\lambda_1^+ \varepsilon}$. So, with probability greater than $\widehat{\varpi}_1 \overset{\text{def}}{=\!=} (1 - e^{-\varepsilon\mu_2^-})e^{-\lambda_1^+ \varepsilon}$, $i_{\widehat{\theta}_k} = j_{\widehat{\theta}_k} = 0$.

2. If $j_{\theta_k} = 0$, then $i_{\theta_k} = j_{\theta_k} = 0$, then with probability $\widehat{\varpi}_2 = 1$, $i_{\widehat{\theta}_k} = j_{\widehat{\theta}_k} = 0$, where $\widehat{\theta}_k < \theta_k + \varepsilon$. So, with probability greater then $\widetilde{\varpi}_1 = \min\{\widehat{\varpi}_1, \widehat{\varpi}_2\} = \widehat{\varpi}_1$, $i_{\widehat{\theta}_k} = j_{\widehat{\theta}_k} = 0$.

And in both situations, with probability greater then $\widetilde{\varpi}_1 = e^{-\lambda_1^+ \delta} e^{-\lambda_2^+ \delta}$, at the time $\widetilde{\theta}_k < \theta_k + \varepsilon + \delta$, $i_{\widetilde{\theta}_k} = j_{\widetilde{\theta}_k} = 0$. And with probability greater then $\varpi_0 = e^{-\delta\lambda_1^-} e^{-\delta\lambda_2^-}$, the process Y_t is in the stay $\{i_t = j_t = 0\}$ until the time $\widehat{\theta}_k + \delta$.

Now, consider the process $\widehat{Y}_t$ at the time $\widehat{\theta}_k$. Both its elements can be in two states. Anew consider the situations:

1. $\widehat{i}_{\widehat{\theta}_k} = \widehat{j}_{\widehat{\theta}_k} = 0$. With probability $\widetilde{\varpi}_1 = 1$, at the time $\widetilde{\theta}_k' = \widehat{\theta}_k < \widehat{\theta}_k + \delta$, $\widehat{i}_{\widetilde{\theta}_k'} = \widehat{j}_{\widetilde{\theta}_k'} = 0$.

2. $\widehat{\imath}_{\widehat{\theta}_k} = 0$, $\widehat{\jmath}_{\widehat{\theta}_k} = 1$. With probability greater then $\bar{\varpi}_2 = e^{-\delta\lambda_1^+}(1 - e^{-\delta\mu_2^-})$, at the time $\widetilde{\theta}'_k < \widehat{\theta}_k + \delta$, $\widehat{\imath}_{\widetilde{\theta}'_k} = \widehat{\jmath}_{\widetilde{\theta}_k} = 0$.

3. $\widehat{\imath}_{\widehat{\theta}_k} = 1$, $\widehat{\jmath}_{\widehat{\theta}_k} = 0$. With probability greater then $\bar{\varpi}_3 = e^{-\delta\lambda_2^+}(1 - e^{-\delta\mu_1^-})$, at the time $\widetilde{\theta}'_k < \widehat{\theta}_k + \delta$, $\widehat{\imath}_{\widetilde{\theta}'_k} = \widehat{\jmath}_{\widetilde{\theta}_k} = 0$.

4. $\widehat{\imath}_{\widehat{\theta}_k} = 1$, $\widehat{\jmath}_{\widehat{\theta}_k} = 1$. With probability greater then $\bar{\varpi}_4 = (1 - e^{-\delta\mu_1^+}(1 - e^{-\delta\mu_1^-})$, at the time $\widetilde{\theta}'_k < \theta_k + \delta$, $\widehat{\imath}_{\widetilde{\theta}'_k} = \widehat{\jmath}_{\widetilde{\theta}'_k} = 0$.

Thus, with probability greater then $\varpi \stackrel{\text{def}}{=} \min_{i,j}\{\widehat{\varpi}_i, \bar{\varpi}_j\}$, at the time $\widetilde{\theta}'_k$, $\widehat{\imath}_{\widetilde{\theta}'_k} = \widehat{\jmath}_{\widetilde{\theta}'_k} = \widehat{\imath}_{\widetilde{\theta}'_k} = \widehat{\jmath}_{\widetilde{\theta}'_k} = 0$.

From Remark 2 it follows that with probability greater then $\kappa_{0,0}$, at the time of the next after $\widetilde{\theta}'_k$ change of the first components of the processes Y_t and $\widehat{Y}_t$, the event $Y_t = \widehat{Y}_t$ will be happen. This is a coupling epoch. The schema of the construction of successful coupling is done. We skip here the proof that $\mathbf{E}\,\tau < \infty$.

3   Main Result

Theorem 1. *If conditions (3) are satisfied, then for any initial state X_0 of the process X_t, it is possible to calculate the numbers $\alpha > 0$ and $\mathcal{K} = \mathcal{K}(\alpha, \lambda_m^-, \lambda_m^+, \mu_m^-, \mu_m^+)$ such that $\|\mathcal{P}_t - \mathcal{P}\|_{TV} \leq \mathcal{K}e^{-\alpha t}$ for all $t \geq 0$.*

Corollary 1. *Recall that the availability factor $A_2(t)$ is $A_2(t) = \mathbf{P}\{X_t \notin \{((1,\cdot),(1,\cdot))\}\} = 1 - \mathcal{P}_t((1,\cdot),(1,\cdot))$. Thus, in the conditions of Theorem 1, the inequality $|A_2(t) - A_2(\infty)| \leq \mathcal{K}e^{-\alpha t}$ is true.* ▷

Proof. The idea of this proof is the algorithm of computing the values $\alpha > 0$ and $\mathcal{K}$. It is impossible here give the full proof. But above (Sect. 2.3) the schema of the construction of successful coupling for two versions of studied process X_t is given.

With probability greater then with probability greater then $\widehat{\kappa} \stackrel{\text{def}}{=} \kappa_{0,0} \times \varpi \times \varpi_0$, after any time θ_k, the coincidence of the processes Y_t and $\widehat{Y}_t$ can be occurred at the time $\theta_m \stackrel{\text{def}}{=} \min\{\theta_i : \theta_i > \widetilde{\theta}'_k\}$.

So, the coupling epoch $\tau(X_0, \widehat{X}_0)$ is the time $\theta_k + \varepsilon + \delta$ and + one (residual) period of work.

Thus, θ_k is a geometrical summa of double (work+repair) periods. And τ is summa of θ_k and two (conditional) periods (work or repair). Therefore, $\tau \prec \sum_1^{\nu+2}(\xi_k^- + \zeta_k^-)$, where $\mathbf{P}\{\nu > m\} \leq (1 - \widehat{\kappa})$. Hence, it can estimate for $\alpha < \widehat{\kappa}$

$$\mathbf{E}\,e^{\alpha\tau} \leq e^{\alpha(\varepsilon+\delta)}\mathbf{E}\,e^{\alpha(\zeta_1^- + \xi_1^-)}\sum_{i=1}^{\infty}\left(\mathbf{E}\,e^{\alpha(\zeta_1^- + \xi_1^-)}\right)^i(1 - \widehat{\kappa})^i \leq \mathcal{K} < \infty \text{ for some}$$

(small) $\alpha > 0$.

Now return to the inequality (4) and use Markov inequality:

$$
\begin{aligned}
|\mathbf{P}\{X_t \in \mathcal{A}\} - \mathbf{P}\{\widehat{X}_t \in \mathcal{A}\}| &\leq \mathbf{P}\{\tau > t\} \\
&= \mathbf{P}\{\exp(\alpha\tau) > \exp(\alpha t)\} \leq \frac{\mathbf{E}\,e^{\alpha\tau}}{e^{\alpha t}} \leq \mathcal{K}e^{-\alpha t}.
\end{aligned} \tag{6}
$$

As inequality (6) is uniform by the initial states of the processes X_t and $\widehat{X}_t$, this inequality is true in the situation where the process $\widehat{X}_t$ has an initial distribution equal to the stationary distribution $\mathcal{P}$. Therefore,

$$
\sup_{S \in \mathcal{B}(\mathcal{X})} |\mathbf{P}\{X_t \in S\} - \mathcal{P}(S)| = \|\mathcal{P}_t(S) - \mathcal{P}(S)\|_{TV} \leq \frac{\mathbf{E}\,e^{\alpha\tau}}{e^{\alpha t}} = \mathcal{K}e^{-\alpha t}.
$$

4   Conclusion

The construction of successful coupling presented here is simplified. So, the bounds obtained according to the described scheme are rather rough.

However, using the proposed approach with the analysis of *all possible situations suitable for possible constructions of coupling epoch*, as well as the possibility to take into account some specific characteristics of the considered system, and using of different initial states of the process X_t allow to significantly improve the proposed estimate.

Acknowledgement. The author thanks E.Yu. Kalimulina, V. V. Kozlov and A.Yu. Veretennikov for valuable recommendations and help in the preparation of the article. The work is supported by RFBR (project No 17-01-00633 A). The reported study was funded by Presidium of RAS according to the research project by Program I.30.

References

1. Afanasyeva, L.G., Tkachenko, A.V.: On the convergence rate for queueing and reliability models described by regenerative processes. J. Math. Sci. **218**(2), 119–36 (2016)
2. Asmussen, S.: Applied Probability and Queues, 2nd edn. Springer, New York (2003). https://doi.org/10.1007/b97236
3. Gnedenko, B.V., Belyayev, Y.K., Solovyev, A.D.: Mathematical Methods of Reliability Theory. Academic Press, Cambridge (2014)
4. Gnedenko, B.V., Kovalenko, I.N.: Introduction to Queuing Theory. In: Mathematical Modeling, Birkhaeuser Boston, Boston (1989)
5. Griffeath, D.: A maximal coupling for Markov chains. Zeitschrift für Wahrscheinlichkeitstheorie und Verwandte Gebiete **31**(2), 95–106 (1975)
6. Doeblin, W.: Exposé de la théorie des chaînes simples constantes de Markov à un nombre fini d'états. Rev. Math. de l'Union Interbalkanique **2**, 77–105 (1938)
7. Doob, J.L.: Stochastic Processes. Wiley, Hoboken (1953)
8. Kalimulina, E.Y.: Analysis of unreliable open queueing network with dynamic routing. In: Vishnevskiy, V.M., Samouylov, K.E., Kozyrev, D.V. (eds.) DCCN 2017. CCIS, vol. 700, pp. 355–367. Springer, Cham (2017). https://doi.org/10.1007/978-3-319-66836-9_30

9. Kalimulina, E.Y.: Rate of convergence to stationary distribution for unreliable Jackson-type queueing network with dynamic routing. In: Vishnevskiy, V.M., Samouylov, K.E., Kozyrev, D.V. (eds.) DCCN 2016. CCIS, vol. 678, pp. 253–265. Springer, Cham (2016). https://doi.org/10.1007/978-3-319-51917-3_23

10. Kato, K.: Coupling Lemma and Its Application to The Security Analysis of Quantum Key Distribution. Tamagawa University Quantum ICT Research Institute Bulletin, vol. 4, no. 1, pp. 23–30 (2014)

11. Thorisson, H.: Coupling. In: Accardi, L., Heyde, C.C. (eds.) Probability Towards 2000, vol. 128. Springer, New York (2000). https://doi.org/10.1007/978-1-4612-2224-8_19

12. Veretennikov, A., Butkovsky, O.A.: On asymptotics for Vaserstein coupling of Markov chains. Stoch. Process. Appl. **123**(9), 3518–3541 (2013)

13. Veretennikov, A.Y., Zverkina, G.A.: Simple proof of Dynkin's formula for single-server systems and polynomial convergence rates. Markov Process. Relat. Fields **20**, 479–504 (2014)

14. Zverkina, G.: On strong bounds of rate of convergence for regenerative processes. In: Vishnevskiy, V.M., Samouylov, K.E., Kozyrev, D.V. (eds.) DCCN 2016. CCIS, vol. 678, pp. 381–393. Springer, Cham (2016). https://doi.org/10.1007/978-3-319-51917-3_34

15. Zverkina G.: About some extended Erlang-Sevast'yanov queueing system and its convergence rate (English and Russian versions). https://arxiv.org/abs/1805.04915. Fundamentalnaya i Prikladnaya Matematika, 2018, No 22, issue 3 - in print

16. Stoyan, D.: Qualitative Eigenschaften und Abschtzungen stochastischer Modelle. Berlin (1977)

AQM Mechanism with the Dropping Packet Function Based on the Answer of Several PI^α Controllers

Adam Domański[1], Joanna Domańska[2], Tadeusz Czachórski[2], Jerzy Klamka[2], Dariusz Marek[1], and Jakub Szyguła[1(✉)]

[1] Institute of Informatics, Silesian University of Technology,
Akademicka 16, 44-100 Gliwice, Poland
jakub.szygula@polsl.pl
[2] Institute of Theoretical and Applied Informatics, Polish Academy of Sciences,
ul. Bałtycka 5, 44-100 Gliwice, Poland

Abstract. In this paper the performance of AQM mechanism based on three PI^α controllers and the impact of traffic self-similarity on network utilization are investigated with the use of discrete event simulation modelling. The queue is divided into several thresholds. Each segment of the queue is controlled by a different PI^α mechanism. We analyze in tests the length of the queue and the number of rejected packets. The results obtained by the proposed approach are compared to the results obtained for AQM mechanism based on single PI^α controller.

Keywords: AQM · Congestion control ·
Non-integer order PI^α controller

1 Introduction

The most important factor of the TCP/IP network traffic control is the rejection of packets arriving to an IP router to be queued and send then forward. At first, packets are queued following FIFO algorithm and rejected only when the whole buffer space used to queue the packets was already occupied. Since many years, the recommended by IETF active queue management (AQM) where packets are rejected following a certain algorithm, enhances the efficiency of transfers [20] and cooperates better with TCP congestion window mechanism in adapting the flows intensity to the congestion of the network [2].

In the classic RED algorithms (the basic AQM mechanism) the incoming packet is dropped according to the given by a predefined function. Usually, this function is linear and depends on the queue length [8,11,15].

Our previous works proposed to base the probability function on the answer of the PI^α controller [5–7,10,13]. The considered models were based on the controller with the non-integer integrate/derivative orders.

In this article we reconsider this problem by extending the controller to include variable parameters. Similarly to algorithm DSRED [22] (the well-known

The original version of this chapter was revised: the acknowledgement section has been removed. The correction to this chapter is available at
https://doi.org/10.1007/978-3-030-21952-9_30

P. Gaj et al. (Eds.): CN 2019, CCIS 1039, pp. 400–412, 2019.
https://doi.org/10.1007/978-3-030-21952-9_29

variant of the RED algorithm) we divided the queue length into the three separate segments. For each segment we choose a different set of parameters (controller PI^α coefficients and integrate/derivative orders). The first two controllers are *weak* i.e. for high traffic load the bulk of packets are dropped due to maximal queue size exceeding. The third controller is *strong*: its main task is to counteract the buffer overloading. Such choice of controllers enables the incremental increase of controller response as a result of growth of the traffic load.

The remainder of the paper is organized as follows: Sect. 2 gives basic notions on active queue management and presents the DSRED algorithm, Sect. 3 presents briefly theoretical basis for PI^α controller. Section 4 discusses numerical results. Some conclusions are given in Sect. 5.

2   The RED and DSRED Algorithms

The RED algorithm was the solution which fundamentally changed the principles of discarding packets in a router queue. In the case of passive queue management newly incoming packets are dropped only when the buffer is totally full. In the case of RED queue packets are rejected earlier - when the queue length exceeds a planned level. The authors of the RED algorithm: Sally Floyd and Van Jacobson [15] suggested that the destiny of this type of mechanism is to cooperate with transport protocols and congestion control mechanisms based on the positive acknowledgment.

Its performance is based on a drop function giving the probability that a packet is rejected. In RED drop function there are two thresholds: Min_{th} and Max_{th}. The argument avg of this function is a weighted moving average queue length. If $avg < Min_{th}$, all packets are admitted. If $Min_{th} < avg < Max_{th}$, then dropping probability p increases linearly:

$$p = p_{max} \frac{avg - Min_{th}}{Max_{th} - Min_{th}}$$

The value p_{max} corresponds to a probability of packet rejection in the case of $avg = Max_{th}$. If $avg > Max_{th}$ then all packets are dropped. Efficient operation of the RED mechanism is dependent on the proper selection of its parameters. There were several works studying the impact of various parameters on the RED performance.

Many variations of the RED mechanism were developed to improve its performance. They can be classified according to the modification of the method of control variable or dropping packet function calculation and according to how to configure and set the parameters of the algorithm.

One of the possibilities is to increase the thresholds number in the queue. In the algorithm DSRED (Double-Slope RED) [22], the bufor is divided into four sections. Three thresholds K_l, K_m and K_h (usually $K_m = (K_l + K_h)/2$) and parameter γ determine two slopes of this drop function:

$$p(avg) = \begin{cases} 0 & \text{if } avg < K_l \\ \alpha(avg - K_l) & \text{if } K_l \leq avg < K_m \\ 1 - \gamma + \beta(avg - K_m) & \text{if } K_m \leq avg < K_h \\ 1 & \text{if } K_h \leq avg \leq N \end{cases}$$

where

$$\alpha = \frac{2(1 - \gamma)}{K_h - K_l}, \qquad \beta = \frac{2\gamma}{K_h - K_l}.$$

The double slope function makes the algorithm more elastic (more parameters to fix); gentle at the beginning (for low congestion) drop function enhances throughput and reduces queue waiting times. The advantages of this algorithm authors presented in [4] (Fig. 1).

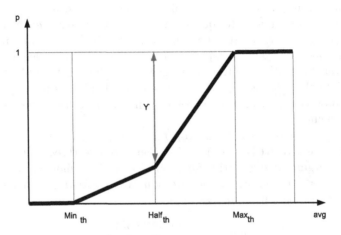

Fig. 1. The probability function of rejection the packet for the DSRED mechanism [4]

3   AQM Mechanism Based on Non-integer Order PI^α Controller.

Our papers [5–7,10] describe how to use the response from PI^α(non-integer integral order) to calculate the probability of packet loss. It is described by a formula:

$$p_i = max\{0, -(K_P e_k + K_I \Delta^\alpha e_k) \tag{1}$$

where K_P, K_I are tuning parameters, e_k is the error in current slot $e_k = Q_k - Q$, i.e. the difference between current queue Q_k and desired queue Q.

For standard PI controller (for $\alpha = -1$ and $\beta = 1$) the packet dropping probability is defined as follows:

$$p_i = max\{0, -(K_p e_i + K_i \sum_{j=i}^{0} e_j)\} \tag{2}$$

In this approach, the dropping probability depends on three parameters: the coefficients for the proportional and integral terms (K_p, K_i) and integrals (α) orders.

The Fractional Order Derivatives and Integrals (FOD/FOI) definitions unify the notions of derivative and integral to one differintegral definition. The most popular formulas to calculate differintegral numerically are Grunwald-Letnikov (GrLET) formula and Riemann-Liouville formulas (RL) [3,16,18].

Differintegral is a combined differentiation/integration operator. The q-differintegral of function f, denoted by $\Delta^q f$ is the fractional derivative (for $q > 0$) or fractional integral (if $q < 0$). If $q = 0$, then the q-th differintegral of a function is the function itself.

In the case of discrete systems (in the active queue management, packet drop probabilities are determined at discrete moments of packet arrivals) there is only one definition of differ-integrals of non-integer order. This definition is a generalization of the traditional definition of the difference of integer order to the non-integer order and it is analogous to a generalization used in Grunwald-Letnikov (GrLET) formula.

For a given sequence $f_0, f_1, ..., f_j, ..., f_k$

$$\Delta^q f_k = \sum_{j=0}^{k} (-1)^j \binom{q}{j} f_{k-j} \tag{3}$$

where $q \in R$ is generally a non-integer fractional order, f_k is a differentiated discrete function and $\binom{q}{j}$ is generalized Newton symbol defined as follows:

$$\binom{q}{j} = \begin{cases} 1 & \text{for } j = 0 \\ \dfrac{q(q-1)(q-2)..(q-j+1)}{j!} & \text{for } j = 1, 2, ... \end{cases} \tag{4}$$

Articles [5,7,10] show that using the non-integer order PI^α controller as AQM mechanism is more efficient in network congestion control than standard RED mechanism and improves the router performance. The approach proposed in the article divides the queue length into several segments and for each of them use a different set of controller coefficients. This solution should result in more flexible behavior of AQM mechanism independently of the network load or long-range dependence of the network traffic.

4   Packet Dropping Scheme Based on the Answer of the Three PI^α Controllers

The AQM algorithms drop packets following a dropping packet function. The choice of the proper coefficients of this function is not easy. These parameters may differ and depend on the network traffic profile.

This problem also exists in the case of the dropping packets functions based on the answers of the PI^α controllers. Our previous works show significant influence of the traffic parameters (intensity and self-similarity) on the choice of the optimal controller parameters. The AQM mechanism should change its parameters during operation as a result of traffic load. One of the possibilities is to change the parameters of the controller as a function of the queue occupancy.

This article presents the DSRED-like solution. We divide the queue length iwith the use of thresholds. Each segment of the queue is controlled by a different PI^α mechanism.

For queue length between 0 and 180 (packets) we use only one controller. For queue length from 180 to 220 the probability of packet dropping is a sum of answers of the first and second controller. When the queue occupancy exceeds 220 the probability is the sum of responses of all three controllers.

The packet dropping probability may be defined as follows:

$$p(q) = \begin{cases} p_1(q) & \text{if} \quad q < 180 \\ p_1(q) + p_2(q) & \text{if} \quad 180 \leq q < 220 \\ p_1(q) + p_2(q) + p_3(q) & \text{if} \quad 220 \leq q \end{cases}$$

where

p_1 - answer of the first controller,
p_2 - answer of the second controller,
p_2 - answer of the third controller.

All presented in this article results were obtained using the simulation model. The simulations were done using the Simpy Python packet. To accelerate the calculations the PI^α module was written in C language. During the tests, we analyzed the following parameters of the AQM transmission: the length of the queue and the number of rejected packets. The input traffic intensity $\lambda = 0.5$ was considered independently of the Hurst parameter. During the tests we changed the Hurst parameter of the input traffic within the range from 0.5 to 0.90. We use a fast algorithm for generating approximate sample paths for a fGn process, first introduced in [17]. After each trace generation the Hurst parameter was estimated with the use of popular self-similarity parameter estimators [9,12,14]: the R/S statistic, aggregated variance, periodogram as well known methods with a significant history of use in estimating LRD and wavelet based method, local Whittle's estimator as newer techniques. Traditional Hurst parameter estimators can be really biased [1,21]. Additionally, the different implementations of the same method may give varying results [19]. Only Hurst parameter estimator based on wavelets can be treated as unbiased and robust [21].

The Table 1 presents the estimations for sample generated trace with the assumed Hurst parameter. These results show that the assumed and estimated Hurst parameters are not the same. The obtained results changed for subsequent generated samples and differed depending on the method of estimating the Hurst parameter. For all differences in results, the dependence of the increase in the estimated Hurst parameter with the increase in the assumed parameter is clearly visible.

Table 1. Hurst parameter estimates for IITiS data traces

	H = 0.5	H = 0.6	H = 0.7	H = 0.8	H = 0.9
Estimator	Estimated Hurst parameter				
R/S method	0.6289	0.6638	0.7338	0.7486	0.7666
Aggregate variance method	0.5710	0.6710	0.7805	0.8785	0.9521
Periodogram method	0.5278	0.6383	0.7601	0.8735	0.9589
Whittle method	0.6889	0.7485	0.8021	0.8429	0.8565
Wavelet-based method	0.5872	0.6859	0.7893	0.8759	0.9337

The service time represents the time of a packet treatment and dispatching. In packet-switched networks it is the time required to transmit information. We have used discrete-time model, hence we have assumed that service-time distribution is geometric (which corresponds to Poisson traffic in case of continuous time models). The distribution of service time μ changed during the test.

The high traffic load was considered for parameter $\mu = 0.25$. The average traffic load we obtained for $\mu = 0.5$. Small network traffic was considered for parameter $\mu = 0.75$

The PI^α controllers coefficients and setpoints presents Table 2. The impact of controller parameters on the behavior of the AQM mechanism and packet dropping probability were described in [5,13]. In presented solution first and second controllers drop the some packets but mostly the queue size crosses the third threshold. When the queue size exceeded third threshold, the third controller begin to work. The third (strong) controller protects the queue against exceeding the maximum size.

Table 2. PI^α controllers coefficients

	K_p	K_i	α	Setpoint
1	0.0001	0.00040	−0.4	100
2	0.0001	0.00015	−0.5	180
3	0.0001	0.00035	−0.6	220

The distributions of the queue length present Figs. 2 and 3. The Tables 3, 4 and 5 present the detailed results. The results consider the high traffic load ($\mu = 0.25$). The Table 3 presents the results for the first controller. In our solution this controller works for the queue occupancy between 0 and 180. The controller parameters were chosen to maximize the queue length. However, the majority of packets are dropped by PI^α mechanism, several packets are dropped due to maximum queue length exceeding. The number of packets dropped by the queue increases with the Hurst parameter. The Table 4 presents the controller that starts when queue length exceeds 180. This controller is weak. The most packets are dropped by the queue. The third controller (Table 5) is very strong. All packets are discarded from the queue by controller mechanism. Obtained results confirmed the assumptions of the controllers behavior.

Table 3. PI^α controller, $\mu = 0.25$, $K_p = 0.0001$, $K_i = 0.0004$, $\alpha = -0.4$, setpoint $= 100$

Hurst	Avg. queue length	Packet drop by	
		PI^α	Queue
0.50	270.42	2492385	10134
0.60	268.90	2467277	30821
0.70	264.08	2374004	124992
0.80	246.98	2115155	384445
0.90	203.94	1744739	875155

Table 4. PI^α controller, $\mu = 0.25$, $K_p = 0.0001$, $K_i = 0.00015$, $\alpha = -0.5$, setpoint $= 100$

Hurst	Avg. queue length	Packet drop by	
		PI^α	Queue
0.50	296.95	1350549	1149714
0.60	295.94	1339802	1157844
0.70	292.22	1307309	1190066
0.80	275.13	1162557	1339373
0.90	222.05	875029	1735620

The proposed solution sums the behavior of all three presented above controllers. The packet dropping probability increases with assumed thresholds.

Tables 6, 7 and 8 present obtained results for different traffic intensity. The Table 6 presents the overloaded network. Although two first controllers drop most packets, the queue length exceeds the third threshold. The advantage of this solution is a small reaction of the first and second PI^α in the case of highly variable traffic. For $H = 0.90$ the most packets are dropped by third PI^α.

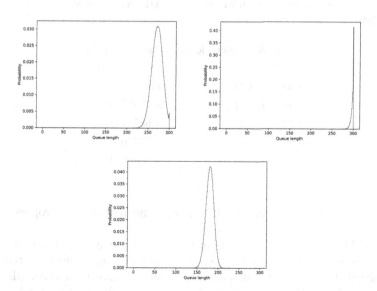

Fig. 2. Distribution of queue length for high traffic load ($\mu = 0.25$) and H = 0.5, left: $K_p = 0.0001$, $K_i = 0.00015$, $\alpha = -0.5$, right: $K_p = 0.0001$, $K_i = 0.0004$, $\alpha = -0.4$, center: $K_p = 0.0001$, $K_i = 0.00035$, $\alpha = -0.6$

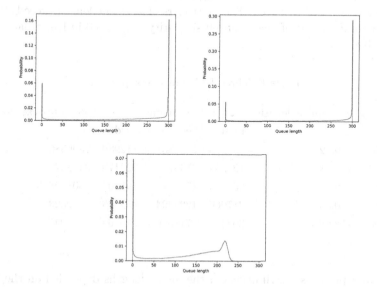

Fig. 3. Distribution of queue length for high traffic load ($\mu = 0.25$) and H = 0.9, left: $K_p = 0.0001$, $K_i = 0.00015$, $\alpha = -0.5$, right: $K_p = 0.0001$, $K_i = 0.0004$, $\alpha = -0.4$, center: $K_p = 0.0001$, $K_i = 0.00035$, $\alpha = -0.6$

Table 5. PI^{α} controller, $\mu = 0.25$, $K_p = 0.0001$, $K_i = 0.00035$, $\alpha = -0.6$, set-point $= 100$

Hurst	Avg. queue length	Packet drop by	
		PI^{α}	Queue
0.50	179.73	2499983	0
0.60	179.11	2498358	0
0.70	177.20	2498517	0
0.80	169.19	2504336	0
0.90	142.84	2638408	0

Independently of the degree of self-similarity, no packets are dropped by the queue.

The Table 7 presents the results in the case of the average traffic load. The average queue length does not exceed 80 packets (independently of the traffic self-similarity). However, the detailed results suggest that temporarily the queue length exceeds the third thresholds. The number of dropped packet by second and third controller grows with the degree od self-similarity. This phenomenon is caused by high variability of queue occupancy. This variability grows with Hurst parameter.

The average queue length for small network traffic (Table 8) is the largest in the case of $H = 90$. The queue length never exceeds the third threshold. All packets are dropped by two earlier controllers. The queue length exceeds the firt threshold only in case of degree of self-similarity (expressed in Hurst parameter) exceeds 0.8.

Table 6. Three PI^{α} controllers, $\mu = 0.25$

Hurst	Avg. queue length	Packet drop in stage			Sum of packet loss
		First	Second	Third	
0.50	199.42	62275	2324331	112980	2499586
0.60	199.78	126334	2148758	223179	2498271
0.70	199.32	251743	1758947	491208	2501898
0.80	190.33	317066	1229494	955963	2502523
0.90	158.95	204411	711370	1716164	2631945

Figure 4 presents distributions of the queue lengths depended on the traffic intensity and the degree of self-similarity.

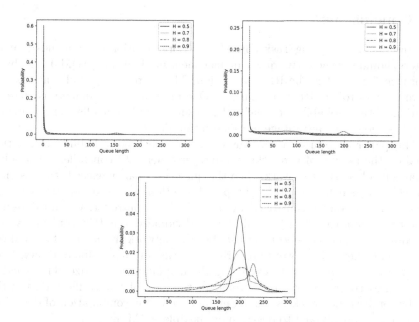

Fig. 4. The influence of degree of traffic self-similarity on queue distribution, three PI^α controllers, $\mu = 0.75$ (left), $\mu = 0.5$ (right), $\mu = 0.25$ (bottom)

Table 7. Three PI^α controllers, $\mu = 0.50$

Hurst	Avg. queue length	Packet drop in stage			Sum of packet loss
		First	Second	Third	
0.50	58.47	33866	0	0	33866
0.60	61.19	86436	62	0	86498
0.70	65.77	208121	20793	2045	230959
0.80	72.98	293810	218707	42700	555217
0.90	78.88	205968	840522	86260	1132750

Table 8. Obtained results for the input traffic intensity $\mu = 0.75$

Hurst	Avg. queue length	Packet drop in stage			Sum of packet loss
		First	Second	Third	
0.50	0.79	0	0	0	0
0.60	1.23	0	0	0	0
0.70	3.07	94	0	0	94
0.80	14.0	48981	53	0	49034
0.90	35.9	368562	1097	0	369659

5   Conclusions

The Internet Engineering Task Force (IETF) organization recommends that IP routers should use the active queue management mechanisms (AQMs). The basic algorithm for AQM is the RED algorithm. There are many modifications and improvements to the RED mechanism. One of these improvements is the calculation of the probability of packet loss using a PI^α controller. Our previous work has shown the advantage of this solution [5,10].

This paper introduces a new way of packet rejecting probability calculation based on the answer of three the non-integer order PI^α controllers. The additional controllers start to work when the queue occupancy exceeds the assumed threshold. The behavior of the proposed solution was also compared to the behavior of the queue controlled by a single PI^α controller. Obtained results show the advantage of such a solution. Individually, the PI^α controllers presented in the article are poorly adjusted to the network traffic. The first and the second controller did not work properly in the case of high traffic intensity. Most packets were dropped due to exceeding the maximum queue size. The reaction of the third controller was too strong. In the case of low traffic intensity the number of discarded packets was redundant. Only the combination of described above controllers allowed to design more flexible AQM mechanism.

Our article presents also the impact of the degree of self-similarity (expressed in the Hurst parameter) on the length of the queue and the number of rejected packets. Obtained results are closely related to the degree of self-similarity. The experiments are carried out for the four types of traffic ($H = 0.5, 0.7, 0.8, 0.9$). Additionally, we evaluate the number of dropped packets in assumed queue segments. This results allowed to select the desired parameters of the controller.

The results described in this article confirm that our approach increases the efficiency of the AQM mechanism based on the PI^α controller. In presented solution we refere mainly to the queue occupancy. In our future work we will focus on mechanisms based on the evaluation of the network traffic parameters and the selection of controller parameters according to the intensity or the self-similarity of the network traffic.

References

1. Abry, P., Veitch, D.: Wavelet analysis of long-range-dependent traffic. IEEE Trans. Inf. Theory **44**(1), 2–15 (1998). https://doi.org/10.1109/18.650984
2. Braden, B., et al.: Recommendations on queue management and congestion avoidance in the internet. Network Working Group - Request for Comments: 2309, IETF (1998)
3. Leszczyński, J., Ciesielski, M.: A numerical method for solution of ordinary differential equations of fractional order. In: Wyrzykowski, R., Dongarra, J., Paprzycki, M., Waśniewski, J. (eds.) PPAM 2001. LNCS, vol. 2328, pp. 695–702. Springer, Heidelberg (2002). https://doi.org/10.1007/3-540-48086-2_77

4. Domańska, J., Domański, A., Czachórski, T.: The drop-from-front strategy in AQM. In: Koucheryavy, Y., Harju, J., Sayenko, A. (eds.) NEW2AN 2007. LNCS, vol. 4712, pp. 61–72. Springer, Heidelberg (2007). https://doi.org/10.1007/978-3-540-74833-5_6

5. Domański, A., Domańska, J., Czachórski, T., Klamka, J.: Self-similarity traffic and AQM mechanism based on non-integer order $PI^\alpha D^\beta$ controller. In: Gaj, P., Kwiecień, A., Sawicki, M. (eds.) CN 2017. CCIS, vol. 718, pp. 336–350. Springer, Cham (2017). https://doi.org/10.1007/978-3-319-59767-6_27

6. Domański, A., Domańska, J., Czachórski, T., Klamka, J., Marek, D., Szyguła, J.: GPU accelerated non-integer order $PI^\alpha D^\beta$ controller used as AQM mechanism. In: Gaj, P., Sawicki, M., Suchacka, G., Kwiecień, A. (eds.) CN 2018. CCIS, vol. 860, pp. 286–299. Springer, Cham (2018). https://doi.org/10.1007/978-3-319-92459-5_23

7. Domański, A., Domańska, J., Czachórski, T., Klamka, J., Szyguła, J.: The AQM dropping packet probability function based on non-integer order $PI^\alpha D^\beta$ controller. In: Ostalczyk, P., Sankowski, D., Nowakowski, J. (eds.) RRNR 2017. LNEE, vol. 496, pp. 36–48. Springer, Cham (2019). https://doi.org/10.1007/978-3-319-78458-8_4

8. Domański, A., Domańska, J., Czachórski, T.: Comparison of AQM control systems with the use of fluid flow approximation. In: Kwiecień, A., Gaj, P., Stera, P. (eds.) CN 2012. CCIS, vol. 291, pp. 82–90. Springer, Heidelberg (2012). https://doi.org/10.1007/978-3-642-31217-5_9

9. Domańska, J., Domański, A., Czachórski, T.: Estimating the intensity of long-range dependence in real and synthetic traffic traces. In: Gaj, P., Kwiecień, A., Stera, P. (eds.) CN 2015. CCIS, vol. 522, pp. 11–22. Springer, Cham (2015). https://doi.org/10.1007/978-3-319-19419-6_2

10. Domańska, J., Domański, A., Czachórski, T., Klamka, J.: The use of a non-integer order PI controller with an active queue management mechanism. Int. J. Appl. Math. Comput. Sci. **26**, 777–789 (2016). https://doi.org/10.1515/amcs-2016-0055

11. Domańska, J., Domański, A., Augustyn, D., Klamka, J.: A RED modified weighted moving average for soft real-time application. Int. J. Appl. Math. Comput. Sci. **24**(3), 697–707 (2014). https://doi.org/10.2478/amcs-2014-0051

12. Domańska, J., Domański, A., Czachórski, T.: Modeling packet traffic with the use of superpositions of two-state MMPPs. In: Kwiecień, A., Gaj, P., Stera, P. (eds.) CN 2014. CCIS, vol. 431, pp. 24–36. Springer, Cham (2014). https://doi.org/10.1007/978-3-319-07941-7_3

13. Domański, A., Domańska, J., Czachórski, T., Klamka, J., Marek, D., Szyguła, J.: The influence of the traffic self-similarity on the choice of the non-integer order PI^α controller parameters. In: Czachórski, T., Gelenbe, E., Grochla, K., Lent, R. (eds.) ISCIS 2018. CCIS, vol. 935, pp. 76–83. Springer, Cham (2018). https://doi.org/10.1007/978-3-030-00840-6_9

14. Estrada-Vargas, L., Torres Roman, D., Toral-Cruz, H.: A study of wavelet analysis and data extraction from second-order self-similar time series. Math. Probl. Eng. 1–14 (2013). https://doi.org/10.1155/2013/102834

15. Floyd, S., Jacobson, V.: Random early detection gateways for congestion avoidance. IEEE/ACM Trans. Netw. **1**(4), 397–413 (1993). https://doi.org/10.1109/90.251892

16. Miller, K., Ross, B.: An Introduction to the Fractional Calculus and Fractional Differential Equations. Wiley, New York (1993)

17. Paxson, V.: Fast, approximate synthesis of fractional Gaussian noise for generating self-similar network traffic. ACM SIGCOMM Comput. Commun. Rev. **27**(5), 5–18 (1997). https://doi.org/10.1145/269790.269792

18. Podlubny, I.: Fractional Differential Equations, vol. 198. Academic Press, San Diego (1999)
19. Ramirez-Pacheco, J., Torres-Román, D., Toral-Cruz, H., Estrada-Vargas, L.: High-performance tool for the test of long-memory and self-similarity. In: Simulation Technologies in Networking and Communications: Selecting the Best Tool for the Test. CRC Press/Taylor & Francis Group, pp. 93–114 (2014). https://doi.org/10.1201/b17650-6
20. Sawicki, M., Kwiecień, A.: Unexpected anomalies of isochronous communication over USB 3.1 Gen 1. Comput. Stand. Interfaces 49, 67–70 (2017). https://doi.org/10.1016/j.csi.2016.08.010
21. Stolojescu, C., Isar, A.: A comparison of some Hurst parameter estimators. In: 13th International Conference on Optimization of Electrical and Electronic Equipment (OPTIM), pp. 1152–1157 (2012). https://doi.org/10.1109/OPTIM.2012.6231802
22. Zheng, B., Atiquzzaman, M.: DSRED: a new queue management scheme for the next generation internet. IEICE Trans. Commun. 242–251 (2000). https://doi.org/10.1093/ietcom/e89-b.3.764

Correction to: AQM Mechanism with the Dropping Packet Function Based on the Answer of Several PI^α Controllers

Adam Domański, Joanna Domańska, Tadeusz Czachórski,
Jerzy Klamka, Dariusz Marek, and Jakub Szyguła

Correction to:
Chapter "AQM Mechanism with the Dropping Packet Function Based on the Answer of Several PI^α Controllers" in: P. Gaj et al. (Eds.): *Computer Networks*, CCIS 1039, https://doi.org/10.1007/978-3-030-21952-9_29

In the originally published version of the chapter 29 the acknowledgement was cancelled. Hence, the acknowledgement section has been removed.

The updated original version of this chapter can be found at
https://doi.org/10.1007/978-3-030-21952-9_29

Author Index

Printed in the United States
by Baker & Taylor Publisher Services